Stochastic stability of differential equations

Monographs and textbooks on mechanics of solids and fluids
editor-in-chief: G. Æ. Oravas

Mechanics: Analysis
editor: V. J. Mizel

1. M. A. Krasnoselskii *et al.*
   Integral operators in spaces of summable functions

2. V. V. Ivanov
   The theory of approximate methods and their application to the numerical solution of singular integral equations

3. A. Kufner *et al.*
   Function spaces

4. S. G. Mikhlin
   Approximation on a rectangular grid

5. D. G. B. Edelen
   Isovector methods for equations of balance

6. R. S. Anderssen, F. R. de Hoog, M. A. Lukas
   The application and numerical solution of integral equations

7. R. Z. Has'minskiĭ
   Stochastic stability of differential equations

Stochastic stability of differential equations

R. Z. Has'minskiĭ

Translated by

D. Louvish

Edited by

S. Swierczkowski

SIJTHOFF & NOORDHOFF 1980
Alphen aan den Rijn, The Netherlands
Rockville, Maryland, USA

Library of Congress Catalog Card Number: 80-83267

ISBN 90 286 0100 7

Original title:
Ustoĭčivost' sistem differencial'nyh uravneniĭ pri slučaĭnyh vozmuščenijah ih parametrov

Printed in The Netherlands

CONTENTS

PREFACE TO THE ENGLISH EDITION

I am very pleased to witness the printing of an English edition of this book by Noordhoff International Publishing. Since the date of the first Russian edition in 1969 there have appeared no less than two specialist texts devoted at least partly to the problems dealt with in the present book (Bunke [4], Morozan [7]). There have also appeared a large number of research papers on our subject. Also worth mentioning is the monograph of Sagirov [1] containing applications of some of the results of this book to cosmology.

In the hope of bringing the book somewhat more up to date we have written, jointly with M.B. Nevel'son, an Appendix containing an exposition of recent results. Also, we have in some places improved the original text of the book and have made some corrections. Among these changes, the following two are especially worth mentioning: A new version of Section 8.4, generalizing and simplifying the previous exposition, and a new presentation of Theorem 7.4.1 rendering correct the reference to this Theorem in Section 8.5.

Finally, there have been added about thirty new titles to the list of references (these are marked with an asterisk). In connection with this we would like to mention the following. In the first Russian edition we tried to give as complete as possible a list of references to works concerning the subject. This list was up to date in 1967. Since then the annual output of publications on stability of stochastic systems has increased so considerably that the task of supplying this book with a totally up to date and complete bibliography became very difficult indeed. Therefore we have chosen to limit ourselves to listing only those titles which pertain directly to the contents of this book. We have also mentioned some more recent papers which were published in Russian, assuming that those will be less known to the western reader.

I would like to conclude this preface by expressing my gratitude to M.B. Nevel'son for his help in the preparation of this new edition of the book.

R.Z. HAS'MINSKIĬ

PREFACE TO THE RUSSIAN EDITION

This monograph is devoted to the study of the qualitative theory of differential equations with random right-hand side. More specifically, we shall consider here problems concerning the behavior of solutions of systems of ordinary differential equations whose right-hand sides involve stochastic processes. Among these the following questions will receive most of our attention.

1. When is each solution of the system defined with probability 1 for all $t > 0$ (i.e., the solution does not "escape to infinity" in a finite time)?

2. If the function $X(t) \equiv 0$ is a solution of the system, under which conditions is this solution stable in some stochastic sense?

3. Which systems admit only bounded for all $t > 0$ (again in some stochastic sense) solutions?

4. If the right-hand side of the system is a stationary (or periodic) stochastic process, under which additional assumptions does the system have a stationary (periodic) solution?

5. If the system has a stationary (or periodic) solution, under which circumstances will every other solution converge to it?

The above problems are also meaningful (and motivated by practical interest) for deterministic systems of differential equations. In that case, they received detailed attention in Ljapunov [1], Malkin [2], Krasovskiĭ [1], Pliss [1], Krasnosel'skiĭ [1], LaSalle and Lefschetz [1], and others.

Problems 3-5 have been thoroughly investigated for linear systems of the type $\dot{x} = Ax + \xi(t)$, where A is a constant or time-dependent matrix and $\xi(t)$ a stochastic process. For that case one can obtain not only qualitative but also quantitative results (i.e., the moment, correlation and spectral characteristics of the output process $x(t)$, in terms of the corresponding characteristics of the input process $\xi(t)$). Methods leading to this end are presented e.g., in Pugačev [1], Laning and Battin [1], etc. In view of this, we shall concentrate our attention in the present volume primarily on non-linear systems, and on linear systems whose parameters (the elements of the matrix A) are subjected to random perturbations.

In his celebrated memoir Ljapunov [1] applied his method of auxiliary functions (Ljapunov functions) to the study of stability. His method proved later to be applicable also to many other problems in the qualitative theory of differential equations. Also in this book we shall utilize an appropriate modification of the method of Ljapunov functions when discussing the solutions to the above mentioned problems.

In Chapters I and II we shall study problems 1-5 without making any specific assumptions on the form of the stochastic process on the right side of the special equation. We shall be predominantly concerned with systems of the type $\dot{x} = F(x,t) + \sigma(x,t) \times \xi(t)$ in euclidian ℓ-space. We shall discuss their solutions, using the Ljapunov functions of the truncated system $\dot{x} = F(x,t)$. In this we shall try to impose as few restrictions as possible on the stochastic process $\xi(t)$; e.g., we may require only that the expectation of $|\xi(t)|$ be bounded. It seems convenient to take this approach, first, because sophisticated methods are available for constructing Ljapunov functions for deterministic systems, and second, because the results so obtained will be applicable also when the properties of the process $\xi(t)$ are not completely known, as is often the case.

Evidently, to obtain more detailed results, we shall have to restrict the class of stochastic processes $\xi(t)$ that may appear on the right side of the equation. Thus in Chapters III through VII we shall study the solutions of the equation $\dot{x} = F(x,t) + \sigma(x,t) \times \xi(t)$ where $\xi(t)$ is a white noise, i.e. a Gaussian process such that $\mathbf{E}\xi(t) = 0$, $\mathbf{E}[\xi(s)\xi(t)] = \delta(t - s)$. We have chosen this process, because:

1. In many real situations physical noise can be well approximated by white noise.
2. Even under conditions different from white noise, but when the noise acting upon the system has a finite memory interval τ (i.e., the values of the noise at times t_1 and t_2 such that $|t_2 - t_1| > \tau$ are virtually independent), it is often possible after changing the time scale to find an approximating system, perturbed by white noise.
3. When solutions of an equation are sought in the form of a process, continuous in time and without after-effects, the assumption that the noise in the system is "white" is essential. The investigation is facilitated by the existence of a well developed theory of processes without after-effects (Markov processes).

Shortly after the publication of Kolmogorov's paper [1], which laid the foundations for the modern analytical theory of Markov processes, Andronov, Pontrjagin and Vitt [1] pointed out that actual noise in dynamic systems can be replaced by white noise, thus showing that the theory of Markov processes is a convenient tool for the study of such systems.

Certain difficulties in the investigation of the equation $\dot{x} = F(x,t) + \sigma(x,t)\xi(t)$, where $\xi(t)$ is white noise are caused by the fact that, strictly speaking, "white" noise processes do not exist; other difficulties arise because of the many ways of interpreting the equation itself. These difficulties have been largely overcome by the efforts of Bernšteĭn, Gihman and Itô. In Chapter III we shall state without proof a theorem on the existence and uniqueness of the Markov process determined by an equation with white noise. We shall assume a certain interpretation of this equation. For a detailed proof we refer the reader to Doob [3], Dynkin [3], Gihman and Skorohod [1].

However, we shall consider in Chapter III various other issues in great detail, such as sufficient conditions for a sample path of the process not to "escape to infinity" in a finite time, or to reach a given bounded region with probability 1. It turns out that such conditions are often conveniently formulated in terms of certain auxiliary functions analogous to Ljapunov functions. Instead of the Ljapunov operator (the derivative along the path) one uses the infinitesimal generator of the corresponding Markov process.

In Chapter IV we examine conditions under which a solution of a differential equation where $\xi(t)$ is white noise, converges to a stationary process. We show how this is related to the ergodic theory of dynamic systems and to the problem of stabilization of the solution of a Cauchy problem for partial differential equations of parabolic type.

Chapters V-VIII contain the elements of stability theory of stochastic systems without after-effects. This theory has been created in the last few years for the purpose of studying the stabilization of controlled motion in systems perturbed by random noise. Its origins date from the 1960 paper by Kac and Krasovskiĭ [1] which has stimulated considerable further research. More specifically, in Chapter V we generalize the theorems of Ljapunov's second method; Chapter VI is devoted to a detailed investigation of linear systems, and in Chapter VII we prove theorems on stability and instability in the first approximation. We do this, keeping in view applications to stochastic approximation and certain other problems.

Chapter VIII is devoted to application of the results of Chapters V to VII to optimal stabilization of controlled systems. It was written by the author in collaboration with M.B. Nevel'son. In preparing this chapter we have been influenced by Krasovskiĭ's excellent Appendix IV in Malkin [2].

As far as we know, there exists only one other monograph on stochastic stability. It was published in the U.S.A. in 1967 by

Kushner [4], and its translation into Russian is now ready for print. Kushner's book contains many interesting theorems and examples. They overlap partly with the results of §7, in Chapter III, and §§1-5, in Chapter V, of this book.

Though our presentation of the material is abstract, the reader who is primarily interested in applications should bear in mind that many of the results admit a directly "technical" interpretation. For example, problem 4, stated above, concerning the question of the existence of a stationary solution, is equivalent to the problem of determining when stationary operating conditions can prevail within a given, generally non-linear, automatic control system, whose parameters experience random perturbations and whose input process is also stochastic. Similarly, the convergence of each solution to a stationary solution (see Chapter IV) means that each output process of the system will ultimately "settle down" to stationary conditions.

In order not to deviate from the main purpose of the book, we shall present without proof many facts from analysis and from the general theory of stochastic process. However, in all such cases we shall mention either in the text or in a footnote where the proof can be found. For the reader's convenience, such references will usually be not to the original papers but rather to more accessible textbooks and monographs. On the other hand, in the rather narrow range of the actual subject matter we have tried to give precise references to the original research. Most of the references appear in footnotes.

Part of the book is devoted to the theory of stability of solutions of stochastic equations (§§5-8 of Chapter I, Chapters V-VIII). This appears to be an important subject which has recently been receiving growing attention. The volume of the relevant literature is increasing steadily. Unfortunately, in this area various authors have published results overlapping significantly with those of others. This is apparently due to the fact that the field is being studied by mathematicians, physicists, and engineers, and each of these groups publishes in journals not read by the others. Therefore the bibliography given at the end of this book lists, besides the books and papers cited in the text, various other publications on the stability of stochastic systems known to the author, which appeared prior to 1967[1]. For the reason given above, this list is far from complete, and the author wishes to apologize to authors whose research he might have overlooked.

The book is intended for mathematicians and physicists. It may be of particular interest to those who specialize in mechanics, in particular in the applications of the theory of stochastic pro-

---

[1] Most of this book was prepared in 1966. (The exceptions are §§4.10, 4.11, 5.11, 7.6 and Chapter VIII, all of which were added in 1968). The later developments of the theory are discussed in the Appendix at the end of this book, which was added at the time of preparing this English edition.

cesses to problems in oscillation theory, automatic control and related fields. Certain sections may appeal to specialists in the theory of stochastic processes and differential equations. The author hopes that the book will also be of use to specialized engineers interested in the theoretical aspects of the effect of random noise on the operation of mechanical and radio-engineering systems and in problems relating to the control of systems perturbed by random noise.

To study the first two chapters it is sufficient to have an acquaintance with the elements of the theory of differential equations and probability theory, to the extent generally given in higher technical schools (the requisite material from the theory of stochastic processes is given in the text without proofs).

The heaviest mathematical demands on the reader are made in Chapters III and IV. To read them, he will need an acquaintance with the elements of the theory of Markov processes to the extent given, e.g., in Chapter VIII of Gihman and Skorohod [1].

The reader interested only in the stability of stochastic systems might proceed directly from Chapter II to Chapters V-VII, familiarizing himself with the results of Chapters III and IV as the need arises.

Each of the seven chapters is divided into sections. Within a chapter the lemmas and theorems are numbered by section. For example, the third theorem in §4 of Chapter I is designated in that chapter as Theorem 4.3. It is referred to in later chapters as Theorem 1.4.3.

The origin of this monograph dates back to some fruitful conversations which the author had with N.N. Krasovskiĭ. In the subsequent research, here described, the author has used the remarks and advice offered by his teachers A.N. Kolmogorov and E.B. Dynkin, to whom he is deeply indebted.

This book also owes much to the efforts of its editor, M.B. Nevel'son, who not only took part in writing Chapter VIII and indicated several possible improvements, but also placed some of his yet unpublished examples at the author's disposal. We are grateful to him for this assistance. We also would like to thank V.N. Tutubalin, V.B. Kolmanovskiĭ and A.S. Holevo for many critical remarks, and to R.N. Stepanova for her work on the preparation of the manuscript.

R.Z. HAS'MINSKIĬ

BASIC NOTATION

$I_T = \{t: 0 \leqslant t < T\}$: set of points t such that $0 \leqslant t < T$, p. 4

$I = I_\infty$, p. 4

$U_R = \{x: |x| < R\}$, p. 4

E_ℓ: euclidean ℓ-space, p. 2

$E = E_\ell \times T$, p. 2

$\mathbf{L}$: class of functions $f(t)$ absolutely integrable on every finite interval, p. 4

$\mathbf{C}_2$: class of functions $V(t,x)$ twice continuously differentiable with respect to x and once continuously differentiable with respect to t, p. 81

$\mathbf{C}_2^0(U)$: class of functions $V(t,x)$ twice continuously differentiable with respect to $x \in U$ and once continuously differentiable with respect to $t \in I$ everywhere except possibly at the point $x = 0$, p. 157

$\mathbf{C}$: class of functions $V(t,x)$ absolutely continuous in t and satisfying a local Lipschitz condition, p. 7

$\mathbf{C}_0$: class of functions $V(t,x) \in \mathbf{C}$ satisfying a global Lipschitz condition, p. 7

$\mathfrak{A}$: σ-algebra of Borel sets in the initial probability space, p. 1

$\mathfrak{B}$: σ-algebra of Borel sets in euclidean space, p. 50

$V_R = \inf_{t \geqslant t_0, x \geqslant R} V(t,x)$, p. 8

$V^{(\delta)} = \sup_{t \geqslant t_0, |x| < \delta} V(t,x)$, p. 31

CHAPTER I

BOUNDEDNESS IN PROBABILITY AND STABILITY OF STOCHASTIC PROCESSES DEFINED BY DIFFERENTIAL EQUATIONS

1. Brief review of prerequisites from probability theory

Let $\Omega = \{\omega\}$ be a space with a distinguished family of subsets $\mathfrak{A}$ such that, for any finite or countable sequence of sets $A_i \in \mathfrak{A}$, the intersection $\bigcap_i A_i$, union $\bigcup_i A_i$ and complement $\bar{A}_i$ (with respect to Ω) are also in $\mathfrak{A}$. Suppose moreover that $\Omega \in \mathfrak{A}$. A family of subsets possessing these properties is known as a σ-algebra. If a probability measure $\mathbf{P}$ is defined on the σ-algebra $\mathfrak{A}$ (i.e. $\mathbf{P}$ is a non-negative countably additive set function on $\mathfrak{A}$ such that $\mathbf{P}(\Omega) = 1$), then the triple $(\Omega, \mathfrak{A}, \mathbf{P})$ is called a probability space and the sets in $\mathfrak{A}$ are called random events. (For more details, see Doob [3], Dynkin [3], Loève [1].)

The following standard properties of measures will be used without any further reference:

1. If $A \in \mathfrak{A}$, $B \in \mathfrak{A}$, $A \subset B$, then $\mathbf{P}(A) \leqslant \mathbf{P}(B)$.

2. For any finite or countable sequence A_n in $\mathfrak{A}$,

$$\mathbf{P}\left(\bigcup_n A_n\right) \leqslant \sum_n \mathbf{P}(A_n).$$

3. If $A_n \in \mathfrak{A}$ and $A_1 \subset A_2 \subset \dots \subset A_n \subset A_{n+1} \subset \dots$, then

$$\mathbf{P}\left(\bigcup_n A_n\right) = \lim_{n\to\infty} \mathbf{P}(A_n).$$

4. If $A_n \in \mathfrak{A}$ and $A_1 \supset A_2 \supset A_3 \supset \dots \supset A_n \supset \dots$, then

$$\mathbf{P}\left\{\bigcap_n A_n\right\} = \lim_{n\to\infty} \mathbf{P}(A_n).$$

Proofs of these properties may be found in texts on measure theory, such as Gnedenko [1], §8; or Gihman and Skorohod, pp. 69-70.

A random variable is a function $\xi(\omega)$ on Ω which is $\mathfrak{A}$ measurable and almost everyhere finite[1]. In this book we shall consider only random variables which take on values in euclidean ℓ-space E_ℓ i.e., such that $\xi(\omega) = (\xi_1(\omega),\ldots,\xi_\ell(\omega))$ is a vector in E_ℓ $(\ell = 1,2,\ldots)$. A vector-valued random variable $\xi(\omega)$ may be defined by its joint distribution function $F(x_1,\ldots,x_\ell)$, that is, by specifying the probability of the event $\{\xi_1(\omega) < x_1;,,,;\xi_\ell(\omega) < x_\ell\}$. Given any vector $x \in E_\ell$ or a $k \times \ell$ matrix $\sigma = ((\sigma_{ij}))$ $(i = 1,\ldots,k; j = 1,\ldots,\ell)$ we shall denote, as usual,

$$|x| = (x_1^2 + \ldots + x_\ell^2)^{\frac{1}{2}}, \qquad \|\sigma\| = \left(\sum_{i=1}^{k}\sum_{j=1}^{\ell}\sigma_{ij}^2\right)^{\frac{1}{2}}.$$

Then we have the well-known inequalities $|\sigma x| \leqslant \|\sigma\|\,|x|$, $\|\sigma_1\sigma_2\| \leqslant \|\sigma_1\|\,\|\sigma_2\|$.

The expectation of a random variable $\xi(\omega)$ is defined to be the integral

$$\mathbf{E}\xi = \int_\Omega \xi(\omega)P(d\omega),$$

provided the funtion $|\xi(\omega)|$ is integrable.

Let $\mathfrak{B}$ be a σ-algebra of Borel subsets of a closed interval $[s_0,s_1] = T$, $\mathfrak{B}\times\mathfrak{A}$ the minimal σ-algebra of subsets of $T\times\Omega$ containing all subsets of the type $\{t \in \Delta, \omega \in A\}$, where $\Delta \in \mathfrak{B}$, $A \in \mathfrak{A}$. A function $\xi(t,\omega) \in E_\ell$ is called a measurable stochastic process (random function) defined on T with values in E_ℓ if it is $\mathfrak{B}\times\mathfrak{A}$-measurable and $\xi(t,\omega)$ is a random variable for each $t \in T$. For fixed ω, we shall call the function $\xi(t,\omega)$ a trajectory or sample function of the stochastic process. In the sequel we shall consider only separable stochastic processes, i.e., processes whose behavior for all $t \in T$ is determined up to an event of probability zero by its behavior on some dense subset $\Lambda \in T$. To be precise, a process $\xi(t,\omega)$ is said to be separable if, for some countable dense subset Λ of T, there exists an event A of probability 0 such that for each closed subset $C \subset E_\ell$ and each open subset $\Delta \subset T$ the event

$$\{\xi(t_j,\omega) \in C; t_j \in \Lambda \cap \Delta\}$$

implies the event

$$A \cup \{\xi(t,\omega) \in C; t \in \Delta\}.$$

A process $\xi(t,\omega)$ is stochastically continuous at a point $s \in T$ if for each $\varepsilon > 0$

---

[1] Sometimes (see Chapter III), but only when this is explicitly mentioned, we shall find it convenient to consider random variables which can take on the values $\pm\infty$ with positive probability.

$$\lim_{t \to s} \mathbf{P}\{|\xi(t,\omega) - \xi(s,\omega)| > \varepsilon\} = 0.$$

The definitions of right and left stochastic continuity are analogous.

It can be proved (see Doob [3], Chapter II, Theorem 2.6) that for each process $\xi(t,\omega)$ which is stochastically continuous throughout T, except for a countable subset of T, there exists a separable measurable process $\tilde{\xi}(t,\omega)$ such that for every $t \in T$

$$\mathbf{P}\{\xi(t,\omega) = \tilde{\xi}(t,\omega)\} = 1.$$

If $\xi(t,\omega)$ is a measurable stochastic process, then for fixed ω the function $\xi(t,\omega)$, as a function of t, is almost surely[2] Lebesgue-measurable. If, moreover, $\mathbf{E}\xi(t,\omega) = m(t)$ exists, then $m(t)$ is Lebesgue-measurable, and the inequality

$$\int_A \mathbf{E}|\xi(t,\omega)|dt < \infty$$

implies that the process $\xi(t,\omega)$ is almost surely integrable over A (Doob [3], Chapter II, Theorem 2.7).

On the σ-algebra $\mathfrak{B} \times \mathfrak{U}$ there is defined the direct product $\mu \times \mathbf{P}$ of the Lebesgue measure μ and the probability measure $\mathbf{P}$. If some relation holds for $(t,\omega) \in A$ and $\mu \times \mathbf{P}(\bar{A}) = 0$, the relation will be said to hold for almost all t,ω. Let $A_1,\ldots,A_n$ be Borel sets in E_ℓ, and $t_1,\ldots,t_n \in T$; the probabilities

$$\mathbf{P}(t_1,\ldots,t_n,A_1,\ldots,A_n) = \mathbf{P}\{\xi(t_1,\omega) \in A_1,\ldots,\xi(t_n,\omega) \in A_n\}$$

are the values of the n-dimensional distributions of the process $\xi(t,\omega)$. Kolmogorov has shown that any compatible family of distributions $\mathbf{P}(t_1,\ldots,t_n,A_1,\ldots,A_n)$ is the family of the finite-dimensional distributions of some stochastic process.

The following theorem of Kolmogorov will play an important role in the sequel.

THEOREM 1.1. *If α, β, k are positive numbers such that whenever $t_1, t_2 \in T$,*

$$\mathbf{E}|\xi(t_2,\omega) - \xi(t_1,\omega)|^\alpha < k|t_1 - t_2|^{1+\beta}$$

and $\xi(t,\omega)$ is separable, then the process $\xi(t,\omega)$ has continuous sample functions.

Let $\xi(t,\omega)$ be a stochastic process defined for $t \geqslant t_0$. The process is said to satisfy the law of large numbers if for each $\varepsilon > 0$, $\delta > 0$ there exists a $T > 0$ such that for all $t > T$

---

[2] [We shall generally use the phrases "almost sure", "almost surely" instead of "with probability one" — Trans.].

$$\mathbf{P}\left\{\left|\frac{1}{t}\int_{t_0}^{t_0+t}\xi(s,\omega)ds - \frac{1}{t}\int_{t_0}^{t_0+t}\mathbf{E}\xi(s,\omega)ds\right| > \delta\right\} < \varepsilon. \tag{1.1}$$

A stochastic process $\xi(t,\omega)$ satisfies the *strong* law of large numbers if

$$\mathbf{P}\left\{\frac{1}{t}\int_{t_0}^{t_0+t}\xi(s,\omega)ds - \frac{1}{t}\int_{t_0}^{t_0+t}\mathbf{E}\xi(s,\omega)ds \xrightarrow[t\to\infty]{} 0\right\} = 1. \tag{1.2}$$

The most important characteristics of a stochastic process are its expectation $m(t) = \mathbf{E}\xi(t,\omega)$ and covariance matrix

$$K(s,t) = \operatorname{cov}(\xi(s),\xi(t))$$
$$= ((\mathbf{E}[(\xi_i(s) - m_i(s))(\xi_j(t) - m_j(t))])).$$

In particular, all the finite-dimensional distributions of a Gaussian process can be reconstructed from the function $m(t)$ and $K(s,t)$. A Gaussian process is stationary if

$$m(t) = \text{const}, \qquad K(s,t) = K(t - s). \tag{1.3}$$

A stochastic process $\xi(t,\omega)$ satisfying condition (1.3) is said to be stationary in the wide sense. The Fourier transform of the matrix $K(\tau)$ is called the spectral density of the process $\xi(t,\omega)$. It is clear that the spectral density $f(\lambda)$ exists and is bounded if the function $\|K(\tau)\|$ is absolutely integrable.

2. Dissipative systems of differential equations

In this section we prove some theorems from the theory of differential equations that we shall need later. We begin with a few definitions.

Let I_T denote the set $0 < t < T$, $I = I_\infty$, $E = E_\ell \times I$; U_R the ball $|x| < R$ and $\overline{U}_R$ its complement in E_ℓ. If $f(t)$ is a function defined on I, we write $f \in \mathbf{L}$ if $f(t)$ is absolutely integrable over every finite interval. The same notation $f \in \mathbf{L}$ will be retained for a stochastic function $f(t,\omega)$ which is almost surely absolutely integrable over every finite interval.

Let $F(x,t) = (F_1(x,t),\ldots,F_\ell(x,t))$ be a Borel-measurable function defined for $(x,t) \in E$. Let us assume that for each $R > 0$ there exist functions $M_R(t) \in \mathbf{L}$ and $B_R(t) \in \mathbf{L}$ such that

$$|F(x,t)| \leqslant M_R(t) \tag{2.1}$$

$$|F(x_2,t) - F(x_1,t)| \leqslant B_R(t)|x_2 - x_1| \tag{2.2}$$

for $x, x_i \in U_R$.

We shall say that a function $x(t)$ is a solution of the equation

$$\frac{dx}{dt} = F(x,t), \tag{2.3}$$

satisfying the initial condition

$$x(t_0) = x_0 \qquad (t_0 \geq 0) \tag{2.4}$$

on the interval $[t_0,t_1]$, if for all $t \in [t_0,t_1]$

$$x(t) = x_0 + \int_{t_0}^{t} F(x(s),s)ds. \tag{2.5}$$

In cases where solutions are being considered under varying initial conditions, we shall denote this solution by $x(t,x_0,t_0)$.

The function $x(t)$ is evidently absolutely continuous, and at all points of continuity of $F(x,t)$ it also satisfies equation (2.3).

THEOREM 2.1 *If conditions* (2.1) *and* (2.2) *are satisfied, then the solution* $x(t)$ *of problem* (2.3),(2.4) *exists and is unique in some neighbourhood of* t_0. *Suppose moreover that for every solution* $x(t)$ *(if a solution exists) and some function* τ_R *which tends to infinity as* $R \to \infty$, *we have the following "a priori estimate":*

$$\inf\{t : t \geq t_0; |x(t)| > R\} \geq \tau_R. \tag{2.6}$$

Then the solution of the problem (2.3),(2.4) *exists and is unique for all* $t \geq t_0$ *(i.e., the solution can be indefinitely continued*[3] *for* $t \geq t_0$*).*

PROOF. We may assume without loss of generality that the function $M_R(t)$ in (2.1) satisfies the inequality

$$|M_R(t)| > 1. \tag{2.7}$$

Therefore we can find numbers R and $t_1 > t_0$ such that $|x_0| \leq R/2$ and

$$\Phi(t_0,t_1) = \int_{t_0}^{t_1} M_R(s)ds \exp\left\{\int_{t_0}^{t_1} B_R(s)ds\right\} = \frac{R}{2}. \tag{2.8}$$

Applying the method of successive approximations to equation (2.5) on the interval $[t_0,t_1]$,

[3] [LaSalle and Lefschetz [1] use the term "defined in the future". — Trans.].

$$x^{(n+1)}(t) = x_0 + \int_{t_0}^{t} F(x^{(n)}(s),s)ds, \qquad x^0(t) \equiv x_0,$$

and using (2.1),(2.2) and (2.8), we get the estimates

$$|x^{(1)}(t) - x_0| \leq \int_t^t M_R(s)ds \leq \frac{R}{2},$$

$$|x^{(n+1)}(t) - x^{(n)}(t)| \leq \int_{t_0}^{t} B_R(s)|x^{(n)}(s) - x^{(n-1)}(s)|ds.$$

Together with (2.8), these imply the inequality

$$|x^{(n+1)}(t) - x^{(n)}(t)| \leq \int_{t_0}^{t} M_R(s)ds \frac{\left[\int_{t_0}^{t} B_R(s)ds\right]^n}{n!}. \tag{2.9}$$

It follows from (2.9) that $\lim_{n\to\infty} x^{(n)}(t)$ exists and that it satisfies equation (2.5). The proof of uniqueness is similar.

Now consider an arbitrary $T > t_0$ and choose R so that, besides the relations $|x_0| < R/2$ and (2.8), we also have $\tau_{R/2} > T$. Then by (2.6), it follows that $x(t_1) \leq R/2$ and thus the solutions can be continued to a point t_2 such that $\Phi(t_1,t_2) = R/2$. Repeating this procedure, we get $t_n \geq T$ for some n, since the functions $M_R(t)$ and $L_R(t)$ are integrable over every finite interval. This completes the proof.

If the function $M_R(t)$ is independent of t and its rate of increase in R is at most linear, i.e.,

$$|F(x,t)| \leq c_1|x| + c_2, \tag{2.10}$$

we get the following estimate for the solution of problem (2.3), (2.4), valid for $t \geq t_0$ and some $c_3 > 0$:

$$|x(t)| \leq |x_0|c_3e^{c_1(t-t_0)}.$$

We omit the proof now, since we shall later prove a more general theorem. But if condition (2.10) fails to hold, the solution will generally "escape to infinity" in a finite time. (As for example, the solution $x = (1 - t)^{-1}$ of the problem $dx/dt = x^2$, $x(0) = 1$). Since condition (2.10) fails to cover many cases of practical importance, we shall need a more general condition implying that the solution can be indefinitely continued. We present first some definitions.

The Ljapunov operator associated with equation (2.3) is the

operator d^0/dt defined by

$$\frac{d^0 V(x,t)}{dt} = \overline{\lim_{h \to +0}} \frac{1}{h}[V(x(t+h,x,t),t+h) - V(x,t)]. \quad (2.11)$$

It is obvious that if $V(x,t)$ is continuously differentiable with respect to x and t, then for almost all t the action of the Ljapunov operator

$$\frac{d^0 V}{dt} = \frac{\partial V}{\partial t} + \sum_{i=1}^{\ell} \frac{\partial V}{\partial x_i} F_i(x,t) = \frac{\partial V}{\partial t} + \left(\frac{\partial}{\partial x} V, F\right) \quad (2.12)$$

is simply a differentiation of the function V along the trajectory of the system (2.3).

In his classical work [1], Ljapunov discussed the stability of systems of differential equations by considering non-negative functions for which d^0V/dt satisfies certain inequalities.

These functions will be called Ljapunov functions.

In §§5,6,8, and also in Chapters V to VII we shall apply Ljapunov's ideas to stability problems for random perturbations.

In this section and the next we shall use Ljapunov functions to find conditions under which the solution can be indefinitely continued for $t > 0$ to a bounded solution. All Ljapunov functions figuring in the discussion will be henceforth assumed to be absolutely continuous in t, uniformly in x in the neighbourhood of every point. Moreover we shall assume a Lipschitz condition with respect to x:

$$|V(x_2,t) - V(x_1,t)| < B|x_2 - x_1| \quad (2.13)$$

in the domain $U_R \times I_T$, with a Lipschitz constant which generally depends on R and T. As an abbreviation for this situation we shall write $V \in \mathbf{C}$. If the function V satisfies condition (2.13) with a constant B not depending on R and T, we shall write $V \in \mathbf{C}_0$.

If $V \in \mathbf{C}$ and the function $y(t)$ is absolutely continuous, then it is easily verified that the function $V(y(t),t)$ is also absolutely continuous. Hence, for almost all t,

$$\frac{d^0 V(x,t)}{dt} = \frac{d}{dt} V(x(t),t)\big|_{x(t)=x},$$

where $x(t)$ is the solution of equation (2.3). We shall use this fact frequently without further reference.

THEOREM 2.2[4]. *Assume that there exists a Ljapunov function $V \in \mathbf{C}$ defined on the domain $E_\ell \times \{t > t_0\}$ such that for some $c_1 > 0$*

[4] General conditions for every solution to be indefinitely continuable have been obtained by Okamura and are described in LaSalle and Lefschetz [1]. These results imply Theorem 2.2.

$$V_R = \inf_{(x,t)\in \overline{U}_R \times \{t>t_0\}} V(x,t) \to \infty \quad as \quad R \to \infty, \tag{2.14}$$

$$\frac{d^0 V}{dt} \leqslant c_1 V, \tag{2.15}$$

and let the function F satisfy conditions (2.1), (2.2).

Then the solution of problem (2.3), (2.4) *can be indefinitely continued for $t \geqslant t_0$.*

The proof of this theorem employs the following well-known lemma, which will be used repeatedly.

LEMMA 2.1. *Let the function $y(t)$ be absolutely continuous for $t \geqslant t_0$, and let the derivative dy/dt satisfy the inequality*

$$\frac{dy}{dt} < A(t)y + B(t) \tag{2.16}$$

for almost all $t \geqslant t_0$, where $A(t)$ and $B(t)$ are almost everywhere continuous functions integrable over every finite interval. Then for $t > t_0$

$$y(t) < y(t_0)\exp\left\{\int_{t_0}^{t} A(s)ds\right\} + \int_{t_0}^{t} \exp\left\{\int_{s}^{t} A(u)du\right\}B(s)ds. \tag{2.17}$$

PROOF. It follows from (2.16) that for almost all $t \geqslant t_0$

$$\frac{d}{dt}\left[y(t)\exp\left\{-\int_{t_0}^{t} A(s)ds\right\}\right] < B(t)\exp\left\{-\int_{t_0}^{t} A(s)ds\right\}.$$

Integration of this inequality yields (2.17).

PROOF OF THEOREM 2.2. It follows from (2.15) that for almost all t we have $dV(x(t),t)/dt \leqslant c_1 V(x(t),t)$. Hence, by Lemma 2.1, it follows that for $t > t_0$

$$V(x(t),t) \leqslant V(x_0,t_0)\exp\{c_1(t - t_0)\}.$$

If τ_R denotes a solution of the equation

$$V(x_0,t_0)\exp\{c_1(\tau_R - t_0)\} = V_R,$$

then condition (2.6) is obviously satisfied. Thus all assumptions of Theorem 2.1 are now satisfied. This completes the proof.

Let us now consider conditions under which the solutions of equation (2.3) are bounded for $t > 0$. There exist in the literature various definitions of boundedness. We shall adopt here only one which is most suitable for our purposes, referring the reader for more details to Yoshizawa [1], LaSalle and Lefschetz [1], and Demidovič [1].

The system (2.3) is said to be dissipative for $t > 0$ if there exists a positive number $R > 0$ such that for each $r > 0$, beginning from some time $T(r,t_0) \geq t_0$, the solution $x(t,x_0,t_0)$ of problem (2.3),(2.4), $x_0 \in U_r$, $t_0 > 0$, lies in the domain U_R. (Yoshizawa [1] calls the solutions of such a system equi-ultimately bounded.)

THEOREM 2.3[5]. *A sufficient condition for the system* (2.3) *to be dissipative is that there exist a nonnegative Ljapunov function* $V(x,t) \in \mathbf{C}$ *on* E *with the properties*

$$V_R = \inf_{(x,t)\in\overline{U}_R\times I} V(x,t) \longrightarrow \infty, \qquad as\ R \to \infty, \tag{2.18}$$

$$\frac{d^0V}{dt} < -cV \qquad (c = \text{const} > 0). \tag{2.19}$$

PROOF. It follows from Lemma 2.1 and from (2.19) that for $t > t_0$, $x_0 \in U_r$,

$$V(x(t),t) \leq V(x_0,t_0)e^{-c(t-t_0)} \leq e^{-c(t-t_0)} \sup_{|x_0|<r} V(x_0,t_0).$$

Therefore $V(x(t),t) < 1$ for $t > T(t_0,r)$. This inequality and (2.18) imply the statement of the theorem.

REMARK 1. The converse theorem is also valid: Yoshizawa [1] proves that for each system which is dissipative in the above sense there exists a nonnegative function V with properties (2.18), (2.19), provided $F(x,t)$ satisfies a Lipschitz condition in every bounded subset of E.

REMARK 2. It is easy to show that the conclusion of Theorem 2.3 remains valid if it is merely assumed that (2.19) holds in a domain $\overline{U}_R$ for some $R > 0$, and in the domain U_R the functions V and d^0V/dt are bounded above. To prove this, it is enough to apply Lemma 2.1 to the inequality

$$\frac{d^0V}{dt} < -cV + c_1,$$

which is valid under the above assumptions for some positive constant c_1 and for $(x,t) \in E$.

In the sequel we shall need a certain frequently used estimate; its proof may be found, e.g., in Bellman [1].

GRONWALL-BELLMAN LEMMA. *Let* $u(t)$ *and* $v(t)$ *be nonnegative functions and let* k *be a positive constant such that for* $t \geq s$

---

[5] See Yoshizawa [1].

$$u(t) \leqslant k + \int_s^t u(t_1)v(t_1)dt_1.$$

Then for $t \geqslant s$

$$u(t) \leqslant k\exp\left\{\int_s^t v(t_1)dt_1\right\}.$$

3. Stochastic processes as solutions of differential equations

Let $\xi(t,\omega)$ $(t \geqslant 0)$ be a separable measurable stochastic process with values in E_k, and let $G(x,t,z)$ $(x \in E_\ell,\ t \geqslant 0,\ z \in E_k)$ be a Borel-measurable function of (x,t,z) satisfying the following conditions:

1. There exists a stochastic process $B(t,\omega) \in \mathbf{L}$ such that for all $x_i \in E_\ell$

$$|G(x_2,t,\xi(t,\omega)) - G(x_1,t,\xi(t,\omega))| \leqslant B(t,\omega)|x_1 - x_2|. \tag{3.1}$$

2. The process $G(0,t,\xi(t,\omega))$ is in $\mathbf{L}$, i.e., for every $T > 0$,

$$\mathbf{P}\left\{\int_0^T |G(0,t,\xi(t,\omega))|dt < \infty\right\} = 1. \tag{3.2}$$

We shall show presently that under these assumptions the equation

$$\frac{dx}{dt} = G(x,t,\xi(t,\omega)) \tag{3.3}$$

with initial condition

$$x(t_0) = x_0(\omega) \tag{3.4}$$

determines a new stochastic process in E_ℓ for $t \geqslant t_0$.

THEOREM 3.1. *If conditions* (3.1) *and* (3.2) *are satisfied, then problem* (3.3), (3.4) *has a unique solution* $x(t,\omega)$, *determining a stochastic process which is almost surely absolutely continuous for all* $t \geqslant t_0$. *For each* $t \geqslant t_0$, *this solution admits the estimate*

$$|x(t,\omega) - x_0(\omega)|$$
$$\leq \int_{t_0}^{t} |G(x_0(\omega),s,\xi(s,\omega))|ds \exp\left\{\int_{t_0}^{t} B(s,\omega)ds\right\}. \quad (3.5)$$

The proof is analogous to that of Theorem 2.1.

EXAMPLE. Consider the linear system

$$\frac{dx}{dt} = A(t,\omega)x + b(t,\omega), \qquad x(0) = x_0(\omega).$$

If $\|A(t,\omega)\|$, $|b(t,\omega)| \in L$, then it follows from Theorem 3.1 that this system has a solution which is a continuous stochastic process for all $t > 0$.

The global Lipschitz condition (3.1) fails to hold in many important applications. Most frequently the following local Lipschitz condition holds: For each $R > 0$, there exists a stochastic process $B_R(t,\omega) \in L$ such that if $x_i \in U_R$, then

$$|G(x_2,t,\xi(t,\omega)) - G(x_1,t,\xi(t,\omega))|$$
$$\leq B_R(t,\omega)|x_2 - x_1|. \quad (3.6)$$

As we have already noted in Section 2, condition (3.6) does not prevent the sample function escaping to infinity in a finite time, even in the deterministic case. However, we have the following theorem which is a direct corollary of Theorem 2.1.

THEOREM 3.2. *Let $\tau(R,\omega)$ be a family of random variables such that $\tau(R,\omega) \uparrow \infty$ almost surely as $R \to \infty$. Suppose that these random variables satisfy almost surely for each solution $x(t,\omega)$ of problem* (3.3), (3.4) *(if a solution exists) the following inequality*

$$\inf\{t: |x(t,\omega)| \geq R\} \geq \tau(R,\omega). \quad (3.7)$$

Assume moreover that conditions (3.2) *and* (3.6) *are satisfied. Then the solution of problem* (3.3), (3.4) *is almost surely unique and it determines an absolutely continuous stochastic process for $t \geq t_0$ (indefinitely continuable for $t \geq t_0$).*

Assume that the function G in equation (3.3) depends linearly on the third variable, i.e.,

$$\frac{dx}{dt} = F(x,t) + \sigma(x,t)\xi(t,\omega). \quad (3.8)$$

(Here σ is a $k \times \ell$ matrix, ξ a vector in E_k and k a positive integer.) Then the solution of equation (3.8) can be indefinitely continued if there exists a Ljapunov function of the truncated system

$$\frac{dx}{dt} = F(x,t). \tag{3.9}$$

Let us use $d^{(1)}/dt$ to denote the Ljapunov operator of the system (3.8), retaining the notation d^0/dt for the Ljapunov operator of the system (3.9).

THEOREM 3.3. *Let* $\xi(t,\omega) \in \mathbf{L}$ *be a stochastic process,* F *a vector and* σ *a matrix satisfying the local Lipschitz condition* (2.13), *where* $F(0,t) \in \mathbf{L}$ *and*

$$\sup_{E_\ell \times \{t>t_0\}} \|\sigma(x,t)\| < c_2. \tag{3.10}$$

Assume that a Ljapunov function $V(x,t) \in \mathbf{C}_0$ *of the system* (3.9) *exists with*

$$V_R = \inf_{\overline{U}_R \times \{t>t_0\}} V(x,t) \to \infty \quad as \quad R \to \infty, \tag{3.11}$$

$$\frac{d^0 V}{dt} < c_1 V. \tag{3.12}$$

Then the solution of problem (3.8), (3.4) *exists and determines an absolutely continuous stochastic process for all* $t \geqslant t_0$.

To prove this theorem we need the following lemma.

LEMMA 3.1. *If* $V(x,t) \in \mathbf{C}_0$, *then for almost all* t *the following relation holds almost surely:*

$$\frac{d^{(1)}V(x,t)}{dt} \leqslant \frac{d^0 V(x,t)}{dt} + B\|\sigma(x,t)\|\,|\xi(t,\omega)|, \tag{3.13}$$

where B *is the constant in condition* (2.13).

PROOF. It can be easily verified that the difference $x(t + h,\omega,x,t) - x(t + h,x,t)$ between solutions of equations (3.8) and (3.9) with the initial condition $x(t) = x$, satisfies for almost all t,ω the inequality

$$|x(t + h,\omega,x,t) - x(t + h,x,t)| \leqslant h\|\sigma(x,t)\|\,|\xi(t,\omega)| + o(h) \qquad (h \to 0).$$

This inequality, together with (2.13), implies (3.13).

PROOF OF THEOREM 3.3. We shall show that the assumptions of Theorem 3.2 are satisfied. Since conditions (3.2) and (3.6) are obviously satisfied, it will suffice to prove (3.7). Let $x(t,\omega)$ be a solution of problem (3.8), (3.4). It follows from the assumptions of the theorem and from Lemma 3.1 that the function $V(x(t,\omega),t)$ is absolutely continuous, and for almost all t,ω

$$\frac{dV(x(t,\omega),t)}{dt} \leqslant \frac{d^0V(x(t,\omega),t)}{dt} + B\|\sigma(x(t,\omega),t)\|\,|\xi(t,\omega)|$$

$$\leqslant c_1V(x(t,\omega),t) + Bc_2\,\xi(t,\omega)\ .$$

Combining this with Lemma 2.1 we get that almost surely

$$V(x(t,\omega),t)$$

$$\leqslant e^{c_1(t-t_0)}\left[V(x_0(\omega),t_0) + Bc_2\int_{t_0}^{t}|\xi(s,\omega)|ds\right]. \qquad (3.14)$$

Let $\tau_R(\omega)$ denote a solution of the equation

$$e^{c_1(\tau_R-t_0)}\left[V(s_0(\omega),t_0) + Bc_2\int_{t_0}^{\tau_R}|\xi(s,\omega)|ds\right] = V_R. \qquad (3.15)$$

It now follows from the relation $\xi(t,\omega)\in\mathbf{L}$ and from (3.11) that $\tau_R\uparrow\infty$ almost surely as $R\to\infty$. (3.7) follows now from (3.14) and (3.15). Thus all assumptions of Theorem 3.2 are satisfied.

REMARK 1. If the relation $|\xi(t,\omega)|^{(1+\varepsilon)/\varepsilon}\in\mathbf{L}$ holds for some $\varepsilon > 0$, condition (3.10) can be slightly weakened and replaced by the condition

$$\|\sigma(x,t)\|^{1+\varepsilon} \leqslant c_3V(x,t). \qquad (3.16)$$

To prove this, we need only use Young's inequality

$$|ab| < \frac{|a|^p}{p} + \frac{|b|^q}{q} \qquad \left(\frac{1}{p} + \frac{1}{q} = 1,\ p,q > 0\right) \qquad (3.17)$$

for estimating $\|\sigma\|\,|\xi|$. In particular, if for each $T > 0$, there is a constant c such that the process $\xi(t,\omega)$satisfies the condition

$$\mathbf{P}\{\sup_{0\leqslant t\leqslant T}|\xi(t,\omega)| < c\} = 1,$$

then it is enough to require that inequality (3.16) holds for sufficiently small $\varepsilon > 0$.

REMARK 2. The conditions of Theorem 3.3 guarantee that the solutions of equation (3.8) are indefinitely continuable, uniformly in the following sense: For all initial conditions $x_0(\omega)$ which satisfy the relation

$$\mathbf{P}\{|x_0(\omega)| < K\} = 1 \qquad (3.18)$$

for some K, one can find a family of random variables $\tau(R,\omega)$ satisfying condition (3.7). Since

$$\mathbf{P}\{\max_{0\geqslant t\geqslant T} |x(t,\omega,x_0(\omega))| > R\} \leqslant \mathbf{P}\{\tau_R < T\},$$

this implies in particular that for every $\varepsilon > 0$, $T > 0$ and $K > 0$ there exists an $R > 0$ such that

$$\mathbf{P}\{\max_{0\geqslant t\geqslant T} |x(t,\omega,x_0(\omega))| > R\} > \varepsilon$$

for all $x_0(\omega)$ satisfying condition (3.18).

EXAMPLE 1. In the one-dimensional case with the Ljapunov function $V(x,t) = |x| + 1$ we get the following result. If $F \in \mathbf{C}$, $\sigma \in \mathbf{C}$, σ satisfies condition (3.10), while $\xi(t,\omega), F(0,t) \in \mathbf{L}$, then a sufficient condition for the solutions of problem (3.8), (3.4) to be indefinitely continuable is that $(\operatorname{sign} x)F(x,t) < c(|x| + 1)$ for some $c > 0$.

EXAMPLE 2. Consider the equation

$$x'' + f(x)x' + g(x) = \sigma(x,x')\xi(t,\omega). \tag{3.19}$$

This equation describes the process "at the output" of many radio systems driven by a stochastic process. In particular, for $f(x) = x^2 - 1$, $g(x) = x$ and $\sigma(x,x') = 1$, the process at the output is that of a system described by a van der Pol equation. Let the function $f(x)$ be bounded below and assume that

$$|\sigma(x,x')| < c_1, \qquad \left|\frac{g(x)}{x}\right| < c_2.$$

Then

$$V(x,y) = (x^2 + y^2)^{\frac{1}{2}} \in \mathbf{C}_0$$

is obviously a Ljapunov function for the system

$$\frac{dx}{dt} = y, \qquad \frac{dy}{dt} = - f(x)y - g(x).$$

Moreover V satisfies conditions (3.11) and (3.12). Applying Theorem 3.3, we see that the process in equation (3.19) exists for all $t \geqslant t_0$ provided that $\xi(t,\omega) \in \mathbf{L}$.

4. Boundedness in probability of stochastic processes defined by systems of differential equations

A stochastic process $\xi(t,\omega)$ $(t \geq 0)$ is said to be bounded in probability if the random variables $|\xi(t,\omega)|$ are bounded in probability uniformly in t, i.e.,

$$\sup_{t>0} \mathbf{P}\{|\xi(t,\omega)| > R\} \to 0 \quad \text{as} \quad R \to \infty.$$

We shall say that a random variable $x_0(\omega)$ is in the class A_{R_0} if

$$\mathbf{P}\{|x_0(\omega)| < R_0\} = 1. \tag{4.1}$$

The system (3.3) will be called dissipative if the random variables $|x(t,\omega,x_0,t_0)|$ are bounded in probability, uniformly in $t \geq t_0$ whenever $x_0(\omega) \in A_R$ for some $R > 0$.

It is readily seen that this definition agrees with that of a deterministic dissipative system (see Section 2).

THEOREM 4.1. *Let $V(x,t) \in \mathbf{C}_0$ be a non-negative Ljapunov function, defined on the domain E which satisfies condition* (3.11) *and the condition*

$$\frac{d^0V}{dt} \leq -c_1V \qquad (c_1 = \text{const} > 0). \tag{4.2}$$

Let F and σ satisfy a local Lipschitz condition (2.13), *and let σ also satisfy condition* (3.10).

Then the system (3.8) *is dissipative for every stochastic process $\xi(t,\omega)$ such that*

$$\sup_{t>0} \mathbf{E}|\xi(t,\omega)| < \infty. \tag{4.3}$$

Before proving this theorem, we shall prove a lemma which yields a convenient form of Čebyšev's inequality.

LEMMA 4.1. *Let $V(x,t)$ be a nonnegative function and $\eta(t,\omega)$ a stochastic process such that $\mathbf{E}V(\eta(t,\omega),t)$ exists.*

Then

$$\mathbf{P}\{|\eta(t,\omega)| > R\} \leq \frac{\mathbf{E}V(\eta(t,\omega),t)}{\inf_{\overline{U}_R\times\{s>t_0\}} V(x,s)}. \tag{4.4}$$

The proof follows from the following chain of inequalities:

$$V(\eta(t,\omega),t) \geqslant \int_{|\eta(t,\omega)|>R} V(\eta(t,\omega),t)\mathbf{P}(d\omega)$$

$$\geqslant \inf_{\overline{U}_R\times\{s>t_0\}} V(x,s)\mathbf{P}\{|\eta(t,\omega)| > R\}.$$

PROOF OF THEOREM 4.1. Let $x(t,\omega)$ be a solution of problem (3.8), (3.4). Then the function $V(x(t,\omega),t)$ is differentiable for almost all t,ω. By Lemma 3.1 and by (4.2),

$$\frac{dV(x(t,\omega),t)}{dt} \leqslant \frac{d^0V(x(t,\omega),t)}{dt} + Bc_2|\xi(t,\omega)|$$

$$\leqslant - c_1V(x(t,\omega),t) + Bc_2|\xi(t,\omega)|.$$

Hence, by Lemma 2.1,

$$V(x(t,\omega),t) \leqslant V(x_0(\omega),t_0)e^{c_1(t-t_0)}$$

$$+ Bc_2\int_{t_0}^{t} e^{c_1(s-t)}|\xi(s,\omega)|ds.$$

Calculating the expectation of both sides of this inequality and using (4.3), we see that the variable $\mathbf{E}V(x(t,\omega),t)$ is bounded uniformly for $t \geqslant t_0$ and for all $x_0(\omega)$ satisfying condition (4.1). Together with (4.4), this implies the theorem.

REMARK 1. It is clear from Remark 1 to Theorem 2.3 that the existence of a function V satisfying conditions (3.11), (4.2) is not only sufficient but also necessary for the system (3.8) to be dissipative for each stochastic process $\xi(t,\omega)$ satisfying (4.3).

REMARK 2. If for some $\varepsilon > 0$

$$\sup_{t>0} \mathbf{E}|\xi(t,\omega)|^{(1+\varepsilon)/\varepsilon} < \infty,$$

then using (3.17) one easily shows that one may replace condition (3.10) in the formulation of Theorem 4.1 by condition (3.16). Another modification of this theorem is obtained by requiring that condition (4.2) only holds in some U_R, where $R > 0$, and that V and d^0V/dt are bounded in the domain U_R (see Remark 2 to Theorem 2.3, and also Has'minskiĭ [5]).

REMARK 3. It is not hard to conclude from the assumptions of Theorem 4.1 that the growth of the function $V(x,t)$ is at most linear, i.e.,

$$V(x,t) < c_1|x| + c_2.$$

If moreover there exist positive constants c_3 and c_4 such that

$$V(x,t) > c_3|x| - c_4, \tag{4.5}$$

then it follows from the proof of the theorem that

$$\sup_{t>0} \mathbf{E}|x(t,\omega)| < \infty.$$

The following theorem generalizes this observation.

THEOREM 4.2. *Let the functions V, F and σ satisfy the assumptions of Theorem 4.1 and assume moreover that V satisfies also (4.5). Suppose further that for some $\alpha > 1$*

$$\sup_{t>0} \mathbf{E}|\xi(t,\omega)|^{\alpha} < \infty. \tag{4.6}$$

Then every solution $x(t,\omega)$ of problem (3.8), (3.4) satisfies the inequality

$$\sup_{t>0} \mathbf{E}|x(t,\omega)|^{\alpha} < \infty.$$

Moreover, there exist constants c and $T = T(R_0,t_0)$ such that for every initial condition $x_0(\omega)$ which satisfies the equality (4.1) for some R_0, the solution $x(t,\omega)$ satisfies for all $t > T(R_0,t_0)$ the following inequality:

$$\mathbf{E}|x(t,\omega)|^{\alpha} < c.$$

PROOF. Consider the Ljapunov function $W(x,t) = [V(x,t)]^{\alpha}$. The assumptions of the theorem, Lemma 3.1 and (3.17) imply that

$$\frac{dW(x(t,\omega),t)}{dt} \leqslant - c_5 W(x(t,\omega),t) + c_6 [V(x(t,\omega),t)]^{\alpha-1} |\xi(t,\omega)|$$

$$\leqslant - c_7 W(x(t,\omega),t) + c_8 |\xi(t,\omega)|^{\alpha}$$

for some $c_5 > 0,\ldots,c_8 > 0$. Further, as in the proof of Theorem 4.1, we see that

$$\sup_{t>0} \mathbf{E}W(x(t,\omega),t) < \infty.$$

The first part of the theorem follows now from this inequality and the inequality

$$W(x,t) \geqslant c_9 |x|^{\alpha} - c_{10},$$

which is a consequence of (4.5). The proof of the second part is analogous.

By considering various narrower classes of stochastic processes $\xi(t,\omega)$, we can derive various dissipativity conditions under less stringent restrictions on the Ljapunov functions. The following theorem is an example.

THEOREM 4.3. *Let the process* $\xi(t,\omega)$ *be such that for some* $c_1 > 0$, $c_2 > 0$, $A > 0$ *and all* $0 \leqslant s \leqslant t$

$$\mathrm{E}\exp\left\{c_1\int_s^t |\xi(u,\omega)|du\right\} \leqslant A\exp\{c_2(t - s)\}. \tag{4.7}$$

Assume that there exists a non-negative function $V(x,t) \in \mathbf{C}_0$ *defined on E, satisfying condition* (3.11) *and the conditions*

$$\sup_{t>0} V(0,t) < \infty, \qquad \frac{d^0V}{dt} < c,$$

$$\frac{d^0V}{dt} < - c_2 - \varepsilon \quad \text{for} \quad |x| > R_0 \quad \text{and some} \quad \varepsilon > 0,$$

$$\lim_{R\to\infty} \;\; \sup_{x_i \in \overline{U}_R, t \geqslant 0} \frac{|V(x_2,t) - V(x_1,t)|}{|x_2 - x_1|} = B_1.$$

Further let F and σ *satisfy condition* (2.13) *and the condition* $\|\sigma\| \leqslant K$, *where* $B_1K < c_1$. *Then the system* (3.8) *is dissipative.*

PROOF. Let $V(x,t)$ be a function satisfying the assumptions of our theorem. Assume moreover that $R > R_0$ is large enough, so that for $|x_i| > R$ we have

$$|V(x_2,t) - V(x_1,t)| < \frac{c_1}{K}|x_2 - x_1|.$$

Set $W(x,t) = \exp\{V(x,t)\}$. It follows from the assumptions of the theorem that for almost all $t \geqslant t_0$ and for (t,ω) such that $|x(t,\omega)| > R$ we have

$$\frac{dW(x(t,\omega),t)}{dt} \leqslant W\left[\frac{d^0V}{dt} + c_1|\xi(t,\omega)|\right]$$

$$\leqslant W[-(c_2 + \varepsilon) + c_1|\xi(t,\omega)|].$$

Since $V \in \mathbf{C}_0$ and both V and d^0V/dt are bounded for $|x| > R$, this implies that there exist constants c_3, c_4, such that the following estimate is valid for almost all t,ω:

$$\frac{dW(x(t,\omega),t)}{dt} \leqslant W[-(c_2 + \varepsilon) + c_1|\xi(t,\omega)|] + c_3 + c_4|\xi(t,\omega)|.$$

Applying Lemma 2.1, we see that almost surely

$$W(x(t,\omega),t) \leqslant W(x_0(\omega),t_0)\exp\left\{\int_{t_0}^{t} (-c_2 - \varepsilon + c_1|\xi(s,\omega)|)ds\right\}$$

$$+ \int_{t_0}^{t} \exp\left\{\int_{s}^{t} (-c_2 - \varepsilon + c_1|\xi(u,\omega)|)du\right\}$$

$$\times (c_3 + c_4|\xi(s,\omega)|)ds$$

$$\leqslant (W(x_0(\omega),t_0) + c_5)\exp\left\{\int_{t_0}^{t} (-c_2 + c_1|\xi(s,\omega)|)ds\right\}$$

$$+ c_6\int_{t_0}^{t} \exp\left\{\int_{s}^{t} (-c_2 - \varepsilon + c_1|\xi(u,\omega)|)du\right\}ds.$$

Calculating the expectation of each of the sides of this inequality and using (4.1) and (4.7), we easily see that $\mathbf{E}W(x(t,\omega),t) < c_7$. This together with (4.4) implies the assertion of Theorem 4.3.

The following example shows that the assertion of Theorem 4.1 fails to hold if we replace condition (4.3) in this theorem by the condition

$$\sup \mathbf{E}|\xi(t,\omega)|^{\alpha} < \infty \qquad (\alpha < 1).$$

EXAMPLE 1. Let $x(t,\omega)$ be a solution in E_ℓ of the problem

$$\frac{dx}{dt} = -x + \xi(t,\omega), \qquad x(1) = 0. \tag{4.8}$$

Define the stochastic process $\xi(t,\omega)$ by

$$\xi(t) = \begin{cases} 2^{k/a}\exp\{2^k - t + \gamma(2^k - \tau_k)\} & \text{for } t \in [\tau_k, \tau_k + 2^{-k}], \\ 0 \quad \text{otherwise.} \end{cases} \tag{4.9}$$

Here $\tau_1, \tau_2, \ldots$ are independent random variables such that τ_k is distributed on the interval $[2^{k-1}, 2^k - 2^{-k}]$ with density

$$p_k(s) = \lambda_k \exp\{-\gamma(2^k - s)\}, \tag{4.10}$$

where $\gamma > 0$ and λ_k is determined by the normalization requirement (it is clear that $\lambda_k \to \gamma$ as $k \to \infty$).

From (4.8)-(4.10) we readily get the estimate

$$x(2^k,\omega) = \int_1^{2^k} \exp(s - 2^k)\xi(s,\omega)ds$$

$$> \int_{2^{k-1}}^{2^k} \exp(s - 2^k)\xi(s,\omega)ds$$

$$= \int_{\tau_k}^{\tau_k+2^{k-1}} 2^{k/\alpha} e^{\gamma(2^k-\tau_k)} ds$$

$$\geqslant 2^{k\left(\frac{1}{\alpha}-1\right)} \to \infty \quad \text{as} \quad k \to \infty$$

which holds almost surely.

On the other hand, if $\gamma > \alpha/(1-\alpha)$, $2^{k-1} \leqslant t \leqslant 2^k$, then

$$\mathbf{E}|\xi(t,\omega)|^\alpha \leqslant \lambda_k \int_{t-2^{-k}}^{t} e^{-\gamma(2^k-s)} 2\, e^{\alpha(2^k-t)+\alpha\gamma(2^k-s)} ds$$

$$\leqslant \lambda_k e^{-(2^k-t)[\gamma(1+\alpha)-\alpha]} < \infty.$$

Thus, if (4.3) does not hold, we cannot assert that the system (3.8) is dissipative, even when the unperturbed system (3.9) is an asymptotically stable linear system.

The next example will show that condition (4.7) in Theorem 4.3 cannot be replaced by (4.3), or even by the stronger condition that $\mathbf{E}|\xi(t,\omega)| \to 0$ as $t \to \infty$.

EXAMPLE 2. Consider in E_ℓ the problem

$$\frac{dx}{dt} = -\operatorname{sign}x.\ln(|x| + 1) + \eta(t,\omega), \qquad x(0) = 0, \tag{4.11}$$

where $\eta(t,\omega)$ is a stochastic step process satisfying the conditions

$$\eta(t,\omega) = \begin{cases} 2(2^{n+1} - \tau_n)\ln(2^{n+1} - \tau_n) & \text{for } \tau_n < t < \tau_n + 1 \\ & (n = 2,3,4,\ldots), \\ 0 & \text{otherwise.} \end{cases}$$

Let $\tau_2,\ldots,\tau_n,\ldots$ be independent random variables, τ_n being distributed on the interval $[2^n, 2^{n+1} - 1]$ with density

$$p_n(s) = \frac{c_n}{(2^{n+1} - s)\ln(2^{n+1} - s)}, \quad c_n = (\ln\ln(2^n - 2) - \ln\ln 2)^{-1}.$$

It is clear that for $2^n \leqslant t \leqslant 2^{n+1}$ we have

$$\mathbf{E}\eta(t,\omega) = \mathbf{E}|\eta(t,\omega)|$$

$$\leqslant \int_{t-1}^{t} 2p_n(s)(2^{n+1} - s)\ln(2^{n+1} - s)ds = 2c_n$$

and therefore $\mathbf{E}|\eta(t,\omega)| \to 0$ as $t \to \infty$.

Let $\tilde{x}(t,\omega)$ denote the solution of equation (4.11) satisfying the initial condition $\tilde{x}(\tau_n,\omega) = 0$. Then, by the uniqueness of the solution of the Cauchy problem for equation (4.11) and by the definition of the process $\eta(t)$, we have the inequality

$$x(2^{n+1},\omega) \geqslant \tilde{x}(2^{n+1},\omega) = \int_{\tau_n}^{2^{n+1}} [\eta(t,\omega) - \ln(\tilde{x}(t,\omega) + 1)]dt$$

$$\geqslant 2(2^{n+1} - \tau_n)\ln(2^{n+1} - \tau_n) \tag{4.12}$$

$$- (2^{n+1} - \tau_n)\ln[2(2^{n+1} - \tau_n)\ln(2^{n+1} - \tau_n)].$$

By the definition of $p_n(s)$,

$$\mathbf{P}\{2^{n+1} - \tau_n > n\} > \frac{1}{2}.$$

Hence, using (4.12), we see that for sufficiently large n

$$\mathbf{P}\left\{x(2^{n+1},\omega) > \frac{n}{2}\ln n\right\} > \frac{1}{2}.$$

This means that the system (4.11) is non-dissipative.

It is readily seen that the Ljapunov function $V(x) = |x|$ for this system satisfies

$$\frac{d^0V}{dt} = - \ln(V + 1) \to - \infty \quad \text{as} \quad |x| \to \infty.$$

Using the same method, one easily constructs examples of non-dissipative systems satisfying all assumptions of Theorem 4.1 except (4.2), instead of which we have

$$\frac{d^0V}{dt} < - \Phi(V), \tag{4.13}$$

where $\Phi(V)$ is any function such that

$$\int_1^\infty \frac{dx}{x\Phi(x)} = \infty. \tag{4.14}$$

(Examples of this type were constructed by the author in [5].) It is as yet an open question whether condition (4.2) in Theorem 4.1 may be replaced by condition (4.13) with a function $\Phi(V)$ such that the integral (4.14) is convergent. We do not even know the answer to the following more specific question: Do there exist non-dissipative systems of the type

$$\frac{dx}{dt} = \begin{cases} -\operatorname{sign} x|x|^{\alpha} + \xi(t,\omega) & \text{for} \quad |x| > 1, \\ -x + \xi(t,\omega) & \text{for} \quad |x| \leqslant 1, \end{cases}$$

where $0 < \alpha < 1$ and $\xi(t,\omega)$ satisfies condition (4.3)?

We now apply the theorems of this section to one-dimensional systems.

EXAMPLE 3. Consider equation (3.8) in E_ℓ, and assume that $|\xi(t,\omega)| < c$ almost surely. Assume further that the necessary smoothness conditions hold and that $|\sigma| \leqslant k$. Set $V(x) = |x|$, $c_1 = k + \varepsilon$, $c_2 = c(k + \varepsilon)$. Condition (4.7) is obviously valid if the constants c_1, c_2 are chosen in this way. If moreover

$$\frac{d^0 V}{dt} = \operatorname{sign} x F(x,t) \leqslant -ck - \varepsilon_1 \qquad (|x| > R_0) \tag{4.15}$$

for some $\varepsilon_1 > 0$ and all sufficiently large $|x|$, then also all the other assumptions of Theorem 4.3 hold. Thus we may conclude from this theorem the following corollary.

COROLLARY. A sufficient condition for equation (3.8) to be dissipative in E_ℓ is the existence of positive constants c, k, ε_1 such that (4.15) and

$$\mathbf{P}\{|\xi(t,\omega)| \leqslant c\} = 1, \qquad |\sigma| \leqslant k$$

hold.

On the other hand, it is clear that if

$$F(x,t) > -ck$$

holds for all $x \in R_0$, then the equation is non-dissipative for $\sigma = k$, $\xi(t,\omega) \equiv c$.

EXAMPLE 4. Suppose that for some positive constant

$$\frac{F(x,t)}{x} < -c_1, \tag{4.16}$$

whenever $|x| > R_0$ and assume that the process $\xi(t,\omega)$ satisfies condition (4.3). Considering the Ljapunov function

$$V(x,t) = \begin{cases} |x| - R_0, & |x| > R_0, \\ 0 & |x| \leqslant R_0, \end{cases}$$

we see that all assumptions of Theorem 4.1 are satisfied. Thus, relations (4.16) and (4.3) are sufficient conditions for the one-dimensional system (3.8) to be dissipative.

Note that the above Ljapunov function satisfies inequality (4.5). Thus, applying Theorem 4.2, we get the following result: If condition (4.16) is satisfied and condition (4.6) holds for some $\alpha > 1$, then the solution $x(t,\omega)$ of problem (3.8), (3.4) in E_ℓ has a bounded α-th moment.

EXAMPLE 6[6]. Let us again consider the equation

$$x'' + f(x)x' + g(x) = \sigma(x,x')\xi(t,\omega). \tag{4.17}$$

Assume that the process $|\xi(t,\omega)|$ has a bounded expectation, that the coefficients σ are bounded from above and from below and that there exists an $x_0 > 0$ such that

$$0 < c_1 < \frac{g(x)}{x} < c_2, \qquad 0 < c_3 < f(x) < c_4 \tag{4.18}$$

for $|x| > x_0$. Then the process defined by the system (4.17) is dissipative. To prove this consider the system

$$x' = y, \qquad y' = -f(x)y - g(x) + \sigma(x,y)\xi(t,\omega) \tag{4.19}$$

which is equivalent to (4.17), and set

$$F(x) = \int_0^x f(t)dt, \qquad G(x) = \int_0^x g(t)dt,$$

$$W(x,y) = (F - \gamma x)y + G(x) + \int_0^x f(t)(F(t) - \gamma t)dt + 1 + \frac{y^2}{2},$$

$$V(x,y) = [W(x,y)]^\alpha - c \quad \text{for} \quad [W(x,y)]^\alpha > c,$$

$$V(x,y) = 0 \quad \text{for} \quad [W(x,y)]^\alpha \leqslant c.$$

Regarding the function W was a quadratic form in y and using (4.18), we easily see that for a certain $\gamma > 0$ we have that $W \to \infty$ as $r = (x^2 + y^2)^{\frac{1}{2}} \to \infty$. Next, we can choose an $\alpha > 0$ in such a way that $V(x,y) \in \mathbf{C}_0$. Using the equality

$$\frac{d^0 W}{dt} = -[\gamma y^2 + g(F - \gamma x)]$$

and (4.18) we see that, for sufficiently small $\gamma > 0$ and $\beta > 0$,

---

[6] The author's exposition of this example in [5] contains an error. The following corrected version is due to Nevel'son.

condition (4.2) holds whenever $r > r_0$. Hence it follows that for a suitable choice of c inequality (4.2) is valid for $V(x,y)$ everywhere. It now follows from Theorem 4.1 that our process is dissipative.

For the general system (3.3) one can prove the following result which is analogous to a theorem of Demidovič [1] for the deterministic case.

THEOREM 4.4. *Let the following conditions hold:*

1. $\mathbf{E}|G(0,t,\xi(t,\omega))| < c < \infty \qquad (t \geqslant t_0)$.

2. *There exists a symmetric positive definite matrix* D $((d_{ij}))$ *such that the Jacobian* $J(x,t,z) = \left\{\left[\frac{\partial G}{\partial x}(x,t,z)\right]\right\}$, *symmetrized by the matrix* D, *is negative definite uniformly in* x, t *and* z, *i.e., all roots of the symmetric matix* $DJ + J^*D$ *satisfy the inequality* $\lambda(x,z,t) < -\lambda_U < 0$. *Then the system* (3.3) *is dissipative.*

PROOF. Set $V(x) = (Dx,x)^{\frac{1}{2}}$. Obviously,

$$\frac{dV(X(t,\omega))}{dt} = (\text{grad}V,G) = \frac{(DG(X(t,\omega),t,\xi(t,\omega)),X(t,\omega))}{V(X(t,\omega))} .$$

It follows from the assumptions of the theorem and from the fundamental lemma in Demidovič [1] that

$$(DG(x,t,\zeta) - DG(0,t,z),x) < -\lambda_0(x,x,).$$

Thus we get the inequality

$$\frac{dV(X(t,\omega))}{dt} \leqslant -\lambda_0 \frac{(X(t,\omega),X(t,\omega))}{V(X(t,\omega))}$$

$$+ \frac{(DG(0,t,\xi(t,\omega),X(t,\omega)))}{V(X(t,\omega))}$$

$$\leqslant -c_1V + c_2\, G(0,t,\xi(t,\omega)) .$$

By Lemma 2.1, this inequality implies the desired conclusion.

5. Stability[7]

In this section we shall study conditions ensuring the stability of a particular solution $y = y(t,\omega)$ of the equation

---

[7] Almost sure stability has been considered by Kozin [1] and Caughey and Gray [1] for less general systems. Mean and mean square stability has been considered by Bertram and Sarachik [1], Malahov [1] and others.

$$\frac{dx}{dt} = G(x,t,\xi(t,\omega)). \tag{5.1}$$

Following the usual procedure of introducing new variables, equal to the deviations of the corresponding coordinates of the "perturbed" motion from their "unperturbed" values, we see that we only need to consider the stability of the solution $x(t) \equiv 0$ of an equation of type (5.1) in which the function G satisfies the condition

$$G(0,t,\xi(t,\omega)) \equiv 0. \tag{5.2}$$

Even in the deterministic case the concept of stability of the trivial solution $x(t) \equiv 0$ can be given various meanings. For example, one distinguishes between local stability and stability in the large, also between asymptotic and nonasymptotic stability. The diversity is even greater in the presence of "randomness". We shall not list here all the possible definitions, but we shall confine ourselves to those which are in our view of greatest practical interest. Accordingly, we introduce the following definitions.

The solution $x(t) \equiv 0$ is said to be

1. *(Weakly) stable in probability* (for $t \geqslant t_0$) if, for every $\varepsilon > 0$ and $\delta > 0$, there exists an $r \geqslant 0$ such that if $t > t_0$ and $|x_0| < r$, then

$$\mathbf{P}\{|x(t,\omega,t_0,x_0)| > \varepsilon\} < \delta. \tag{5.3}$$

2. *(Weakly) asymptotically stable in probability*[8] if it is stable in probability and, for each $\varepsilon > 0$, there exists an $r = r(\varepsilon)$ such that for $t \to \infty$

$$\mathbf{P}|\{x(t,\omega,t_0,x_0)| > \varepsilon\} \to 0, \quad \text{if} \quad |x_0| < r.$$

3. *p-stable*, if for each $\varepsilon > 0$, there exists an $r > 0$ such that

$$\mathbf{E}|x(t,\omega,t_0,x_0)|^p < \varepsilon \qquad (p > 0),$$

whenever $t \geqslant t_0$ and $|x_0| < r$.

4. *Asymptotically p-stable*, if it is p-stable and for sufficiently small values of

$$\mathbf{E}|x(t,\omega,t_0,x_0)|^p \to 0 \quad \text{as} \quad t \to \infty.$$

---

[8] Throughout this chapter we shall consider stability and asymptotic stability in the weak sense (compare Chapter V, where stability in the strong sense will be discussed).

5. *Stable in probability in the large* if it is stable in probability and if furthermore for every x, $\varepsilon > 0$ and $\delta > 0$, there exists a $T = T(x_0,\varepsilon,\delta)$ such that (5.3) is valid for all $t > T$. A similar definition obtains for asymptotic stability in probability and p-stability in the large.

6. *Exponentially p-stable* if there exist constants $A > 0$ and $\alpha > 0$ such that

$$\mathbf{E}|x(t,\omega,x_0,t_0)|^p \leqslant A|x_0|^p \exp\{-\alpha(t - t_0)\}.$$

7. *Almost surely stable in any of the above senses* if almost all sample functions i.e. all, except those from some set of probability 0, are stable in the appropriate sense.

It follows from Čebyšev's inequality that (asymptotic) p-stability of the trivial solution for any value of p implies its (asymptotic) p-stability for every smaller value of p and stability in probability. On the other hand, one can easily show by an example that a solution could be (asymptotically) p-stable for some p and not (asymptotically) p-stable for $p_1 > p$ (see below, Section 6).

The case most often discussed in the literature is asymptotic p-stability for $p = 2$. Henceforth we shall refer to it as mean square stability.

Unless certain restrictive assumptions are made concerning a given system, it is not likely that non-trivial and effective stability conditions can be found. For example, in Bertram and Sarachik [1] stability conditions are given in terms of a Ljapunov function $V(x,t) \geqslant 0$ such that $\mathbf{E}\dot{V}(x,t) < 0$ where $\dot{V}$ denotes the derivative in the sense of (5.1)). However, in order to calculate the expectation $\mathbf{E}V(x,t)$ one must solve the system (5.1) with a suitable initial condition, and this limits the practical use of the criterion.

Here we shall limit ourselves to stability conditions for systems of the type

$$\left.\begin{aligned} &\frac{dx}{dt} = F(x,t) + \sigma(x,t)\xi(t,\omega), \\ &F(0,t) \equiv 0, \qquad \sigma(0,t) \equiv 0 \end{aligned}\right\}. \tag{5.4}$$

Sufficient conditions for stability will be given in terms of the existence of a Ljapunov function for the truncated system

$$\frac{dx}{dt} = F(x,t). \tag{5.5}$$

We shall assume throughout this section that all Ljapunov functions under consideration are positive definite uniformly in t, i.e.,

$$\inf_{t>0,|x|>r} V(x,t) = V_r > 0 \quad \text{for} \quad r > 0. \tag{5.6}$$

We set

$$B = \sup_{t>0, x_i \in E_\ell} \frac{|V(x_2,t) - V(x_1,t)|}{|x_2 - x_1|}.$$

THEOREM 5.1. *Suppose that there exists a Ljapunov function* $V(x,t) \in \mathbf{C}_0$ *for the system* (5.5) *satisfying condition* (5.6) *and the condition*

$$V(0,t) \equiv 0,$$

$$\frac{d^0 V}{dt} \leqslant -c_1 V, \qquad \|\sigma\| \leqslant c_2 V \tag{5.7}$$

($c_1, c_2 > 0$ are constants).

Suppose moreover that the process $|\xi(t,\omega)|$ *satisfies the law of large numbers* (1.1) *and the condition*

$$\sup_{t>0} \mathbf{E}|\xi(t,\omega)| < \frac{c_1}{Bc_2}. \tag{5.8}$$

Then the trivial solution of the system (5.4) *is asymptotically stable in probability in the large. If the process* $|\xi(t,\omega)|$ *satisfies the strong law of large numbers* (1.2), *while all the other assumptions remain unchanged, then the solution* $x = 0$ *is almost surely asymptotically stable in the large.*

PROOF. By Lemma 3.1, it follows from (5.7) that

$$\frac{dV(x(t,\omega),t)}{dt} \leqslant -c_1 V(x(t,\omega),t) + Bc_2|\xi(t,\omega)|V.$$

We may assume without loss of generality that $t_0 = 0$. Applying Lemma 2.1, we get the estimate

$$V(x(t,\omega),t) \leqslant V(x_0,0)\exp\left\{\int_0^t (Bc_2|\xi(s,\omega)| - c_1)ds\right\}$$

$$\leqslant V(x_0,0)\exp\left\{Bc_2\left(\frac{1}{t}\int_0^t |\xi(s,\omega)|ds - \frac{c_1}{Bc_2}\right)t\right\}. \tag{5.9}$$

Now let $\varepsilon > 0$ and $\delta > 0$ be arbitrary. Using (5.8) and the fact that the process $|\xi(t,\omega)|$ satisfies the law of large numbers, we see that there exists a number $T > 0$ such that for $t \geqslant T$

$$\mathbf{P}\left\{\frac{1}{t}\int_0^t |\xi(s,\omega)|ds > \frac{c_1}{Bc_2}\right\} < \varepsilon. \tag{5.10}$$

We now choose a large enough number $M > 1$ such that

$$\mathbf{P}\left\{Bc_2\int_0^T |\xi(s,\omega)|ds > \ln M\right\} < \varepsilon. \tag{5.11}$$

Finally, we choose r small enough so that, for $|x_0| < r$,

$$V(x_0,0)M < V_\delta. \tag{5.12}$$

It follows now from the inequalities (5.9)-(5.12), considered separately for $t < T$ and $t \geqslant T$, that for $|x_0| < r$ and all $t \geqslant 0$

$$\mathbf{P}\{|x(t,\omega)| > \delta\} \leqslant \mathbf{P}\{V(x(t,\omega),t) > V_\delta\} \leqslant \varepsilon.$$

Hence, using the relation

$$\mathbf{P}\left\{\frac{1}{t}\int_0^t |\xi(s,\omega)|ds > \frac{c_1}{Bc_2}\right\} \to 0 \quad \text{as} \quad t \to \infty,$$

we get the first part of the theorem.

The proof of the second part is analogous.

THEOREM 5.2. *Suppose that there exists a Ljapunov function $V(x,t) \in \mathbf{C}_0$ for the system* (5.5), *satisfying condition* (5.7) *and the inequality*

$$V(x,t) > c|x|, \qquad (c > 0). \tag{5.13}$$

Assume that the process $\xi(t,\omega)$ is such that for some positive constants k_1, k_2 and $t > 0$

$$\mathbf{E}\exp\left\{k_1\int_0^t |\xi(s,\omega)|ds\right\} \leqslant \exp\{k_2 t\}, \tag{5.14}$$

where the constants k_i, c_i, B satisfy the inequality

$$Bk_2c_2 \leqslant k_1c_1. \tag{5.15}$$

Then the solution $x(t) \equiv 0$ of the system (5.4) *is p-stable for $p \leqslant k_1/Bc_2$. If the strict inequality is valid,*

$$Bk_2c_2 < k_1c_1, \tag{5.16}$$

then the solution is exponentially p-stable for $p \leqslant k_1/Bc_2$.

PROOF. The proof is based on inequality (5.9). Raising both sides of this inequality to the power k_1/Bc_2 and then calculating the expectation of both sides, we see, using (5.13), that

$$c^{k_1/Bc_2}\mathbf{E}|\xi(t,\omega)|^{k_1/Bc_2}$$

$$\leq \mathbf{E}[V(x(t,\omega),t)]^{k_1/Bc_2}$$

$$\leq [V(x_0,0)]^{k_1/Bc_2}\mathbf{E}\exp\left\{k_1\int_0^t|\xi(s,\omega)|ds - \frac{c_1k_1}{Bc_2}t\right\}.$$

This, together with the inequalities (5.14)-(5.16), implies the assertion.

REMARK 1. It is clear from the proof of Theorem 5.1 that we can somewhat weaken the requirement that the process $|\xi(t,\omega)|$ should satisfy the law of large numbers. However, this condition cannot be completely dropped. This can be seen from the example

$$\frac{dx}{dt} = (-a + \xi)x,$$

where $a > 0$ and the random variable ξ can take arbitrarily large positive values. However small the expectation $\mathbf{E}|\xi|$ may be, the solution $x(t,\omega) = x_0\exp\{(-a + \xi)t\}$ of this equation tends to infinity with probability $p = \mathbf{P}\{\xi > a\}$.

The same example shows that condition (5.14) of Theorem 5.2 cannot be essentially weaked.

REMARK 2. Theorems 5.1 and 5.2 which furnish conditions for the occurrence of a stable equilibrium are not local in nature since conditions (5.7) imposed on the Ljapunov functions must hold everywhere, not only in the neighborhood of the origin. It is not hard to devise examples of stochastic processes $\xi(t,\omega)$ for which all assumptions of Theorems 5.1 and 5.2 hold locally, but the origin is nevertheless unstable. This is the case, for example, for the equation

$$\frac{dx}{dt} = \frac{-x + x^3}{1 + |x|^3} + x\xi(t,\omega),$$

where the process $\xi(t,\omega)$ vanishes everywhere except on intervals of length $\Delta_k \to 0$, on which it is equal to $1/\Delta_k^2$. Scattering the intervals Δ_k at random and sufficiently sparsely over the t-axis, we can ensure that the law of large numbers and condition (5.8) will hold. Nevertheless, $x(t) \to \infty$ almost surely if $x_0 > 0$.

This example shows that the existence of a local Ljapunov

function is not sufficient for stability in probability. It is also clear that these systems do not satisfy the analog of Ljapunov's theorem on stability in the first approximation. As we shall see later, the situation changes radically if $\xi(t,\omega)$ is assumed to be a process with "independent values".

6. Stability of randomly perturbed deterministic systems

The following problem has been considered by several authors. Let $x \equiv 0$ be a stable solution, in some sense, of the equation

$$\frac{dx}{dt} = F(x,t) \qquad (F(0,t) \equiv 0). \tag{6.1}$$

Will the solution of this system remain in a given neighbourhood of the origin for all $t \geqslant t_0$ if the right-hand side $F(x,t)$ is perturbed, say, by sufficiently small random forces? More precisely, along with the system (6.1), we consider the system

$$\frac{dx}{dt} = F(x,t) + R(x,t) \tag{6.2}$$

and call the solution $x(t) \equiv 0$ of (5.1) stable under continually acting perturbations if, for each $\varepsilon > 0$, there exists a $\delta > 0$ such that if

$$|x_0| < \delta, \qquad |R(x,t)| < \delta,$$

then the solution $x(t,x_0,t_0)$ of the system (6.2) satisfies the inequality $|x(t,x_0,t_0)| < \varepsilon$ for all $t \geqslant t_0$. It is known (Malkin [2]) that a sufficient condition for stability under continually acting perturbations is that the trivial solution of the system (6.1) be asymptotically stable uniformly in x_0,t_0.

Sometimes, however, assumptions of this kind might be too restrictive. This is the case, for instance, when the right-hand side of the system (6.1) is subjected to random perturbations which are small only on the average, but sometimes, even if only rarely, experience quite significant "overshoots" which begin at a random time and extend over a period which is not necessarily short. It is clear that then restrictions on F only in the neighborhood of the point $x = 0$ will not imply stability of the trivial solution, since the solution may sometimes extend far beyond the origin. The only meaningful definition of stability in such a situation is that at any fixed time the sample function should lie in the neighbourood of the origin with sufficiently high probability.

We now present the rigorous definition.

Along with equation (6.1), we consider

$$\frac{dx}{dt} = F(x,t) + R(x,t,\omega), \tag{6.3}$$

where the function $R(x,t,\omega)$ is such that equation (6.3) satisfies the existence and uniqueness theorems of Section 2. We also assume that the stochastic process

$$\eta(t,\omega) = \sup_{x} |R(x,t,\omega)|$$

has finite expectation. The solution $x \equiv 0$ of the system (6.1) will be called stable for $t \geq t_0$, under continually acting random perturbations which are small *on the average* (briefly: *stable under small random perturbations*) if the solution of equation (6.3) satisfying the initial condition $x(t_0,\omega) = x_0$ tends to zero in probability uniformly for $t \geq t_0$ as

$$|x_0| + \sup_{t \geq t_0} \mathsf{E}\eta(t,\omega) \to 0. \tag{6.4}$$

In other words, the solution $x \equiv 0$ of the system (6.4) is stable under random perturbations if, for each $\varepsilon > 0$ and $\Delta > 0$, there exists a $\gamma > 0$ such that, when

$$|x_0| + \sup_{t \geq t_0} \mathsf{E}\eta(t,\omega) < \gamma,$$

then the following inequality holds for $t \geq t_0$:

$$\mathbf{P}\{|x(t,\omega)| > \Delta\} < \varepsilon.$$

THEOREM 6.1. *Suppose that there exists a Ljapunov function $V(x,t) \in \mathbf{C}_0$ on E with the following properties:*

1. $V(0,t) \equiv 0$, $V_\delta > 0$ *for* $\delta > 0$.
2. *For each* $\delta > 0$, *there exists a* $c_\delta > 0$ *such that*

$$\frac{d^0V}{dt} < -c_\delta V \tag{6.5}$$

holds in the domain $\{|x| > \delta\} \times \{t > t_0\}$.

Then the solution $x \equiv 0$ *of equation* (6.1) *is stable under small random perturbations for* $t \geq t_0$.

PROOF. It follows from the assumptions of the theorem that in the domain $|x| > \delta$

$$\frac{dV(x(t,\omega),t)}{dt} \leq \frac{d^0V}{dt} + c\eta(t,\omega). \tag{6.6}$$

Set $V^{(\delta)} = \sup V(x,t)$. Then $V^{(\delta)} \to 0$ as $\delta \to 0$. Moreover, the assumptions of the theorem imply that $d^0V/dt < 0$ for $x \neq 0$. In view of this inequality and (6.5), it follows that for all x and almost all ω

$$\frac{dV}{dt} \leqslant - c_\delta V + c\eta(t,\omega) + c_\delta V^{(\delta)}. \tag{6.7}$$

Applying Lemma 2.1 and then taking the expectation of both sides of the inequality (see proof of Theorem 4.1), we easily see that

$$\mathbf{E}V(x(t,\omega),t) \leqslant V(x_0,t_0)e^{-c_\delta(t-t_0)} + \frac{c}{c_\delta} \sup_{t>t_0} \mathbf{E}\eta(t,\omega) + V^{(\delta)}. \tag{6.8}$$

Now let $\varepsilon > 0$ and $\Delta > 0$ be arbitrary. Taking δ, $|x_0|$ and $\mathbf{E}\eta(t,\omega)$ sufficiently small, we easily get the inequality

$$\mathbf{E}V(x(t,\omega),t) \leqslant \varepsilon \sup_{t\geqslant t_0, |x|>\delta} V(x,t). \tag{6.9}$$

This inequality and Lemma 4.1 imply the required assertion.

REMARK 1. Let the point $x = 0$ be exponentially stable for the system (6.1), i.e., assume that the solution $x(t,x_0,t_0)$ of the system (6.1) with initial condition $x(t_0) = x_0$ admits the estimate $|x(t,x_0,t_0)| < B|x_0|\exp\{ - \alpha(t - t_0)\}$, where $B,\alpha > 0$ are constants independent of x_0 and t_0. Then (see Krasovskii [1], p. 72) there exists a function $W(x,t)$ for the system (6.1) such that

$$c_1|x|^2 < W(x,t) < c_2|x|^2,$$

$$\frac{d^0W}{dt} \leqslant - c_3|x|^2, \qquad \left|\frac{\partial W}{\partial x}\right| < c_4|x|,$$

provided $\|\partial F/\partial x\|$ is bounded in E. It follows from these estimates that the function $V(x,t) = [W(x,t)]$ satisfies all the assumptions of Theorem 6.1. Thus, if the point $x = 0$ is exponentially stable for the system (6.1), then it is also stable under small random perturbations.

REMARK 2. If the function $V(x,t)$ satisifes the assumptions of Theorem 6.1 and moreover $V(x,t) > c_1|x|$ for some $c_1 > 0$, then it evidently follows from the proof of Theorem 6.1 that the system (6.1) is stable under small random perturbations in a stronger sense. Indeed, we have then

$$\sup_{t>0} \mathbf{E}|x(t,\omega)| \to 0 \quad \text{as} \quad |x_0| + \sup_{t>0} \mathbf{E}\eta(t,\omega) \to 0.$$

The preceding remark shows that the above type of stability (mean stability) holds when the unperturbed system is exponentially stable. It is readily shown that in the latter case we also have mean square stability with respect to random perturbations which are small in mean square, i.e.,

$$\sup_{t>0} \mathbf{E}|x(t,\omega)|^2 \to 0 \quad \text{as} \quad |x_0| + \sup_{t>0} \mathbf{E}\eta^2(t,\omega) \to 0.$$

The derivation of further, more general criteria for mean square stability presents no difficulties.

REMARK 3. It is clear from Example 2 in Section 4 that condition (6.5) cannot be replaced by the condition: $d^0V/dt < -c_\delta$ in the domain $|x| > \delta$. By slightly modifying Example 1 of that section one readily shows that even a linear asymptotically stable system may be unstable under perturbations such that only $\sup_{t>0} \mathbf{E}[\eta(t,\omega)]^\alpha \to 0$, if $\alpha < 1$. Thus, even the "best-behaved" stable systems may lose their stability if $\mathbf{E}\eta(t,\omega)$ does not tend to zero (we are not considering here the case of white noise, when $\mathbf{E}\eta(t,\omega)$ does not exist; for this case, see Chapter V).

The assumptions of Theorem 6.1 may be slightly weakened by further restricting the range of admissible random perturbations. It seems that in most applications it is sufficient to consider random perturbations of the type

$$R(x,t,\omega) = \sigma(x,t)\xi(t,\omega). \tag{6.10}$$

We shall say that the solution $x(t) \equiv 0$ of the system (6.1) is stable under small random perturbations of type (6.10) if, for each $\varepsilon > 0$ and $\Delta > 0$, there exists a $\varkappa > 0$ such that, whenever

$$|x_0| + \sup_{x,t} \|\sigma(x,t)\| < \varkappa, \tag{6.11}$$

then the following inequality holds for $t > t_0$:

$$\mathbf{P}\{|x(t,\omega)| > \Delta\} < \varepsilon.$$

(This definition is natural, since the matrix $\sigma(x,t)$ characterizes the intensity of the random perturbations at the point (x,t).)

THEOREM 6.2. *Let $V(x,t) \in \mathbf{C}_0$ be a Ljapunov function in E satisfying assumption* 1 *of Theorem* 6.1 *and assumption* 2 *with inequality* (6.5) *replaced by*

$$\frac{d^0V}{dt} < -c_\delta.$$

Assume further that the process $\xi(t,\omega)$ satisfies the following *condition: For each $\varepsilon > 0$, there exists a $\gamma > 0$ such that for $t_0 \leqslant s \leqslant t$*

$$\mathrm{E}\exp\left\{\gamma\int_s^t |\xi(u,\omega)|du\right\} \leqslant e^{\varepsilon(t-s)}. \tag{6.12}$$

Then the solution $x \equiv 0$ of equation (6.1) *is stable for $t \geqslant t_0$ under random perturbations of type* (6.10).

PROOF. We set

$$W(x,t) = \exp\{V(x,t)\} - 1, \qquad W^{(\delta)} = \sup_{U_\delta\times\{t>t_0\}} W(x,t).$$

Proceeding as in the proof of (6.7) we get the estimate

$$\frac{dW}{dt} \leqslant W(-c_\delta + \gamma|\xi(t,\omega)|) + \gamma|\xi(t,\omega)| + c_\delta W^{(\delta)}, \tag{6.13}$$

valid for every $\delta > 0$ and $\gamma > 0$, provided inequality (6.11) holds for $x \leqslant x_0(\gamma)$. Using Lemma 2.1, we infer from (6.13) that

$$W(x(t,\omega),t) \leqslant W(x_0,t_0)\exp\left\{\int_{t_0}^t (-c_\delta + \gamma|\xi(s,\omega)|)ds\right\} \tag{6.14}$$

$$+ c_\delta W^{(\delta)}\int_{t_0}^t \exp\left\{\int_s^t (-c_\delta + \gamma|\xi(u,\omega)|)du\right\}ds$$

$$+ \gamma\int_{t_0}^t \exp\left\{\int_s^t (-c_\delta + \gamma|\xi(u,\omega)|)du\right\}|\xi(s,\omega)|ds.$$

Let ε be any number such that $0 < \varepsilon < \frac{1}{2}$. It follows from (6.12) that we can choose a number $\gamma_0(\varepsilon)$ such that, for $\gamma < \gamma_0(\varepsilon)$,

$$\mathrm{E}\exp\left\{\gamma\int_s^t |\xi(u,\omega)|du\right\} \leqslant \exp\{\varepsilon c_\delta(t - s)\}. \tag{6.15}$$

Hence, using the fact that $W^{(\delta)} \to 0$ as $\delta \to 0$ (which follows directly from the assumptions of the theorem) we see that we can choose, first $\delta(\varepsilon)$, and then $\varkappa(\varepsilon)$ and $\gamma_0(\varepsilon)$, so that inequality (6.15) will hold, and moreover

$$\mathbf{E}\left[W(x_0,t_0)\exp\left\{\int_{t_0}^t (-c_\delta + \gamma|\xi(u,\omega)|)du\right\}\right.$$

$$\left. + c_\delta W^{(\delta)}\int_{t_0}^t \exp\left\{\int_s^t (-c_\delta + \gamma|\xi(u,\omega)|)du\right\}ds\right] < \varepsilon. \tag{6.16}$$

Next, from the equalities

$$J = \gamma\int_{t_0}^{t} \exp\left\{\int_{s}^{t} (-c_\delta + \gamma|\xi(u,\omega)|)du\right\}|\xi(s,\omega)|ds$$

$$= \exp\left\{\int_{t_0}^{t} (-c_\delta + \gamma|\xi(u,\omega)|)du\right\} - 1$$

$$+ c_\delta\int_{t_0}^{t} \exp\left\{\int_{s}^{t} (-c_\delta + \gamma|\xi(u,\omega)|)du\right\}ds$$

and from (6.15) it follows that

$$\mathbf{E}J \leqslant \exp\{-c_\delta(1-\varepsilon)(t-t_0)\} - 1$$

$$+ c_\delta\int_{t_0}^{t} \exp\{-c_\delta(1-\varepsilon)(t-s)\}ds$$

$$\leqslant \frac{2\varepsilon}{1-\varepsilon} \leqslant 4\varepsilon. \qquad (6.17)$$

It follows from (6.14), (6.16) and (6.17) that for all $t > t_0$ we have $\mathbf{E}W(x(t,\omega),t) < 5\varepsilon$, whenever $|x_0| + \sup_{t>t_0} \mathbf{E}\eta(t,\omega) < \varkappa(\varepsilon)$. Now it suffices to apply Lemma 4.1 in order to derive the assertion of the theorem.

EXAMPLE. In the one-dimensional case we may use the Ljapunov function $V(x) = x$ to get the following result. Let $F(x,t)/x < -c_\delta$ hold in the domain $|x| > \delta$ for some $c_\delta > 0$. Then the point $x = 0$ is stable under small random perturbations. But if, instead, we assume that $\operatorname{sign}x.F(x,t) < -c_\delta$ for $|x| > \delta$, then the point $x = 0$ is stable under small random perturbations of type (6.10), provided $\xi(t,\omega)$ satisfies condition (6.12).

7. Estimation of a certain functional of a Gaussian process

One sees from Theorem 6.2 that the following estimate plays a major role in the theory of stability of stochastic systems:

$$\mathbf{E}\exp\left\{k_1\int_{t_0}^{t_1} |\xi(s,\omega)|ds\right\} \leqslant \exp\{k_2(t_1 - t_0)\}, \quad t_1 \geqslant t_0. \qquad (7.1)$$

Stability of the system under continually acting random perturbations requires that for each $\varepsilon > 0$ there exists a $\gamma > 0$ such that for $t_1 > t_0$

$$\mathbf{E}\exp\left\{\gamma\int_{t_0}^{t_1}|\xi(s,\omega)|ds\right\} \leqslant \exp\{\varepsilon(t_1 - t_0)\}. \tag{7.2}$$

In this section we shall derive simple conditions for these estimates to hold for Gaussian processes.

We shall consider a Gaussian stochastic process $\xi(t,\omega)$ in E_ℓ, i.e., a process all of whose finite-dimensional distributions are Gaussian. Assume in addition that this process is measurable and that its kernel $K(s,t)$ is continuous[9].

As usual, we define the *trace* of a square matrix $A = ((a_{ij}))$ by

$$\mathrm{tr}A = \sum_{i=1}^{\ell} a_{ii}.$$

THEOREM 7.1[10]. *Assume that the Gaussian process $\xi(t,\omega)$ with zero expectation satisfies*

$$\mathrm{tr}K(s,s) = \mathbf{E}|\xi(s,\omega)|^2 \leqslant c_1, \tag{7.3}$$

$$\int_0^\infty \|K(s,u)\|du = \int_0^\infty \|K(u,s)\|du \leqslant c_2, \tag{7.4}$$

for some $c_i > 0$ and all $s > 0$.

Then the following estimate holds:

$$\mathbf{E}\exp\left\{k_1\int_{t_0}^{t_1}|\xi(s,\omega)|ds\right\} < \exp\left\{k_1\left(\sqrt{c_1} + \frac{k_1 c_2}{2}\right)(t_1 - t_0)\right\}. \tag{7.5}$$

LEMMA 7.1. *The process $\xi(t,\omega)$, $t_0 \leqslant t \leqslant t_1$, can be expanded in a series*

$$\xi(t,\omega) = \sum_{k=1}^{\infty} \sqrt{\lambda_k}\varphi_k(t)\xi_k(\omega), \tag{7.6}$$

which is almost surely convergent for every t and satisfies Parseval's identity

---

[9] We denote (see Section 1)

$$K(s,t) = ((K^{ij}(s,t))) = \mathrm{cov}(\xi(s),\xi(t)).$$

[10] Theorem 7.1 generalizes a result of Šur [1].

$$\int_{t_0}^{t_1} |\xi(t,\omega)|^2 dt = \sum_{k=1}^{\infty} \lambda_k \xi_k^2. \tag{7.7}$$

Here $\varphi_k(t)$ *and* λ_k *are the normalized eigenfunctions and eigenvalues, of the integral equation*

$$\int_{t_0}^{t_1} K(t,\tau)\varphi(\tau)d\tau = \lambda\varphi(t); \tag{7.8}$$

ξ_k *are independent Gaussian random variables with zero expectation and unit variance.*

PROOF. The proof follows easily from the expansion of the process $\xi(t,\omega)$ in the series of eigenfunctions of equation (7.8). The formulas for the Fourier coefficients

$$\sqrt{\lambda_k}\xi_k(\omega) = \int_{t_0}^{t_1} \xi(t,\omega)\varphi_k(t)dt,$$

the orthogonality of the $\varphi_k(t)$ and the fact that the process is Gaussian imply that the random variables ξ_k are independent. Identity (7.7) follows from the completeness of the system of eigenfunctions $\varphi_k(t)$. (For more details see, e.g., Pugačev [1], Gihman and Skorohod [1], Chapter 5, Section 2.)

LEMMA 7.2. *The expectation of the functional*

$$\exp\left\{\alpha\int_{t_0}^{t_1} |\xi(t,\omega)|^2 dt\right\}$$

exists for all $t_0 < t_1$ *and for sufficiently small positive* α. *Moreover we have the representation*

$$\mathbf{E}\exp\left\{\alpha\int_{t_0}^{t_1} |\xi(t,\omega)|^2 dt\right\} = \prod_{k=1}^{\infty} \frac{1}{\sqrt{1 - 2\alpha\lambda_k}}. \tag{7.9}$$

If also conditions (7.3), (7.4) *hold, then for all* $t_0 < t_1$

$$\mathbf{E}\exp\left\{\alpha\int_{t_0}^{t_1} |\xi(t,\omega)|^2 dt\right\} \leqslant \exp\left\{\frac{\alpha c_1}{1 - 2\alpha c_2}(t_1 - t_0)\right\}.$$

PROOF. By (7.7),

$$\sum_{k=1}^{\infty} \lambda_k = \mathbf{E}\int_{t_0}^{t_1} |\xi(u,\omega)|^2 du \leqslant c_1(t_1 - t_0). \tag{7.10}$$

Consequently $\lambda_{max} = \max_{1 \leqslant k < \infty} \lambda_k$ exists; we may assume without loss of generality that $\lambda_{max} = \lambda_1$.

The identity (7.9) now follows easily from (7.7) and the following relation, valid for $\alpha < 1/(2\lambda_1)$:

$$\mathbf{E}\exp\{\alpha\lambda_k\xi_k^2(\omega)\} = \frac{1}{\sqrt{1 - 2\alpha\lambda_k}} .$$

Let us now prove that $\lambda_1 = \lambda_{max} \leqslant c_2$. In fact, we easily infer from (7.4) that

$$\lambda_1 = \int_{t_0}^{t_1} \lambda_1(\varphi_1(s),\varphi_1(s))ds = \int_{t_0}^{t_1}\int_{t_0}^{t_1} (K(s,t)\varphi_1(t),\varphi_1(s))dsdt$$

$$\leqslant \int_{t_0}^{t_1}\int_{t_0}^{t_1} \|K(s,t)\| |\varphi_1(t)| |\varphi_1(s)| dsdt \tag{7.11}$$

$$\leqslant \int_{t_0}^{t_1}\int_{t_0}^{t_1} \|K(s,t)\| \frac{|\varphi_1(t)|^2 + |\varphi_1(s)|^2}{2} dsdt \leqslant c_2.$$

Next, using the elementary inequality $1 + \gamma < e^\gamma$ $(\gamma > 0)$, we get the estimate

$$\prod_{k=1}^{\infty} \frac{1}{\sqrt{1 - 2\alpha\lambda_k}} = \prod_{k=1}^{\infty} \left(1 + 2\alpha\lambda_k + \frac{4\alpha^2\lambda_k^2}{1 - 2\alpha\lambda_k}\right)^{\frac{1}{2}}$$

$$\leqslant \exp\left\{\alpha\left(1 + \frac{2\alpha\lambda_1}{1 - 2\alpha\lambda_1}\right) \sum_{k=1}^{\infty} \lambda_k\right\} = \exp\left\{\frac{\alpha}{1 - 2\alpha\lambda_1} \sum_{k=1}^{\infty} \lambda_k\right\} .$$

Hence, in view of (7.10) and (7.11), we get the second assertion of the lemma.

PROOF OF THEOREM 7.1. Applying the inequality

$$a < \frac{\alpha}{2} a^2 + \frac{1}{2\alpha} \qquad (\alpha > 0), \tag{7.12}$$

we get

$$\mathbf{E}\exp\left\{k_1\int_{t_0}^{t_1} |\xi(s,\omega)| ds\right\}$$

$$\leqslant \exp\left\{\frac{k_1}{2\alpha} (t_1 - t_0)\right\}\mathbf{E}\exp\left\{\frac{k_1\alpha}{2}\int_{t_0}^{t_1} |\xi(s,\omega)|^2 ds\right\} .$$

Hence, it follows by Lemma 7.2 that for all $\alpha < 1/k_1c_2$

$$\mathrm{E}\exp\left\{k_1\int_{t_0}^{t_1}|\xi(s,\omega)|ds\right\} \qquad (7.13)$$

$$\leqslant \exp\left\{\left[\frac{k_1}{2\alpha} + \frac{k_1\alpha c_1}{2(1-\alpha k_1 c_2)}\right](t_1 - t_0)\right\}.$$

Setting $\alpha = \alpha^* = 1/(k_1c_2 + \sqrt{c_1})$, we get (7.5)[11].

This completes the proof of Theorem 7.1.

COROLLARY. *One can easily eliminate the condition* $\mathrm{E}\xi(t,\omega) = 0$, *if one knows that*

$$|\mathrm{E}\xi(t,\omega)| \leqslant c_0. \qquad (7.14)$$

Indeed, it is easy to see that estimates similar to (7.1) and (7.2) hold if condition $\mathrm{E}\xi(t,\omega) = 0$ is replaced by (7.14). More precisely, the following analogue of Theorem 7.1 is valid.

If $\xi(t,\omega)$ is a Gaussian process satisfying conditions (7.3), (7.4) and (7.14), then the following estimate is valid for all $k_1 > 0$ and $-\infty < t_0 < t_1 < \infty$:

$$\mathrm{E}\exp\left\{k_1\int_{t_0}^{t_1}|\xi(s,\omega)|ds\right\} \leqslant \exp\left\{k_1\left(c_0 + \sqrt{c_1} + \frac{k_1c_2}{2}\right)(t_1 - t_0)\right\}. \qquad (7.15)$$

We shall prove one more relation for Gaussian processes which will be needed for the examples in the next section:

LEMMA 7.3. *The process* $\xi(t,\omega)$ *satisfies*

$$\mathrm{E}\exp\left\{\int_{t_0}^{t_1}\xi(t,\omega)dt\right\} \qquad (7.16)$$

$$= \exp\left\{\frac{1}{2}\int_{t_0}^{t_1}\int_{t_0}^{t_1}\hat{K}(s,t)dsdt + \int_{t_0}^{t_1}\mathrm{E}\xi(t,\omega)dt\right\},$$

where $\hat{K}(s,t)$ *is the vector with components* $K^{(1,1)}(s,t), \ldots,$ $K^{(1,1)}(s,t)$.

PROOF. It is easily seen that if η is a Gaussian random variable with zero expectation and variance σ^2, then $\mathrm{E}\exp\eta$ exists; in fact

$$\mathrm{E}\exp\eta = \exp\{\tfrac{1}{2}\sigma^2\}. \qquad (7.17)$$

---

[11] It is clear that when $\alpha = \alpha^*$ the argument of the exponential function in (7.13) attains its minimum.

The assertion of the lemma follows immediately from (7.17), if we use the fact that each component of the vector $\int_{t_0}^{t_1} \xi(s,\omega)ds$ has a Gaussian distribution.

REMARK 1. It follows from Lemma 7.3 that if $\xi(t,\omega)$ is a Gaussian process in E_ℓ with zero expectation whose correlation function is nonnegative, then condition (7.4) is necessary for Theorem 7.1 to hold. Indeed, if

$$\int_0^\infty K(s,t)dt = \infty,$$

then it is clear from (7.16) that the estimate (7.1) does not hold for any k_1 and k_2.

REMARK 2. Assume that the process $\xi(t,\omega)$ is stationary. Then condition (7.3) is automatically satisfied and $c_1 = \mathbf{E}|\xi(t,\omega)|^2$. Condition (7.4) can then be replaced by the following: The spectral density $f(\lambda)$ of the process $\xi(t,\omega)$ is bounded in the norm. In fact, we used condition (7.4) only to prove that the greatest eigenvalue of equation (7.8) is bounded. But for stationary stochastic processes it is known that $\max\lambda_k \leqslant \sup\|f(\lambda)\|$.

REMARK 3. The estimates (7.1) and (7.2) remain valid when the process $\xi(t,\omega)$ has finite memory (i.e., there is a τ such that the evolution process up to time t and after time $t + \tau$ are independent for all t), and

$$\mathbf{E}\exp\{u|\xi(t,\omega)|\} < \varphi(u) < \infty.$$

It seems likely that also in the general case one could give conditions in terms of the rate of growth of the function $\varphi(u)$ as $u \to \infty$ and certain "mixing" conditions for the process, which imply (7.1) and (7.9). Questions essentially similar to these arise when one investigates existence conditions for the limit of the configuration integral in statistical physics.

8. Linear systems[12]

We now apply the results of this chapter to linear systems of the type

$$\frac{dx}{dt} = (A(t) + \eta(t,\omega))x + B(t,\omega). \tag{8.1}$$

[12] See Caughey and Gray [1], Has'minskiĭ [9]. One-dimensional linear systems were studied previously in detail by Rosenbloom [1], Tihonov [1] and others.

Without loss of generality, we may assume that the elements of the square matrix $\eta(t,\omega)$ have zero expectation.

We consider first the homogeneous system

$$\frac{dx}{dt} = (A(t) + \eta(t,\omega))x. \tag{8.2}$$

Assume that the deterministic system

$$\frac{dx}{dt} = A(t)x \tag{8.3}$$

is exponentially stable, i.e., every solution $x(t,x_0,t_0)$ of the system admits the estimate

$$|x(t,x_0,t_0)| \leqslant B|x_0|e^{-\alpha(t-t_0)} \qquad (\alpha > 0) \tag{8.4}$$

for $t > t_0$, where the constants B,α are independent of x_0,t_0.

By a well-known theorem of Malkin ([2], p. 321), this implies that there exists a positive definite quadratic form $(C(t)x,x) = W(t,x)$ such that

$$\frac{d^0W}{dt} \leqslant -\lambda|x|^2 \qquad (\lambda > 0). \tag{8.5}$$

In the sequel we shall frequently use the fact that d^0W/dt is a quadratic form, being the derivative of a quadratic form along the trajectory of the system (8.3). Moreover

$$\frac{d^0W}{dt} = \left(\left(CA + A^*C + \frac{\partial C}{\partial t}\right)x,x\right).$$

To be able to apply Theorem 5.1 to the system (8.2), we must write it in the form (5.4). This is easily done if we set $\sigma(x,t)$ equal to the $\ell \times \ell^2$ matrix

$$\sigma(x,t) = \begin{pmatrix} x_1\; x_2 \quad x_\ell & 00\ldots 00 & \ldots\ldots & 000\ldots 0 \\ 0\quad 0\ldots 0 & x_1x_2\ldots x_\ell 0 & \ldots\ldots & 000\ldots 0 \\ 0\quad 0\ldots 0 & & & \\ 0\quad 0\ldots 0 & 00\ldots 00 & & 0x_1x_2\ldots x_\ell \end{pmatrix},$$

and write $\eta_{ik}(t,\omega) = \xi_{(i-1)\ell+k}(t,\omega)$, where $\xi(t,\omega)$ is a vector in E_{ℓ^2}. Considering the Ljapunov function $V(x,t) = (W(x,t))^{\frac{1}{2}}$ and applying Theorem 5.1, we get the following result.

THEOREM 8.1. *Suppose that the solution of the system* (8.3) *satisfies condition* (8.4). *Then there exists a constant* $c > 0$ *such that the system* (8.2) *is asymptotically stable in probability for every (matrix-valued) stochastic process* $\eta(t,\omega)$ *such that the process* $\|\eta(t,\omega)\|$ *satisfies the law of large numbers and* $\mathbf{E}\|\eta(t,\omega)\| < c$. *If the process* $\|\eta(t,\omega)\|$ *also satisfies the strong law of large numbers, the other assumptions remaining unchanged, then the system* (8.2) *is almost surely asymptotically stable in the large.*

Theorem 5.1 also enables us to estimate the constant c. Using the same Ljapunov function and Theorem 5.2, we easily derive sufficient conditions for the system (8.2) to be p-stable, provided the process $\|\eta(t,\omega)\|$ satisfies condition (5.14). Rather than going into further details let us consider an example.

EXAMPLE 1. Consider the following equation in E_ℓ:

$$\frac{dx}{dt} = (a(t) + \xi(t,\omega))x. \tag{8.6}$$

It has a solution

$$x(t) = x_0 \exp\left\{\int_0^t (a(s) + \xi(s,\omega))ds\right\}. \tag{8.7}$$

Using (8.7) and slightly modifying the proof of Theorem 5.1, we get the following result.

The solution $x(t) \equiv 0$ of equation (8.6) is asymptotically stable in probability for $a(t) \leqslant a_0 < 0$ and unstable for $a(t) \geqslant a_0 > 0$ for every stochastic process $\xi(t,\omega)$ with zero expectation and satisfying the law of large numbers.

Applying Lemma 7.3 to a stationary Gaussian process $\xi(t,\omega)$ with $\mathbf{E}\xi(t,\omega) = 0$ and correlation function $K(t - s)$, we get

$$\mathbf{E}|x(t,\omega)| = |x_0| \exp\left\{p\int_0^t a(s)ds + \frac{p^2}{2}\int_0^t\int_0^t K(u-s)duds\right\}. \tag{8.8}$$

Assume that the function $K(u)$ is absolutely integrable and

$$\int_{-\infty}^{+\infty} K(u)du = f(0) > 0.$$

Then it is obvious that

$$\int_0^t\int_0^t K(u - s)duds = f(0)t + o(t) \qquad (t \to \infty). \tag{8.9}$$

Let $a \leqslant a_0 < 0$. It then follows from (8.8) and (8.9) that the solution $x(t) \equiv 0$ of equation (8.6) is asymptotically p-stable for a Gaussian process $\xi(t,\omega)$ if

$$p < -2a_0/f(0) = p_0$$

and not p-stable if $p > p_0$ and $a(t) = a_0 = \text{const}$.

We now consider the case $a = 0$. It is clear from (8.7) that the solution $x(t) \equiv 0$ is unstable in this case if there exists a function $\alpha(t)$ such that $\alpha(t) \to \infty$ as $t \to \infty$ and the probability

$$\mathbf{P}\left\{\int_0^t \xi(s,\omega)ds > \alpha(t)\right\}$$

does not tend to zero. For example this is so when the central limit theorem is applicable to the integral of the process $\xi(t,\omega)$. Fairly broad conditions under which the central limit theorem is applicable in this situation may be found in Rozanov [1].

In the next theorem we adopt the simplifying assumption that A is a constant stable matrix, i.e., a matrix with eigenvalues λ_i such that $\text{Re}\lambda_i < 0$.

Ljapunov showed that, given a stable matrix, one can determine a positive definite matrix C such that the matrix $CA + A^*C$ is negative definite. Let λ denote the greatest positive number such that

$$((CA + A^*C)x,x) \leqslant -\lambda(Cx,x) \tag{8.10}$$

for all $x \in E_\ell$. It is not difficult to estimate the number λ from below in terms of the eigenvalues of the matrices C and $CA + A^*C = D$. Let $\lambda_{\max}^C$ and $\lambda_{\max}^D$ denote the greatest eigenvalues of the matrices C and D. Then it is evident that $\lambda > -\lambda_{\max}^D/\lambda_{\max}^C > 0$.

THEOREM 8.2. *Let A be a stable $\ell \times \ell$ matrix, C a positive definite matrix satisfying condition* (8.10), *and* $\eta(t,\omega) = ((\eta_{ij}(t,\omega)))$, $i,j = 1,\ldots,\ell$, *a Gaussian process. Suppose that the following conditions hold for the process* $\tilde{\eta}(t,\omega) = C^{\frac{1}{2}}\eta(t,\omega)C^{-\frac{1}{2}}$:

$$\|\mathbf{E}\tilde{\eta}(t,\omega)\| \leqslant c_0, \qquad \mathbf{E}\|\tilde{\eta}(t,\omega) - \mathbf{E}\tilde{\eta}(t,\omega)\|^2 \leqslant c_1,$$

$$\int_{t_0}^{t_1} \|K(u,s)\|du \leqslant c_2$$

(here $K(s,t) = \text{cov}(\tilde{\eta}(s,\omega),\tilde{\eta}(t,\omega))$ *is an* $\ell^2 \times \ell^2$ *matrix).*

Then the trivial solution of the system

$$\frac{dx}{dt} = (A + \eta(t,\omega))x \tag{8.11}$$

is asymptotically p-stable for

$$p < \frac{\lambda - 2(c_0 + \sqrt{c_1})}{2c_2},$$

provided $\lambda > 2(c_0 + \sqrt{c_1})$.

PROOF. We consider the Ljapunov function $V(x) = (Cx,x)$. Using (8.10) and the estimate

$$(C\eta x,x) = (C^{1/2}\eta C^{-1/2}C^{1/2}x, C^{1/2}x)$$

$$< \|C^{-1/2}\eta C^{1/2}\| \, \|C^{1/2}x\|^2 = \|\tilde{\eta}\|(Cx,x),$$

we get

$$\frac{dV(x(t,\omega))}{dt} \leqslant -\lambda V + ((C\eta + \eta^* C)x,x) \leqslant V(-\lambda + 2\|\tilde{\eta}\|). \quad (8.12)$$

Therefore

$$[V(x(t,\omega))]^p \leqslant [V(x_0)]^p \exp\left\{-p\lambda t + 2p\int_0^t \|\eta(s,\omega)\| ds\right\}.$$

Hence, calculating expectations and applying the estimate (7.15), we find that

$$\mathbf{E}[V(x(t,\omega))]^p$$

$$\leqslant [V(x_0)]^p \exp^3 pt(-\lambda + 2c_0 + 2\sqrt{c_1} + 2pc_2)\}. \quad (8.13)$$

Inequality (8.13) implies directly the assertion of the theorem.

REMARK 1. It is clear from Example 1 that in the one-dimensional case condition (8.10) is sufficient for the system to be stable for every value of $c_1 = \sup \mathbf{E}\|\eta(t,\omega)\|^2$, if $c_0 = 0$. It is easy to find examples showing that in the many-dimensional case this is generally not true; noise of sufficiently high intensity may "overcome" the stability. The only general conclusion from Theorem 8.2 is that when the coefficients of an exponentially stable linear system are perturbed by Gaussian processes satisfying (7.3), (7.4) and (7.14) with sufficiently small c_0 and c_1, the resulting system is asymptotically p-stable.

REMARK 2. In the one-dimensional case an unstable system driven by Gaussian noise with zero expectation remains unstable. Again, this property does not carry over to many-dimensional systems. In Chapter VI we shall present examples of unstable deterministic systems which can be "stabilized" by specially selected Gaussian noise with zero expectation.

Applying Theorem 6.1, we see that an exponentially stable linear system is stable under small random perturbations. It follows from Theorem 4.1 that the system (8.1) with $\eta(t,\omega) \equiv 0$ is dissipative for every vector $B(t,\omega)$ with finite expectation, provided condition (8.4) (or (8.10)) is satisfied. Let us devote some attention to the case $\eta(t,\omega) \not\equiv 0$, again assuming for simplicity's sake that the process $\eta(t,\omega)$ is Gaussian and the matrix A constant.

THEOREM 8.3. *Let A and $\eta(t,\omega)$ satisfy the assumptions of Theorem 8.2 and let $B(t,\omega)$ be a stochastic process with values in E_ℓ, independent of $\eta(t,\omega)$ and with bounded second moment. Then the system*

$$\frac{dx}{dt} = (A + \eta(t,\omega))x + B(t,\omega) \tag{8.14}$$

is dissipative if

$$2c_0 + 2\sqrt{c_1} + 2c_2 < \lambda. \tag{8.15}$$

PROOF. Setting $V(x) = (Cx,x)$ and using arguments similar to those which yielded (8.12), we infer by means of (7.12) that

$$\frac{dV(x(t,\omega))}{dt} \leqslant (-\lambda + 2\|\tilde{\eta}\|)V + 2(Cx,B(t,\omega))$$

$$\leqslant (-\lambda + 2\|\tilde{\eta}\| + \alpha)V + \frac{\|C^{1/2}\|^2|B(t,\omega)|^2}{\alpha}$$

(where α is any positive number).

Hence, using Lemma 2.1 and the independence of the processes η and B, we can show by standard arguments that

$$\mathbf{E}V(x(t,\omega))$$

$$\leqslant \mathbf{E}\left\{V(x_0(\omega))\exp\left\{\int_0^t(-\lambda + \alpha + 2\|\tilde{\eta}(s,\omega)\|)ds\right\}\right\}$$

$$+ \frac{\|C^{1/2}\|^2}{\alpha}\int_0^t \mathbf{E}\exp\left\{\int_s^t(-\lambda + \alpha + 2\|\tilde{\eta}(u,\omega)\|)du\right\}\mathbf{E}|B(s,\omega)|^2 ds.$$

If $x_0(\omega)$ satisfies condition (4.1) and the constant α is smaller than $\lambda - 2c_0 - 2\sqrt{c_1} - 2c_2$, then, again using the estimate (7.15), we readily see that for some constant $c_3 > 0$

$$\mathbf{E}V(x(t,\omega)) < c_3.$$

Hence it follows that the process $x(t,\omega)$ is dissipative and its second moment is bounded.

REMARK. The assertion of Theorem 8.3 may also hold for certain dependent processes $\eta(t,\omega)$ and $B(t,\omega)$. Indeed, before calculating the expectation in the proof one can estimate the expression $\exp\left\{2\int_s^t \|\eta(u,\omega)\|du\right\}|B(s,\omega)|^2$, using Young's inequality (3.17).

Of course, to continue the proof it is necessary to assume that $B(t,\omega)$ has bounded moments of order higher than 2. Also, it will be necessary to replace condition (8.15) by a somewhat more stringent condition. We omit the details.

We conclude this chapter by considering one more example.

EXAMPLE 2. Let $\eta_0,\ldots,\eta_n,\ldots$ be a sequence of mutually independent and identically distributed $\ell \times \ell$ random matrices, $\mathbf{E}\eta_k = ((0))$, and let A be a constant matrix. Assume that $x(0) = x_0$ and let the evolution of the system on the interval $t \in [k,k+1]$ be governed by the equation

$$\frac{dx(t,\omega)}{dt} = (A + \eta_k(\omega))x(t,\omega). \tag{8.16}$$

Denote $x_k(\omega) = x(k,\omega)$. It is obvious that

$$x_k(\omega) = e^{A+\eta_{k-1}}e^{A+\eta_{k-2}}\cdots e^{A\ \eta_0}x_0. \tag{8.17}$$

By virtue of (8.17) the question as to whether the system (8.16) is stable or unstable reduces to the question as to whether the norm of a product of random matrices does or does not tend to zero.

But although various authors have investigated the limiting behavior of a product of random matrices, at this stage it does not appear possible to derive convenient conditions for the stability of the above system.

Assume that the matrix A is stable, and let C be a positive definite matrix satisfying condition (8.10). Applying the method used to prove Theorem 8.2, we get the results below.

If

$$\mathbf{E}\|C^{1/2}\eta_k C^{-1/2}\| < \frac{\lambda}{2}, \tag{8.18}$$

then the stochastic process $x(t,\omega)$ determined by the system (8.16) *is almost surely asymptotically stable.*

We shall now show that if condition (8.18) is valid, the process $x(t,\omega)$ is asymptotically p-stable for sufficiently small p, provided

$$f(\alpha) = \mathbf{E}\exp\{\alpha\|\eta_k\|\} < \infty \tag{8.19}$$

for sufficiently small $\alpha > 0$.

LEMMA 8.1. *Let ξ be a positive random variable such that $\mathbf{E}\exp\{\alpha_0\xi\}$ exists for some $\alpha_0 > 0$. Then*

$$\mathbf{E}\exp\{\alpha\xi\} < \exp\left\{\alpha\mathbf{E}\xi + \frac{\alpha^2}{2}(\mathbf{E}\xi^2 + \varphi(\alpha))\right\} \tag{8.20}$$

holds for sufficiently small α with $\varphi(\alpha) = O(\alpha)$ for $\alpha \to 0$.

The proof of this lemma follows from the inequalities

$$\mathbf{E}\exp\{\alpha\xi\} = \sum_{n=0}^{\infty} \frac{\alpha^n \mathbf{E}\xi^n}{n!}$$

$$= 1 + \alpha\mathbf{E}\xi + \frac{\alpha^2}{2}\mathbf{E}\xi^2 + \sum_{n=3}\left(\frac{\alpha}{\alpha_0}\right)^n \frac{\alpha_0^n \mathbf{E}\xi^n}{n!}$$

$$\leqslant 1 + \alpha\mathbf{E}\xi + \frac{\alpha^2}{2}\mathbf{E}\xi^2 + \frac{c\alpha^3}{\alpha_0^3 - \alpha^3},$$

where $\alpha < \alpha_0$, and the inequality $1 + \gamma < \exp\gamma$ $(\gamma > 0)$.

Suppose that condition (8.19) is satisfied. Then it is obvious that for sufficiently small α the matrices $\tilde{\eta}_k = C^{1/2}\eta_k C^{-1/2}$ also satisfy this condition. On the other hand, from (8.12) we infer that the following estimate is valid when $k - 1 < t < k$ (where k is an integer):

$$(Cx(t,\omega),x(t,\omega))^p$$

$$\leqslant (Cx_0,x_0)^p\exp\{-p\lambda(k - 1) + 2p(\|\tilde{\eta}_1\| + \ldots + \|\tilde{\eta}_k\|)\}.$$

Since the random variables $\|\tilde{\eta}_i\|$ are independent, it follows from (8.19) and (8.20) that for sufficiently small p

$$\mathbf{E}(Cx(t,\omega),x(t,\omega))^p$$
$$\leqslant (Cx_0,x_0)^p\exp\{-p\lambda(k - 1) + 2p(\|\tilde{\eta}_1\| + \ldots + \|\tilde{\eta}_k\|)\}.$$

This inequality and (8.18) imply that if conditions (8.18) and (8.19) are satisfied, then the process $x(t,\omega)$ determined by the system (8.16) is asymptotically p-stable for sufficiently small p.

Other conditions for the stability of the system (8.16), based on the fact that the process $x_0,x_1(\omega),\ldots,x_k(\omega),\ldots$ is a Markov chain, will be given in Chapter IV.

CHAPTER II

STATIONARY AND PERIODIC SOLUTIONS OF DIFFERENTIAL EQUATIONS

1. Stationary and periodic stochastic processes. Convergence of stochastic processes

A stochastic process $\xi(t) = \xi(t,\omega)$ $(-\infty < t < \infty)$ with values in E_ℓ is said to be stationary (in the strict sense) if for every finite sequence of numbers $t_1,\ldots,t_n$ the joint distribution of the random variables $\xi(t_1 + h),\ldots,\xi(t_n + h)$ is independent of h. If we replace the arbitrary number h by a multiple of a fixed number ϑ, $h = k\vartheta$ $(k = \pm1,\pm2,\ldots)$, we get the definition of a periodic stochastic process with period ϑ, or a ϑ-periodic stochastic process[1]. Stationary and periodic stochastic processes constitute a mathematical idealization of physical noise acting on linear and nonlinear devices functioning in a medium with unvarying or periodically varying properties.

Let $\xi(t)$ be a stationary stochastic process with finite variance. By the definition of stationarity,

$$\mathbf{E}\xi(t) = m = \text{const}, \qquad \mathbf{D}\xi(t) = \mathbf{D} = \text{const},$$
$$K(s,t) = \operatorname{cov}(\xi(s),\xi(t)) = K(t - s). \tag{1.1}$$

As already mentioned in Chapter I, a process satisfying conditions (1.1) is said to be stationary in the wide sense. An important characteristic of stationary processes is their spectral density (see Section 1.1).

---

[1] There is an enormous literature on the properties of stationary stochastic processes. Among others, we might mention the paper Jaglom [1] and the books Doob [3], Rozanov [1], Ibragimov and Linnik [1]. The properties of periodic processes to be discussed below may be found, e.g., in a paper of Dorogovcev [1] and in Stratonovič [1].

If $\xi(t)$ is a ϑ-periodic stochastic process, then $\mathbf{E}\xi(t) = m(t)$ and $\mathbf{D}\xi(t) = D(t)$ are evidently periodic with the same period, i.e.

$$m(t + \vartheta) = m(t), \qquad \mathbf{D}(t + \vartheta) = \mathbf{D}(t). \tag{1.2}$$

The matrix-valued function $K(s,t)$ satisfies then the condition

$$K(s + \vartheta, t + \vartheta) = K(s,t) \tag{1.3}$$

for all s,t. A process whose moments satisfy (1.2) and (1.3) is said to be periodic in the wide sense.

It is obvious that a stationary process is periodic with arbitrary period. Conversely, a periodic process can be made stationary by a simple transformation (shift of the argument). Indeed, if τ is a random variable uniformly distributed on the interval $[0,\vartheta]$ and independent of the ϑ-periodic stochastic process $\xi(t)$, then the process $\eta(t) = \xi(t + \tau)$ is stationary. To prove this it suffices to observe that for every $t_1,\dots,t_n,A_1,\dots,A_n$ the function $\mathbf{P}\{\xi(t_1 + h) \in A_1,\dots,\xi(t_n + h) \in A_n\}$ is ϑ-periodic with respect to h, and therefore, for every h,

$$\mathbf{P}\{\eta(t_1 + h) \in A_1,\dots,\eta(t_n + h) \in A_n\}$$

$$= \frac{1}{\vartheta}\int_0^{\vartheta} \mathbf{P}\{\xi(t_1 + s + h) \in A_1,\dots,\xi(t_n + s + h) \in A_n\}ds$$

$$= \frac{1}{\vartheta}\int_0^{\vartheta} \mathbf{P}\{\xi(t_1 + s) \in A_1,\dots,\xi(t_n + s) \in A_n\}ds$$

$$= \mathbf{P}\{\eta(t_1) \in A_1,\dots,\eta(t_n) \in A_n\}.$$

It is easily verified that by averaging the moments of the process $\xi(t)$ over the period we obtain the corresponding moments of the process $\eta(t)$. For example,

$$\mathbf{E}\eta(t) = \frac{1}{\vartheta}\int_0^{\vartheta} \mathbf{E}\xi(s)ds,$$

$$\operatorname{cov}(\eta(s),\eta(t)) = \frac{1}{\vartheta}\int_0^{\vartheta} \operatorname{cov}(\xi(s + h),\xi(t + h))dh.$$

It is evident that a deterministic periodic function can be regarded as a periodic stochastic process. After a suitable shift of the argument we get a stationary process.

Let $f(t,x)$ be a Borel-measurable function, ϑ-periodic in t, and $\xi(t)$ a ϑ-periodic stochastic process. It is then readily seen that the process $f(t,\xi(t))$ is also ϑ-periodic. For example, if τ is a random variable uniformly distributed on the interval $[0,2\pi]$, then the process $\xi\sin(t+\tau)$ is stationary for every random variable ξ independent of τ, while the process $\xi\cos t\sin(t+\tau)$ is 2π-periodic. The sample functions of the processes in these examples are periodic. It is easy to construct also examples of periodic processes which almost surely have no periodic sample functions (paths).

In this chapter we shall frequently have to deal with sequences of random variables and with stochastic processes converging in various senses. Therefore let us first recall various definitions of convergence and some results connected with them[2].

A sequence of measures $\{\mu_n\}$ defined in $(E_\ell,\mathfrak{B})$ is said to be weakly convergent to a measure μ if

$$\lim_{n\to\infty}\int_{E_\ell} f(x)\mu_n(dx) = \int_{E_\ell} f(x)\mu(dx)$$

for every continuous and bounded function $f(x)$ on E_ℓ.

A sequence of random variables ξ_n is *weakly convergent* to ξ if the sequence of measures $\mathbf{P}_n(A) = \mathbf{P}\{\xi_n \in A\}$ converges weakly to the measure $\mathbf{P}(A) = \mathbf{P}\{\xi \in A\}$.

A sequence of random variables ξ_n is said to be *weakly compact* if it contains a weakly convergent subsequence. A sufficient condition for a sequence ξ_n to be weakly compact is that the random variables ξ_n be uniformly bounded in probability, i.e.,

$$\sup_n \mathbf{P}\{|\xi_n| > R\} \to 0 \quad \text{as} \quad R \to \infty.$$

A sequence ξ_n is said to *converge in probability* to ξ if $\mathbf{P}\{|\xi_n - \xi| > \delta\} \to 0$ as $n \to \infty$ for each $\delta > 0$.

Given a sequence ξ_n which converges weakly to ξ_0, one can construct a sequence $\tilde{\xi}_n$ $(n = 0,1,2,\ldots)$ in another probability space $(\tilde{\Omega},\tilde{\mathfrak{U}},\tilde{P})$ such that $\tilde{\xi}_n \to \tilde{\xi}_0$ in probability and the variables ξ_n and $\tilde{\xi}_n$ have the same distribution function for every $n \geq 0$. Skorohod [1] has generalized these results to stochastic processes as follows.

THEOREM 1.1. *Let $\xi_n(t,\omega)$ $(n = 1,2,\ldots)$ be a sequence of stochastic processes in E_ℓ such that for every $t_1,\ldots,t_k$ the joint distribution of $\xi_n(t_1),\ldots,\xi_n(t_k)$ is weakly convergent to some limit, and the sequence $\xi_n(t)$ is uniformly stochastically continuous, i.e.,*

$$\sup_{n,|s_1-s_2|<h} \mathbf{P}\{|\xi_n(s_1) - \xi_n(s_2)| > \varepsilon\} \xrightarrow[h\to 0]{} 0. \tag{1.4}$$

Then one can construct a sequence of stochastic processes

---

[2] See Prohorov [1], Skorohod [1], Gihman and Skorohod [1].

$\xi_n(s)$ $(n = 0,1,2,\ldots)$ *in another probability space* $(\tilde{\Omega},\tilde{\mathfrak{A}},\tilde{\mathbf{P}})$ *such that the process* $\tilde{\xi}_0(s)$ *is stochastically continuous,* $\tilde{\xi}_n(s) \to \tilde{\xi}_0(s)$ *in probability for all* s *and the finite-dimensional distributions of the processes* $\xi_n(s)$ *and* $\tilde{\xi}_n(s)$ *coincide for* $n > 0$.

THEOREM 1.2. *A sufficient condition for a sequence of stochastic processes* $\xi_n(t)$ *to contain a subsequence of processes with weakly convergent finite-dimensional distributions is that the sequence satisfies condition* (1.4) *and is uniformly bounded in probability:*

$$\sup_{t,n} \mathbf{P}\{|\xi_n(t)| > R\} \to 0 \quad \text{as} \quad R \to \infty. \tag{1.5}$$

Let the processes $\xi_n(t),\xi(t)$ be continuous on the interval $[a,b]$. Let $\mathcal{C}[a,b]$ denote the space of all continuous functions on $[a,b]$; all the sample functions of the processes $\xi_n(t),\xi(t)$ are almost surely in this class.

A sequence $\xi_n(t)$ is said to be *weakly convergent* to $\xi(t)$ as $n \to \infty$ if for every functional f continuous on $\mathcal{C}[a,b]$

$$\mathbf{E}f(\xi_n(t)) \xrightarrow[n\to\infty]{} \mathbf{E}f(\xi(t)).$$

Prohorov [1] has proved the following theorem.

THEOREM 1.3. *If the finite-dimensional distributions of the processes* $\xi_n(t)$ *are weakly convergent to some limit and there exist* $\alpha > 1$, $k > 0$ *and* $a > 0$ *such that for all* t_1,t_2 *and* n

$$\mathbf{E}|\xi_n(t_2) - \xi_n(t_1)|^a < k|t_2 - t_1|^a,$$

then the sequence of processes $\xi_n(t)$ *is weakly convergent to a process* $\xi(t)$ *whose finite-dimensional distributions coincide with the above-mentioned limit distributions.*

2. Existence conditions for stationary and periodic solutions[3]

An important part of the qualitative theory of differential equations is the study of periodic solutions of systems with periodic right-hand sides.

In a more general setting, this corresponds to the study of existence conditions and properties of periodic and stationary solutions of differential equations whose right-hand side is a

---

[3] Existence conditions for stationary and periodic solutions of differential equations with random right-hand side have been investigated under different assumptions and by other methods by Vorovič [1] and Dorogovcev [1].

periodic or stationary process in t for fixed values of the space variable.

In this section we shall present a general, but ineffective, solution of this problem. In the next section we shall use this solution to derive effective sufficient conditions for the existence of stationary and periodic functions in terms of auxiliary functions.

THEOREM 2.1. *Let* $G(x,z)$ $(x \in E_\ell, z \in E_k)$ *be a function and* $\xi(t)$ *a stationary stochastically continuous process in* E_k*, satisfying conditions* (1.3.1), (1.3.2). *Then there exists a stationary solution of the equation*

$$\frac{dx}{dt} = G(x,\xi(t)) \tag{2.1}$$

which is stationarily related to $\xi(t)$ *if and only if this equation has at least one solution* $y(t,\omega)$ *satisfying the condition*

$$\frac{1}{T}\int_0^T \mathbf{P}\{|y(t,\omega)| > R\}dt \to 0 \quad \text{as} \quad R \to \infty \tag{2.2}$$

uniformly in $T > T_0$ *(or* $T < -T_0$*).*

PROOF. Necessity is obvious, since every stationary solution $y(t,\omega)$ satisfies condition (2.2). To prove sufficiency, we first make the following observation. Solving equation (2.1) with initial condition $x(0) = x_0(\omega)$ by successive approximations, one may readily verify that the random variable $x(t,\omega)$ is measurable with respect to the minimal σ-algebra containing all possible events $\{\xi(s) \in A_1\}$ $(s \in [0,t])$ and $\{x_0(\omega) \in A_2\}$. Here and below, $A_i \in \mathfrak{B}$, where $\mathfrak{B}$ is the σ-algebra of Borel sets in Euclidean space. Therefore, in order to prove the existence of a stationary process $(X(t),\xi(t))$ satisfying equation (2.1) it will suffice to show that there exists a random variable $\eta(\omega)$ such that for all $t > 0$, $A_0, A_1, \ldots, A_m$, $s_1, \ldots, s_m$,

$$\begin{aligned}\mathbf{P}\{\eta(\omega) \in A_0, \xi(s_1) \in A_1, \ldots, \xi(s_m) \in A_m\} \\ = \mathbf{P}\{X(t) \in A_0, \xi(s_1 + t) \in A_1, \ldots, \xi(s_m + t) \in A_m\},\end{aligned} \tag{2.3}$$

where $X(t)$ is the solution of equation (2.1) with initial condition $x(0) = \eta(\omega)$.

Assume for definiteness that condition (2.2) holds with $T > 0$. Let $\tau_k(\omega)$ be a random variable, uniformly distributed on $[0,k]$ and independent of $\xi(t)$ and $y(0,\omega)$. We set $x_0^{(k)}(\omega) = y(\tau_k(\omega),\omega)$ and

$$x_k(t,\omega) = y(t + \tau_k(\omega),\omega), \xi_k(t,\omega) = \xi(t + \tau_k(\omega),\omega).$$

It is obvious that

$$\mathbf{P}\{x_k(t) \in A_0, \xi_k(s_1) \in A_1, \ldots, \xi_k(s_m) \in A_m\}$$

$$= \frac{1}{k}\int_0^k \mathbf{P}\{y(t+s) \in A_0,$$

$$\xi(s_1+s) \in A_1, \ldots, \xi(s_m+s) \in A_m\} ds. \tag{2.4}$$

It follows from (2.4) that for every k the distribution of the process $\xi_k(t)$ is the same as that of the process $\xi(t)$. It also follows from (2.2) that uniformly in $k > 0$,

$$\mathbf{P}\{|x_0^{(k)}(\omega)| > R\} = \frac{1}{k}\int_0^k \mathbf{P}\{|y(t)| > R\} dt \xrightarrow[R\to\infty]{} 0. \tag{2.5}$$

By the stochastic continuity of the process $\xi(t)$ and by (2.5), the family $(x_0^{(k)}(\omega), \xi^{(k)}(t,\omega))$ satisfies conditions (1.4) and (1.5). Applying Theorems 1.1 and 1.2, we see that in some probability space $(\tilde{\Omega}, \tilde{\mathfrak{A}}, \tilde{\mathbf{P}})$ there is a sequence $(\tilde{x}_0^{(k)}(\tilde{\omega}), \tilde{\xi}^{(k)}(t,\tilde{\omega}))$ with the same distribution as $(x_0^{(k)}(\omega), \xi^{(k)}(t,\omega))$, such that some subsequence $(\tilde{x}_0^{(n_k)}(\tilde{\omega}), \tilde{\xi}^{(n_k)}(t,\tilde{\omega}))$ converges in probability to $(\tilde{x}(\tilde{\omega}), \tilde{\xi}(t,\tilde{\omega}))$. Obviously, the finite-dimensional distributions of the processes $\tilde{\xi}(t,\tilde{\omega})$ and $\xi(t,\omega)$ are the same. We can now construct on the original probability space random variables $x(\omega)$ and $x^{(n_k)}(\omega)$ whose joint distribution with $\xi(t,\omega)$ is the same as the joint distribution of

$$\tilde{x}(\tilde{\omega}),\ \tilde{x}_0^{(n_k)}(\tilde{\omega}),\ \tilde{\xi}(t,\tilde{\omega}).$$

We shall prove that (2.3) holds for $\eta(\omega) = x(\omega)$. Let $X_{n_k}(t)$ $(k = 1,2,\ldots)$ denote the solution of equation (2.1) with initial condition $X_{n_k}(0) = x^{(n_k)}(\omega)$. Now conditions (1.3.1), (1.3.2) and the Gronwall-Bellman lemma imply the inequality

$$|X_{n_k}(t) - X(t)| < |x^{(n_k)}(\omega) - x(\omega)| \exp\left\{\int_0^t B(u,\omega) du\right\},$$

and so $X_{n_k}(t) \to X(t)$ in probability for every t. Let f be a continuous bounded function. Then, by what we have proved it follows from (2.4) that for each t and $s_1, \ldots, s_m$,

$$\mathbf{E} f(\xi(s_1+t), \ldots, \xi(s_m+t), X(t))$$

$$= \lim_{k\to\infty} \mathbf{E} f(\xi(s_1+t), \ldots, \xi(s_m+t), X_{n_k}(t))$$

$$= \lim_{k\to\infty} \mathbf{E}f(\xi_{n_k}(s_1 + t),\ldots,\xi_{n_k}(s_m + t),x_{n_k}(t))$$

$$= \lim_{k\to\infty} \frac{1}{n_k}\int_0^{n_k} \mathbf{E}f(\xi(s_1 + t + u),\ldots,\xi(s_m + t + u),y(t + u))du$$

$$= \lim_{k\to\infty} \frac{1}{n_k}\int_0^{n_k} \mathbf{E}f(\xi(s_1 + s),\ldots,\xi(s_m + s),y(s))ds$$

$$= \mathbf{E}f(\xi(s_1),\ldots,\xi(s_m),x(\omega)). \tag{2.6}$$

This implies (2.3), and hence the assertion of the theorem.

The analogous result for a periodic process $\xi(t)$ is given by the following theorem.

THEOREM 2.2. *Let* $G(x,z)$ $(x \in E_1, z \in E_k)$ *be a given function and* $\xi(t)$ *a* ϑ*-periodic stochastically continuous process in* E_k *satisfying conditions* (1.3.1), (1.3.2). *Then there exists a periodic solution of equation* (2.1) *which is periodically related to* $\xi(t)$ *if and only if the equation has at least one solution* $y(t,\omega)$ *satisfying the condition*

$$\frac{1}{k + 1} \sum_{n=0}^{k} \mathbf{P}\{|y(n\vartheta)| > R\} \to 0 \quad \text{as} \quad R \to \infty \tag{2.7}$$

uniformly in $k = 1,2,\ldots$ *(or* $k = -1,-2,\ldots$*).*

The proof is entirely analogous to that of Theorem 2.1. The only difference is that instead of the processes $x_k(t) = y(t + \tau_k)$ one must consider a sequence $Y_k(t) = y(t + \chi_k)$, where χ_k is a random variable independent of $\xi(t)$ and $y(0,\omega)$ such that $P\{\chi_k = n\vartheta\} = 1/(k + 1)$ $(n = 0,1,\ldots,k)$.

As we shall see in the next section, the advantage of condition (2.2) over (2.7) is that it is easier to verify whether (2.2) holds even if no solutions of equation (2.1) are known. Thus, the following lemma may be sometimes useful.

LEMMA 2.1. *Condition* (2.7) *of Theorem* 2.2 *can be replaced by condition* (2.2).

PROOF. The necessity of condition (2.2) is obvious. Let us prove the sufficiency. Let $y(t) = y(t,\omega)$ be a solution of equation (2.1) satisfying condition (2.2). Then for each τ, $z(t) = y(t + \tau)$ is a solution of the equation

$$\frac{dz}{dt} = G(z,\xi(t + \tau)). \tag{2.8}$$

Now let τ be a random variable uniformly distributed on $[0,\vartheta]$ and independent of the process $\xi(t)$. Then, as shown in Section 1,

$\xi(t + \tau)$ is a stationary process. Moreover, the solution $z(t)$ of equation (2.8) satisfies condition (2.2), since

$$\frac{1}{T}\int_0^T \mathbf{P}\{|z(t)| > R\}dt = \frac{1}{\vartheta}\int_0^\vartheta ds \frac{1}{T}\int_0^T \mathbf{P}\{|y(t + s)| > R\}dt$$

$$\leqslant \frac{T + \vartheta}{T}\frac{1}{\vartheta}\int_0^\vartheta ds \frac{1}{T + s}\int_0^{T+s} \mathbf{P}\{|y(u)| > R\}du.$$

Applying Theorem 2.1, we see that equation (2.8) has a solution $Z_1(t,\omega)$ which is a stationary process. It follows from Theorem 1.3.1 that

$$\sup_{0\leqslant t\leqslant\vartheta} |Z_1(t)| \leqslant |Z_1(0)| + \int_0^\vartheta |G(Z_1(0),\xi(s + \tau))|ds$$

$$\times \exp\left\{\int_0^\vartheta B(s + \tau,\omega)ds\right\}.$$

By conditions (1.3.1), (1.3.2) and the stationarity of the process Z_1, the probability of the event

$$\left\{\sup_{s\leqslant t\leqslant s+\vartheta} |Z_1(t)| > R\right\}$$

is independent of s and

$$P\left\{\sup_{s\leqslant t\leqslant s+\vartheta} |Z_1(t)| > R\right\} \to 0 \quad \text{as} \quad R \to \infty. \tag{2.9}$$

It is clear now that the function $y_1(t,\omega) = Z_1(t - \tau(\omega),\omega)$ is a solution of equation (2.1). By (2.9), this solution satisfies condition (2.7). Hence, by Theorem 2.2, it follows that equation (2.1) has a periodic solution. This completes the proof of the lemma.

REMARK. The global Lipschitz condition (1.3.1) is sometimes too restrictive. It can be seen from the proofs of Theorem 2.1 and Lemma 2.1 that this condition is used only to verify (2.9) and the relation

$$X_{n_k}(t) \to X(t) \quad \text{in probability as} \quad k \to \infty. \tag{2.10}$$

These relations hold if the solutions of (2.1) are uniformly indefinitely continuable in the sense of Remark 2 to Theorem 1.3.3 and conditions (1.3.2), (1.3.6) are satisfied.

In fact, by conditions (1.3.2), (1.3.6) and the Gronwall-

Bellman lemma, for every fixed t_0 and all sample functions $X_{n_k}(t,\omega),X(t,\omega)$ satisfying the conditions

$$\sup_{0\leqslant t\leqslant t_0} |X_{n_k}(t)| \leqslant R, \qquad \sup_{0\leqslant t\leqslant t_0} |X(t)| \leqslant R, \tag{2.11}$$

we get the inequality

$$|X_{n_k}(t_0) - X(t_0)| \leqslant |x^{(n_k)}(\omega) - x(\omega)|\exp\left\{\int_0^{t_0} B_R(t,\omega)dt\right\}. \tag{2.12}$$

Let $\varepsilon > 0$ be arbitrary. Since the solutions of equation (2.1) are uniformly indefinitely continuable, the probability of the events (2.11) can be made greater than $1 - \varepsilon/2$ by choosing R sufficiently large. Hence and by considering (2.12) for sufficiently large k we get the inequalities

$$\mathbf{P}\{|X_{n_k}(t_0) - X(t_0)| > \varepsilon\} \leqslant \frac{\varepsilon}{2} + \mathbf{P}\left\{|x^{(n_k)}(\omega) - x(\omega)|\exp\left(\int_0^{t_0} B_R(t,\omega)dt\right) > \frac{\varepsilon}{2}\right\} \leqslant \varepsilon.$$

This proves (2.10). The proof of (2.9) is analogous.

This remark, together with Theorem 1.3.3, implies the following corollaries.

COROLLARY 1. *Let the functions* $F(x,t)$ *and* $\sigma(x,t)$ *and the stochastic process* $\xi(t)$ *be* ϑ*-periodic and satisfy the assumptions of Theorem* 1.3.3. *Assume also that the equation* $dx/dt = F(x,t) + \sigma(x,t)\xi(t)$ *has a solution satisfying condition* (2.2). *Then this equation also has a* ϑ*-periodic solution. Similarly, if* F *and* σ *are independent of* t *and* $\xi(t)$ *is a stationary process, then the above conditions imply the existence of a stationary solution.*

COROLLARY 2. *Conditions* (2.2) *and* (2.7) *are valid if the system* (2.1) *is dissipative. Therefore, if the system* (2.1) *is dissipative,* $\xi(t,\omega)$ *is a stationary (periodic) stochastically continuous process and conditions* (1.3.1), (1.3.2) *are satisfied, and then the system* 2.1 *has a stationary (periodic) solution.*

EXAMPLE. Let $G(x,t)$ be a deterministic function which is ϑ-periodic in t and such that conditions (1.3.1), (1.3.2) are satisfied and the equation

$$\frac{dx}{dt} = G(x,t) \tag{2.13}$$

has at least one bounded solution. It follows from Theorem 2.2 that for some (generally random) initial condition the solution of equation (2.13) is a periodic stochastic process. For $n \leqslant 2$ this follows also from a well-known theorem of Massera (see Pliss [1], p. 186). Of course, this result does not guarantee the existence of a deterministic periodic solution of equation (2.13), since a periodic stochastic process need not have periodic sample functions.

3. Special existence conditions for stationary and periodic solutions

For systems of the special form

$$\frac{dx}{dt} = F(x,t) + \sigma(x,t)\xi(t) \tag{3.1}$$

one can derive effective conditions which are sufficient for the existence of periodic and stationary solutions.

THEOREM 3.1. *Suppose that the vector $F(x,t)$ and the matrix $\sigma(x,t)$ are ϑ-periodic in t and that they satisfy a local Lipschitz condition; let further $F(0,t) \in \mathbf{L}$ and*

$$\sup_{x,t} \|\sigma(x,t)\| < \infty. \tag{3.2}$$

Assume moreover that the truncated system

$$\frac{dx}{dt} = F(x,t)$$

has a Ljapunov function $V(x,t) \in \mathbf{C}_0$ satisfying the following conditions:

1. *$V(x,t)$ is nonnegative, and*

$$\inf_{t>0} V(x,t) \to \infty \quad \text{as} \quad |x| \to \infty.$$

2. *d^0V/dt is bounded above, and $\sup_{t>0} d^0V/dt \to -\infty$ as $|x| \to \infty$.*

Then equation (3.1) *has a ϑ-periodic solution for each ϑ-periodic stochastically continuous process $\xi(t)$ with finite expectation. If F and σ are independent of t and $\xi(t)$ is a stationary process, then the same conditions imply the existence of a stationary solution.*

PROOF. Let $x(t) = x(t,\omega)$ be a solution of equation (3.1) satisfying the condition $x(t_0) = x_0$. Using Condition 1 of the theorem, inequality (3.2) and Lemma 1.3.1, we see that almost surely, for $t > t_0$ and some constant $k > 0$,

$$- V(x_0,t_0) \leqslant V(x(t),t) - V(x_0,t_0)$$

$$\leqslant \int_{t_0}^{t} \frac{d^0V(x(s),s)}{ds}\,ds + k\int_{t_0}^{t} |\xi(s)|ds. \qquad (3.3)$$

We set $k_1 = \sup_{E} d^0V/dt$, $-c_r = \sup_{|x|>r} d^0V/dt$. It follows from the assumptions of the theorem that

$$k_1 < \infty, \quad c_r \to \infty \quad \text{as} \quad r \to \infty. \qquad (3.4)$$

Replacing for $|x(s)| > r$ the function d^0V/ds in (3.3) by the bound $-c_r$ and for $|x(s)| \leqslant r$ by the bound k_1 and then taking expectations, we get

$$- V(x_0,t_0) \leqslant - c_r\int_{t_0}^{t} \mathbf{P}\{|x(s)| > r\}ds + k_1(t - t_0)$$

$$+ k\int_{t_0}^{t} \mathbf{E}|\xi(s)|ds.$$

Hence it follows by (3.4) that for some constant k_2

$$\frac{\int_{t_0}^{t} \mathbf{P}\{|x(s)| > r\}ds}{t - t_0} < \frac{k_2}{c_r} \to 0 \quad \text{as} \quad r \to \infty. \qquad (3.5)$$

Condition (3.5) is equivalent to (2.2). Applying Lemma 2.1 and Corollary 1 of Section 2, we get the first assertion of the theorem. The second assertion is proved in the same way.

REMARK. The assertion of the theorem is valid when the assumption that $\inf_{t>0} V(x,t) \to \infty$ as $|x| \to \infty$ is replaced by the assumption that the solutions of equation (3.1) are uniformly indefinitely continuable for $t > 0$. It is also sufficient to require that the solutions be indefinitely continuable for $t < 0$ and that the following condition holds: The function d^0V/dt is bounded below and $d^0V/dt \to \infty$ as $|x| \to \infty$. (This case reduces to the preceding one if we set $s = -t$.)

EXAMPLE 1. If the system (3.1) is one-dimensional ($x \in E_1$), then, considering the Ljapunov function $V = |x|$, we have $d^0V/dt = F(x,t)\operatorname{sign}x$. Hence Theorem 3.1 and the Remark after it yield the following result.

If F and σ are periodic functions of t such that

$$F \in C_0, \quad \sigma \in C_0, \quad \sup|\sigma| < \infty$$

and either $F(x,t)\operatorname{sign}x \to -\infty$ or $F(x,t)\operatorname{sign}x \to \infty$ as $|x| \to \infty$, then equation (3.1) has a periodic solution in E_1 for every periodic process $\xi(t)$ with bounded expectation. An analogous conclusion holds for stationary solutions as well.

For example, if $f(t)$ is a ϑ-periodic continuous function and $\xi(t)$ a ϑ-periodic process, then the equation $dx/dt = xf(t) + \xi(t)$ always has a periodic solution, provided $\mathbf{E}|\xi(t)| < \infty$ and $f(t)$ does not vanish. On the other hand it is obvious that if $F(x,t) > k > -\infty$ (or $F(x,t) < k < \infty$), then equation (3.1) need not have periodic solutions, since for a suitable choice of σ and ξ the right-hand side of equation (3.1) will have fixed sign. A more general result is given by the following

LEMMA 3.1 *Let $F(x) \in \mathbf{C}$ be a function which satisfies none of the conditions*

$$F(x)\operatorname{sign}x \xrightarrow[|x|\to\infty]{} \pm\infty. \tag{3.6}$$

Then there exists a stationary stochastic process $\xi(t)$ with finite expectation such that the equation

$$\frac{dx}{dt} = F(x) + \xi(t) \tag{3.7}$$

has no stationary solution.

PROOF. As already mentioned, the assertion is obvious if the function $F(x)$ is bounded above or below. If it is neither and conditions (3.6) do not hold, then there exist an infinite sequence of points α_k $(k = 1,2,\ldots)$ and a number c such that $\alpha_k \to \infty$ or $\alpha_k \to -\infty$ as $k \to \infty$, $F(\alpha_k) = c$ and each α_k is a stable equilibrium point of the equation

$$\frac{dx}{dt} = F(x) - c. \tag{3.8}$$

To be more specific, suppose that $\alpha_k \to \infty$. Then the following three cases are possible:

(a) $F(x) \geqslant c$ for $x < x'$,

(b) $F(x) \leqslant c$ for $x < x'$,

(c) there exists a sequence $x'_k \to -\infty$ such that $F(x'_k) = c$ and the x'_k are stable equilibrium points of equation (3.8).

CASE (a). We may assume without loss of generality that $x' = \alpha_1$. We claim that in this case equation (3.7) has no stationary solutions if $\xi(t) = -c + |\eta(t)|$, where $\eta(t)$ is a stationary stochastic process such that for every constant A

$$\mathbf{P}\left\{\sup_{0 \leqslant u \leqslant t} \int_u^{u+1} |\eta(s)|ds > A\right\} \to 1 \quad \text{as } t \to \infty. \tag{3.9}$$

(Condition (3.9) holds for instance if $\eta(t)$ is a Gaussian stationary Markov process governed by the differential operator $d^2/dx - xd/dx$.)

Suppose that there exists a stationary process $x(t)$ satisfying equation (3.7). Since $F(x) \geqslant c$ for $x < \alpha_1$, the function $x(t)$ is monotone increasing for $x(t) < \alpha_1$, and therefore

$$\mathbf{P}\{x(0,\omega) < \alpha_1\} = 0. \tag{3.10}$$

We shall prove that $\mathbf{P}\{\alpha_1 \leqslant x(0,\omega) < \alpha_2\} = 0$. To this end, we first observe that, by construction, the points α_k have the following property: once the sample function $x(t)$ has "hit" the point α_k at some time t_0, it "cannot" go to the left of α_k for $t > t_0$. Hence in this case it follows from equation (3.7) that either $X(t + 1) > \alpha_2$ or

$$X(t + 1) - X(t) \geqslant \int_t^{t+1} |\eta(s)|ds + \min_{x \in [a_1,a_2]} (F(x) - c).$$

Hence follows the relation

$$\{x(0,\omega) \geqslant \alpha_1\} \cap \left\{\sup_{0 \leqslant u \leqslant t-1} \int_u^{u+1} |\eta(s)|ds + \right.$$

$$\left. + \min_{x \in [\alpha_1,\alpha_2]} (F(x) - c) > \alpha_2 - \alpha_1\right\}$$

$$\subset \{x(t,\omega) \geqslant \alpha_2\}. \tag{3.11}$$

By (3.9), (3.10) and (3.11),

$$P\{x(0,\omega) \geqslant \alpha_2\} = \lim_{t \to \infty} \mathbf{P}\{x(t,\omega) \geqslant \alpha_2\} = 1.$$

Similarly, we show that $\mathbf{P}\{x(0,\omega) \geqslant \alpha_k\} = 1$ for every k. This contradiction shows that a stationary solution does not exist. The proof for cases (b) and (c) is similar. (In case (b) one sets $\xi(t) = -c - |\eta(t)|$).

EXAMPLE 2.[4] Suppose that for $|x| > x_0$ and some positive integers n and k the coefficients of the equation

$$x'' + f(x)x' + g(x) = \sigma(x,x')\xi(t,\omega) \tag{3.12}$$

[4] This example is due to Nevel'son.

satisfy the conditions

$$0 < g(x)/x^{2n+1} < c, \qquad 0 < f(x)/x^{2k} < c,$$

and also the conditions

$$|\sigma| < c; \quad g(x)F(x) \to \infty \quad \text{as} \quad |x| \to \infty \ (F(x) = \int_0^x f(t)dt),$$

$$F(x)\text{sign}x > \delta > 0 \quad \text{for} \quad |x| > x_0.$$

Let $\xi(t)$ be a periodic (stationary) stochastic process with finite expectation. Then equation (3.12) has a periodic (stationary) solution. The proof utilizes Theorem 3.1 applied to the system of equations derived from (3.12), where we set $x' = y$, and take the Ljapunov function

$$V(x,y) = \left[\frac{y^2}{2} + (F(x) - p(x))y + G(x) + \int_0^x f(t)(F(t) - p(t))dt + 1\right]^{\alpha} - c_1,$$

with $G(x) = \int_0^x g(t)dt$, $p(x) = \nu\text{arctg}x$, and the positive constants γ, c_1, α so chosen that

$$\min V(x,y) = 0, \qquad V \in \mathbf{C}_0,$$

$$d^0V/dt \to -\infty \quad \text{as} \quad x^2 + y^2 \to \infty.$$

Note that the conditions of this example hold for a Van-der-Pol equation in which $f(x) = x^2 - 1$, $g(x) = x$.

4. Conditions for convergence to a periodic solution

Hitherto we have dealt only with conditions concerning the existence of a periodic (stationary) solution of a differential equation whose right-hand side is a periodic (stationary) process for fixed x. However only those periodic solutions are of practical interest which are stable, in the sense that if the initial conditions lie in a certain class, then the solutions ultimately converge to periodic solutions. In most cases it is sufficient to consider stability for initial conditions which are independent of the right-hand side of the system.

In some cases a periodic solution of a differential equation turns out to be stable in the large, i.e., every solution ultimately converges to a periodic solution. It is clear that if a periodic solution is stable in the large it is unique. These definitions are rather vague, for it is not clear in what sense one should understand the concepts "ultimately" and "convergence to a periodic process." The first of these concepts can be made rigorous as follows.

DEFINITION. A periodic (stationary) solution $x^0(t,\omega)$ of equation (2.1) is stable in a certain sense for initial conditions belonging to a class **K** if for all random variables $x_0(t_0,\omega) \in$ **K**, the solution $x(t,x_0(t_0,\omega),t_0,\omega)$ of equation (2.1) with initial condition $x(t_0) = x_0(t_0,\omega)$ converges to $x^0(t,\omega)$ in that same given sense as $t_0 \to -\infty$.

In accordance with the various types of convergence (see Section 1), we can consider almost sure stability, stability in probability, weak stability, and so on. In this section we shall establish some sufficient conditions for almost sure stability.

The following theorem indicates the connection between the asymptotically stable compact invariant set of a deterministic equation and the periodic (stationary) solutions of the perturbed system obtained when a small stochastic process is superimposed on the deterministic system. To simplify the exposition, we shall confine ourselves to the case in which the invariant set is an equilibrium point, the system of equations is autonomous and the random perturbation stationary.

THEOREM 4.1 *Let y_0 be an asymptotically stable singular point of the system*

$$\frac{dx}{dt} = F(x), \tag{4.1}$$

*where $F(x) \in$ **C**. Let $g(x,z)$ $(x \in E_1, z \in E_k)$ be a bounded Borel-measurable function such that $\|\partial g(x,z)/\partial x\|$ is bounded in a neighborhood of the point x_0, and $\xi(t,\omega)$ a stochastically continuous stationary stochastic process in E_k. Then for all sufficiently small $|\varepsilon|$ the equation*

$$\frac{dx}{dt} = F(x) + \varepsilon g(x,\xi(t,\omega)) \tag{4.2}$$

has a stationary solution which almost surely satisfies the condition

$$\sup_{-\infty<t<\infty} |x(t,\omega) - y_0| < \delta(\varepsilon) \qquad (\delta(\varepsilon) \to 0 \quad \text{as} \quad \varepsilon \to 0).$$

If moreover the point y_0 is asymptotically stable for the system (4.1) in the linear approximation, then a sufficiently small neighborhood of the point y_0 contains a unique stationary solu-

tion of equation (4.2) *which is almost surely stable for every initial condition* $x_0(t_0,\omega)$ *such that for some* $\delta_1(\varepsilon)$

$$\mathbf{P}\{|x_0(t_0,\omega) - y_0| < \delta_1(\varepsilon)\} = 1. \tag{4.3}$$

PROOF. Suppose y_0 is asymptotically stable for the system (4.1) and consider a fixed neighborhood of y_0. If $|\varepsilon|$ and $|x(t_0) - y_0|$ are sufficiently small, then no solution of the system (4.2) can leave this neighborhood for $t > t_0$. This follows directly from the stability of the solution $x(t) \equiv y_0$ of (4.1) with respect to continually acting perturbations (see Malkin [2]). This together with Theorem 2.1 implies the first part of the theorem.

Since the linear system

$$\frac{dz}{dt} = \frac{\partial F}{\partial x}(y_0)z$$

is asymptotically stable and the matrix $\|\partial g/\partial x\|$ is bounded in a neighborhood of y_0, there exists a constant $\partial_1(\varepsilon)$ such that for $|x_i - y_0| < \delta_1(\varepsilon)$, all $t > t_0$ and certain positive constants c and λ,

$$|x^{(2)}(t) - x^{(1)}(t)| < ce^{-\lambda(t-t_0)}, \tag{4.4}$$

where $x^{(i)}(t)$ is a solution of equation (4.2) with initial condition $x^{(i)}(t_0) = x_i$, $i = 1,2$.

Let $X(t,\omega)$ be some stationary solution of equation (4.2) in the $\delta_1(\varepsilon)$-neighborhood of the point y_0, and $Y^{(t_0)}(t,\omega)$ a solution of equation (4.2) satisfying the initial condition $Y^{(t_0)}(t_0,\omega) = x_0(t_0,\omega)$, where $x_0(t_0,\omega)$ satisfies condition (4.3). Setting $x^{(1)} = X(t_0,\omega)$, $x^{(2)} = x_0(t_0,\omega)$ in (4.4), we see that

$$\mathbf{P}\left\{\lim_{t_0\to-\infty} Y^{(t_0)}(t,\omega) = X(t,\omega)\right\} = 1$$

as $t_0 \to -\infty$ which implies the required assertions.

Note that if we set $x^{(1)} = X(t_0,\omega)$ in (4.4) and let $t_0 \to -\infty$, the evolution of the process $X(t,\omega)$ for $t \in (-\infty,s)$ is determined by that of the process $\xi(t,\omega)$ on the same interval. If moreover $g(x,z)$ is invertible as a function of z, the converse is also true. Thus the process $X(t,\omega)$ has the same regularity and mixing properties (see Rozanov [1]) as the process $\xi(t,\omega)$.

THEOREM 4.2. *Let* G *be a function which is* ϑ*-periodic in* t *(independent of* t*) and satisfies the assumptions of Theorem* 1.4.4, *and* $\xi(t,\omega)$ *a* ϑ*-periodic (stationary) stochastic process. Then the equation*

$$\frac{dx}{dt} = G(x,t,\xi(t,\omega))$$

has a unique periodic (stationary) solution which is almost surely stable for any initial conditions such that $\mathbf{P}\{|x_0(t_0,\omega)| < c\} = 1$ *for some* c.

The reader should have no difficulty in proving this theorem, employing the arguments used in the proofs of Theorems 4.1 and 1.4.4.

We conclude this chapter with the following comments.

1. Theorem 2.1, which is the fundamental theorem of this chapter, admits various generalizations. For example, it is not hard to prove a corresponding result for equations with delayed argument (see Kolmanovskiĭ [1]) and for Itô (stochastic) equations (see Chapter III).

In Itô and Nisio [1] similar methods were used to prove an analogous theorem for Itô equations with delay.

2. The problem of the existence and stability of stationary (periodic) solutions is also of interest for partial differential equations. For example, let us consider a simple model problem in the strip $0 < x < 1$, $-\infty < t < \infty$:

$$\left.\begin{aligned} \frac{\partial u}{\partial t} &= a(x)\frac{\partial^2 u}{\partial x^2} + b(x)\frac{\partial u}{\partial x} + c(x)u + f(x,\xi(t,\omega)) \\ &= Lu + f(x,\xi(t,\omega)), \\ &\qquad u(0,t) = u(1,t) = 0. \end{aligned}\right\} \tag{4.5}$$

It is readily shown that if $\xi(t)$ is a continuous stochastic process with finite expectation, then problem (4.5) has a stationary solution in the following sense: There exists a function $u(x,t,\omega)$ satisfying the equation and the boundary conditions of (4.5) for almost all ω, which for each fixed x is a stationary stochastic process stationarily related to $\xi(t,\omega)$.

Let $p(x,t,y)$ denote the Green function of the problem

$$\frac{\partial u}{\partial t} = Lu, \qquad u(0,t) = u(1,t) = 0.$$

Then the above-mentioned stationary solution can be determined from the formula

$$u(x,t,\omega) = \int_{-\infty}^{t} ds \int_{-\infty}^{+\infty} p(x,t-s,y) f(y,\xi(s,\omega))\,dy.$$

It is easy to show that this stationary solution is stable in the sense that every solution of problem (4.5) satisfying the initial condition $u(x,t_0) = \varphi(x,t_0)$ converges almost surely to $u(x,t,\omega)$

as $t_0 \to -\infty$, for every bounded function $\varphi(x,t_0)$.

This model is readily generalized; for example, instead of homogeneous boundary conditions one can consider conditions of the form

$$u(0,t) = \xi_1(t,\omega), \qquad u_1(1,t) = \xi_2(t,\omega),$$

where $\xi_1(t,\omega)$, $\xi_2(t,\omega)$ are stationary and stationarily related stochastic processes.

It is also easy to prove the existence of a stationary solution in case of an unbounded domain, provided the coefficient $c(x)$ in the operator L satisfies the condition $c(x) \leqslant c_0 < 0$. There is an analogous result for periodic solutions.

Apparently far more interesting but not so well investigated is the existence problem for stationary solutions of nonlinear partial differential equations. A few papers have been devoted to the solutions of the equations of hydrodynamics with stochastic coefficients (see the survey article of Kampé de Fériet [1] which includes a detailed bibliography).

3. We have established above certain results concerning the almost sure stability of stationary and periodic solutions. Although it seems that weak stability is rather more frequently encountered, no general conditions for weak stability are presently known. In particular, the following well-known problem seems to be yet unsolved. Let $F(x,t)$ $(x \in E_1)$ be a periodic function such that $F(x,t)\operatorname{sign}x \to -\infty$ as $|x| \to \infty$. Consider the equation $dx/dt = F(x,t) + \xi(t,\omega)$. What restrictions do we have to impose on the periodic process $\xi(t)$ in order to ensure that every solution of this equation defined by an initial condition independent of $\xi(t)$ converges to some periodic solution? It seems probable that this property is shared by quite a broad class of processes $\xi(t)$. For example, it is known that even in the relatively "unfavorable" case of a deterministic process $\xi(t)$ the property always holds (see Pliss [1], Theorem 9.2).

4. The question of stability of stationary and periodic solutions is intimately connected with the investigation of the properties of a stationary (periodic) solution of equation (3.1). Suppose that equation (3.1) has a stationary solution $x(t)$. To simplify matters, assume that F and σ are independent of t and $\xi(t)$ is a stationary process which is ergodic, regular, satisfies a mixing condition, etc. Under what restrictions on F, σ will the process $x(t)$ possess the analogous properties?

In the proof of Theorem 4.1 above we have answered this question only in the simplest case.

CHAPTER III

MARKOV PROCESSES AND STOCHASTIC DIFFERENTIAL EQUATIONS

1. Definition of Markov processes[1]

Consider the following equation in E_ℓ:

$$\frac{dx}{dt} = F(x(s),t), \tag{1.1}$$

where for each t, $F(x(s),t)$ is a functional of the segment of the sample function $x(s)$ for $t - \tau(t) \leqslant s \leqslant t$. It is well-known that the specification of $x(t_0)$ does not determine the solution of this equation for $t > t_0$ if $\tau(t) > 0$. Moreover, the solution depends on the "past" of the sample function. Therefore equation (1.1) is known as an equation with after-effect. As opposed to this, an ordinary differential equation might be called an equation without after-effect.

Markov processes occupy roughly the same position among stochastic processes as do ordinary differential equations among equations with after-effect.

To clarify the meaning of this statement, we first consider the case of discrete time. In what follows it will be assumed that the reader is familiar with the concept of conditional expectation; see Gihman and Skorohod [1]. Let $x_0, \xi_1, \xi_2, \ldots, \xi_n, \ldots$ be random variables in E_ℓ. We define new random variables by the formulae

$$x_{n+1} - x_n = a_{n+1}(x_n) + \sigma_{n+1}(x_n)\xi_{n+1} \quad (n = 0,1,2,\ldots). \tag{1.2}$$

Here the vectors $a_n(x)$ and the matrices $\sigma_n(x)$ are $\mathfrak{B}$-measurable functions in E_ℓ. (Recall that $\mathfrak{B}$ denotes the σ-algebra of Borel sets in E_ℓ.)

---

[1] A more detailed definition of Markov processes may be found in Dynkin [2],[3].

In the general case, when the joint distribution of $x_0, \xi_1, \ldots, \xi_n, \ldots$ is arbitrary, the conditional distribution of x_{n+1}, given x_n, is not the same as the conditional distribution of x_{n+1}, given $x_0, x_1, \ldots, x_n$, and in this sense the sequence $x_0, x_1, \ldots$ may be termed a process with after-effect. But if $x_0, \xi_1, \ldots, \xi_n$ are independent, then it is easily shown that for each $A \in \mathfrak{B}$ and $k > 0$,

$$\mathbf{P}\{x_{n+k} \in A \mid x_0, x_1, \ldots, x_n\} = \mathbf{P}\{x_{n+k} \in A \mid x_n\} \qquad \text{(a.s.)}. \qquad (1.3)$$

(Here and below "a.s." and "P-a.s." will serve as abbreviations for almost surely, i.e., the relation in question is valid with probability 1 relative to the measure **P**.)

Intuitively speaking, (1.3) means that the prediction of the behavior of the sequence $x_{n+1}, x_{n+2}, \ldots$ when x_n is known remains unchanged if the entire "history" of the process is known for $k < n$; in other words, the past has no effect on the future when the present is fixed (the past has no after-effects).

Let $P_n(x, A)$ denote the distribution of the random variable $x + a_n(x) + \sigma_n(x)\xi_n$. Then, for each $A \in \mathfrak{B}$,

$$\mathbf{P}\{x_n \in A \mid x_{n-1}\} = P_n(x_{n-1}, A) \qquad \text{(a.s.)}. \qquad (1.4)$$

It is clear that $P_n(x, A)$ is a Borel-measurable function of x.

A sequence $x_0, x_1, \ldots, x_n, \ldots$ satisfying condition (1.3) for all $A \in \mathfrak{B}$ is called a *Markov chain*.

It can be shown (see Doob [1]) that for every Markov chain there exists a function $P_n(x, A)$, which is a $\mathfrak{B}$-measurable function of x, and which is for every fixed n, x a measure satisfying condition (1.4). This function is the one-step transition probability from x to A at time n. This function immediately generalizes to the transition probability from x at time k to A at time n. This satisfies the relation

$$P(k, x, n, A) = \mathbf{P}\{x_n \in A \mid x_k = x\} \qquad \text{(a.s.)}. \qquad (1.5)$$

It is clear that $P_n(x, A) = P(n, x, n + 1, A)$.

It is possible to construct a Markov chain for any a priori given family of transition probabilities $P_n(x, A)$. This Markov chain satisfies the Chapman-Kolmogorov equation:

$$P(k, x, m, A) = \int P(k, x, n, dy) P(n, y, m, A) \qquad (k < n < m). \qquad (1.6)$$

The sequence (1.2) considered in the example above is a very special case of a Markov chain. Nevertheless, every Markov chain can be represented as a system whose evolution at time $n + 1$ is entirely determined by n, its state at time n, and certain random factors which are independent of the entire past history of the system.

In the continuous-time case we introduce the following definition which is analogous to (1.3).

A stochastic process $x(t,\omega)$ with values in E_ℓ, defined for $t \geqslant 0$ on a probability space $(\Omega,\mathfrak{A},P)$, is called a *Markov process* if, for all $A \in \mathfrak{B}$, $0 \leqslant s < t$,

$$\mathbf{P}\{x(t,\omega) \in A | \mathcal{N}_s\} = \mathbf{P}\{x(t,\omega) \in A | x(s,\omega)\} \qquad \text{(a.s.)}, \qquad (1.7)$$

where $\mathcal{N}_s$ is the σ-algebra of events generated by all events of the form

$$\{x(u,\omega) \in A\} \qquad (u \leqslant s,\ A \in \mathfrak{B}).$$

It can be proved that there exists a function $P(s,x,t,A)$, defined for $0 \leqslant s \leqslant t$, $x \in E$, $A \in \mathfrak{B}$, which is $\mathfrak{B}$-measurable in x for every fixed s,t,A, and which constitutes a measure as a function of the set A, satisfying the condition

$$\mathbf{P}\{x(t,\omega) \in A | x(s,\omega)\} = P\{s,x(s,\omega),t,A\} \qquad \text{(a.s.)}. \qquad (1\ 8)$$

One can also prove that for all x, except possibly those from a set B such that $P\{x(s,\omega) \in B\} = 0$, the Chapman-Kolmogorov equation holds:

$$P\{s,x,t,A\} = \int P(s,x,u,dy)P(u,y,t,A). \qquad (1.9)$$

The function $P\{s,x,t,A\}$ is called the *transition function* of the Markov process. It is usually assumed at this point (and we shall indeed assume this in the sequel) that equation (1.9) is valid for all $x \in E_\ell$, and $P\{s,x,s,E_\ell \setminus x\} = 0$.

Conversely, given a transition function $P(s,x,t,A)$, one can construct a Markov process with an arbitrary initial distribution. In particular, for $t > s$ one can associate with the function $P(s,x,t,A)$ a family $x^{(s,x)}(t,\omega)$ of Markov processes such that the process $x^{(s,x)}(t,\omega)$ "exits" from the point x at time s, i.e.,

$$\mathbf{P}\{x^{(s,x)}(s,\omega) = x\} = 1. \qquad (1.10)$$

Later on we shall often deal with families of Markov processes $x^{(s,x)}(t,\omega)$ of this kind and with the measures generated by them (Markov families, in the terminology of Dynkin [3]).

The transition function $P(s,x,t,A)$ is said to be *temporally homogeneous* (and the corresponding Markov process is called *homogeneous*) if the function $P(s,x,t + s,A)$ is independent of s. It is called *periodic* if $P(s,x,t + s,A)$ is periodic in s.

A temporally homogeneous transition function is effectively a function of a single time variable, and we shall therefore use the notation $P(s,x,t,A) = P(x,t - s,A)$.

With each homogeneous transition function $P(x,t,A)$ we can associate two families of operators; the first is defined on functions and the second on measures:

$$T_t V(x) = \int P\{x,t,dy\}V(y) = \mathbf{E}V(x^{(x)}(t,\omega)),$$

$$S_t\mu(A) = \int \mu(dx)P(x,t,A).$$

As usual, we denote by $\mathbf{C}(E_\ell)$ ($\mathbf{C}(E)$) the space of continuous function on E_ℓ (on E). A transition function for which the operator T_t maps the space $\mathbf{C}(E_\ell)$ into itself is known as a *Feller function*. If $P(x,t,A)$ is also stochastically continuous, i.e., $P(x,t,U_\varepsilon(x)) \to 1$ as $t \to +0$ for each $\varepsilon > 0$ (where $U_\varepsilon(x)$ is the ε-neighbourhood of x), then it is readily seen that $T_t f(x) \to f(x)$ as $t \to +0$ for $f(x) \in \mathbf{C}(E_\ell)$. From (1.9) we get then the relation $T_{t+s} = T_t T_s$ ($s > 0$, $t > 0$), and hence the family T_t is a homogeneous semigroup on $\mathbf{C}(E_\ell)$. Its $\mathbf{C}$-infinitesimal operator A is defined by the standard formula (see Dynkin [3])

$$AV(x) = \lim_{t\to+0} \frac{T_t V(x) - V(x)}{t}. \tag{1.11}$$

The domain of definition D_A of the operator A is the set of functions for which the limit in (1.11) exists uniformly in E_ℓ.

A stochastically continuous transition function is uniquely determined by its infinitesimal operator (see Dynkin [3], Chapter II). Let us show that the case of a non-homogeneous transition function can be reduced to the homogeneous case by extension of the phase space.

To do this, we consider the σ-algebra $\mathfrak{B}_0$ of Borel sets on the real axis and we define the function $Q((s,x),t,\Gamma)$ for $t \geqslant 0$, $s \geqslant 0$, $x \in E_\ell$, $\Gamma = A \times \Delta$ ($A \in \mathfrak{B}$, $\Delta \in \mathfrak{B}_0$) by

$$Q((s,x),t,\Gamma) = P(s,x,s+t,A)\chi_\Delta(s+t).$$

(Here and below $\chi_\Delta(t)$ denotes the indicator function of the set Δ, i.e., the function equal to 1 for $t \in \Delta$ and 0 for $t \notin \Delta$.) The measure Q can be extended by standard methods (see Halmos [1]) to the σ-algebra $\mathfrak{B} \times \mathfrak{B}_0$ generated by the sets Γ of the above form. It is readily seen that the resulting function Q is a homogeneous transition function in the phase space $E = E_\ell \times I$. If $P(s,x,t,A)$ is stochastically continuous, then Q is also stochastically continuous. Q is the transition function of the process

$$Y(t,\omega) = \{x^{(s,x)}(s+t,\omega), s+t\}$$

in $E_{\ell+1}$. The infinitesimal operator $\tilde{A}$ of the semigroup $\tilde{T}_t$ defined by Q is obviously given by

$$\tilde{A}V(s,x) = \lim_{h\to+0} \frac{\int P(s,x,s+h,dy)V(s+h,y) - V(s,x)}{h} = \text{(Contd)}$$

$$\text{(Contd)} \qquad = \lim_{h \to +0} \frac{\mathbf{E}V(s + h, x^{(s,x)}(s + h, \omega)) - V(s,x)}{h} . \qquad (1.12)$$

$P(s,x,t,A)$ is known as a *Feller transition function* if Q is a Feller function, i.e., if the operator

$$\tilde{T}_t V(s,x) = \int P(s,x,s + t,dy)V(s + t,y)$$

transforms functions $V(s,x) \in \mathbf{C}(E)$ into continuous functions.

In order to avoid separate consideration of homogeneous and non-homogeneous processes, we shall henceforth adopt the term *generating operator* [or *generator*] for the operator defined by (1.12).

It follows from the above-mentioned result concerning the homogeneous case that the operator $\tilde{A}$ uniquely determines the function $P(s,x,t,A)$.

An important example of a stochastically continuous Markov process with a Feller transition function is the process $\xi(t,\omega)$ of Brownian motion (Wiener process), whose transition probability has a density with respect to the Lebesgue measure on the real line and moreover

$$P(x,t,A) = \frac{1}{\sqrt{2\pi t}} \int_A \exp\left\{ -\frac{(x - y)^2}{2t} \right\} dy. \qquad (1.13)$$

It follows from (1.13) that the probability

$$\mathbf{P}\{\xi(t + h) - \xi(t) \in A/\xi(t) = x\} = \frac{1}{\sqrt{2\pi h}} \int_A \exp\left(-\frac{x^2}{2} \right) dz$$

is independent of x. This implies that the increments of the process $\xi(t,\omega)$ on non-overlapping time intervals are independent. In addition, it also follows from (1.13) that

$$\mathbf{E}(\xi(t + h) - \xi(t)) = 0; \qquad \mathbf{D}(\xi(t + h) - \xi(t)) = h. \qquad (1.14)$$

It is also not hard to see that

$$\mathbf{E}(\xi(t + h) - \xi(t))^{2n} = \frac{1}{\sqrt{2\pi h}} \int_{-\infty}^{+\infty} e^{-z^2/2h} z^{2n} dz$$

$$= (2n - 1)!!h^n. \qquad (1.15)$$

It follows from (1.15) and Theorem 1.1.1, that the process $\xi(t)$ has continuous sample functions.

For definiteness, we assume that $\xi(0) = 0$. Given the process $\xi(t)$, one can construct other continuous processes by means of the transformation

$$x(t) = s(0) + \int_0^t b(s)ds + \xi\left(\int_0^t \sigma^2(s)ds\right).$$

The reader should have no difficulty in verifying that each of the resulting processes also has independent increments and a Gaussian transition function. Moreover,

$$\mathbf{E}(x(t + h) - x(t)) = \int_t^{t+h} b(s)ds;$$

$$\mathbf{D}(x(t + h) - x(t)) = \int_t^{t+h} \sigma^2(s)ds.$$

As we shall see from the sequel, the Wiener process can be used to construct a far more extensive class of Markov processes with continuous sample functions.

It is known (see Skorohod [2]) that every almost surely continuous process with independent increments in E_ℓ is Gaussian.

2. Stationary and periodic Markov processes

We shall now investigate conditions under which a Markov process $x(t,\omega)$ is stationary. One necessary condition, at any rate (see the definition of Section 2.1), is that for every $A,B \in \mathfrak{B}$ the probabilities of the events $\{x(t) \in A\}$ and $\{x(t) \in A, x(t+h) \in B\}$ are independent of t. Hence, expressing these probabilities in terms of the transition function, we see that the transition function of a stationary process is homogeneous, and for every $h > 0$ the initial distribution $P_0(A) = P\{x(0) \in A\}$ satisfies the equation

$$P_0(A) = \int P_0(dx)P(x,h,A). \tag{2.1}$$

These two conditions are also sufficient for a Markov process to be stationary. Indeed, for $0 < h_1 < \ldots < h_n$ the probability of the event

$$\{x(t) \in A_0; x(t + h_1) \in A_1; \ldots; x(t + h_n) \in A_n\}$$

is

$$\int P_0(dx)\int_{A_0} P(0,x,t,dx_1)\dots\int_{A_n} P(t + h_{n-1},x_n,t + h_n,dx_{n+1}).$$ [2]

It follows from the homogeneity of the transition function and from (2.1) that this probability is independent of t.

In exactly the same way one can show that a Markov process $x(t)$ is ϑ-periodic if and only if its transition function is ϑ-periodic and the function $P_0(t,A) = P\{x(t) \in A\}$ satisfies the equation

$$P_0(s,A) = \int P_0(s,dx)P(s,x,s + \vartheta,A) \equiv P_0(s + \vartheta,A) \qquad (2.2)$$

for every $A \in \mathfrak{B}$.

However, it is not true that for every homogeneous (periodic) transition function there exists a corresponding stationary (periodic) Markov process. For example, there is no stationary process corresponding in the Wiener transition function. To prove this, suppose that such a process exists. Then, it follows by (1.13) and (2.1) that for every set $A \in \mathfrak{B}$ on the real line whose Lebesgue measure $m(A)$ is finite

$$P_0(A) = \int P_0(dx)\int_A \frac{1}{\sqrt{2\pi t}} \exp\left\{ - \frac{(x - y)^2}{2t}\right\}dy$$

$$\leqslant \frac{1}{\sqrt{2\pi t}} m(A).$$

Letting $t \to \infty$, we get $P_0(A) = 0$. Hence it follows that $P_0(E_\ell) = 0$, contradicting $P_0(E_\ell) = 1$.

We shall now determine some further conditions that must hold for a homogeneous (periodic) transition function corresponding to a stationary (periodic) Markov process.

THEOREM 2.1. *A necessary and sufficient condition for the existence of a stationary Markov process with the given temporally homogeneous stochastically continuous Feller transition function $P(x,t,A)$ is that for some point $x \in E_\ell$* [3]

$$\lim_{R\to\infty} \varliminf_{T\to\infty} \frac{1}{T}\int_0^T P(x,t,\overline{U}_R)dt = 0. \qquad (2.3)$$

For the proof we need the following lemma.

LEMMA 2.1. *A stochastically continuous Feller transition function $P(x,t,A)$ is a $\mathfrak{B}\times\mathfrak{B}_0$-measurable function of (x,t) for every $A \in \mathfrak{B}$.*

---

[2] Henceforth we shall omit the limits of integration when the integration is performed over the entire space E_ℓ.

[3] Recall that $\overline{U}_R = \{|x| > R\}$.

PROOF. It will suffice to prove the lemma for closed sets A. If A is closed, we can construct a sequence $f_n(x)$ of monotone decreasing continuous functions converging to $\chi_A(x)$. It is clear that for each $t > 0$

$$T_t f_n(x) \to P(x,t,A) \quad \text{as} \quad n \to \infty.$$

The function $T_t f_n(x)$ is continuous in x and right-continuous in t, as follows from the Feller property and the stochastic continuity of the transition function. Therefore it is Borel-measurable as a function of x,t. Hence the function $P(x,t,A)$ is also measurable.

PROOF OF THEOREM 2.1. (1) NECESSITY. Let $P_0(A)$ be the stationary initial distribution. Then, integrating (2.1) with respect to t from 0 to T and applying Fubini's theorem (see Halmos [1]), we get

$$P_0(\overline{U}_R) = \int P_0(dx) \frac{1}{T}\int_0^T P(x,t,\overline{U}_R)dt.$$

Now suppose that condition (2.3) does not hold. Then

$$\lim_{R\to\infty} \underline{\lim}_{T\to\infty} \frac{1}{T}\int_0^T P(x,t,\overline{U}_R)dt = q(x) > 0.$$

Therefore

$$0 = \lim_{R\to\infty} P_0(\overline{U}_R) \geqslant \int P_0(dx)q(x) > 0.$$

This is a contradiction and therefore (2.3) must be satisfied.

(2) SUFFICIENCY. It follows from condition (2.3) that for some x_0 there exists a sequence $T_n \to \infty$ such that

$$\frac{1}{T_n}\int_0^{T_n} P(x_0,t,U_R)dt \to 0 \quad \text{uniformly in } n \text{ as } \quad R \to \infty. \tag{2.4}$$

Consider the sequence of measures on E_ℓ defined by

$$P_n(A) = \frac{1}{T_n}\int_0^{T_n} P(x_0,t,A)dt.$$

By condition (2.4), this sequence is weakly compact (see Section 2.1). Let P_{n_k} be a subsequence converging weakly to some measure P_0. We claim now that the measure P_0 satisfies condition (2.1) and consequently defines the initial distribution of a stationary Markov process.

Let $f(x) \in \mathbf{C}(E_\ell)$. Since P_{n_k} is weakly convergent to P_0, and the Feller property holds for the transition probability, we have

$$\int P_0(dx)\int P(x,t,dy)f(y)$$

$$= \lim_{n_k\to\infty} \frac{1}{T_{n_k}}\int_0^{T_{n_k}} ds\int P(x_0,s,dx)\int P(x,t,dy)f(y)$$

$$= \lim_{n_k\to\infty} \frac{1}{T_{n_k}}\int_0^{T_{n_k}} dsP(x_0,s + t,dy)f(y)$$

$$= \lim_{n_k\to\infty} \frac{1}{T_{n_k}}\left[\int_0^{T_{n_k}} du\int P(x_0,u,dy)f(y) + \int_{T_{n_k}}^{T_{n_k}+t} du\int P(x_0,u,dy)f(y) - \int_0^t du\int P(x_0,u,dy)f(y)\right]$$

$$= \int P_0(dy)f(y).$$

The resulting relation

$$\int P_0(dx)T_tf(x) = \int P_0(dx)f(x)$$

is equivalent to (2.1), i.e. (2.1) holds. This completes the proof[4].

THEOREM 2.2. *A necessary and sufficient condition for the existence of a ϑ-periodic Markov process with a given ϑ-periodic transition function $P(s,x,t,A)$ is that for some x_0, s_0*

$$\lim_{R\to\infty} \varliminf_{n\to\infty} \frac{1}{n}\sum_{k=1}^{n} P(s_0,x_0,s_0 + k\xi,\overline{U}_R) = 0. \tag{2.5}$$

The proof of this theorem is entirely analogous to that of Theorem 2.1. The only difference lies in the definition of $P_n(A)$

[4] The method of showing the existence of a stationary Markov process employed in the proof of Theorem 2.1 is well-known. It was first used by Krylov and Bogoljubov to prove the existence of an invariant measure for a dynamic system. The method is systematically used for Markov processes, e.g., in Doob [3].

which is now given by

$$P_n(A) = \frac{1}{k_n} \sum_{i=1}^{k_n} P(s_0, x_0, s_0 + k\vartheta, A),$$

where k_n is an increasing sequence of integers such that the sequence of measures $P_n(A)$ is weakly compact.

REMARK. Condition (2.5) of Theorem 2.2 can be replaced by the more easily tested condition

$$\lim_{R\to\infty} \varliminf_{T\to\infty} \frac{1}{T}\int_0^T P(s,x,s + u,\overline{U}_R)du = 0, \tag{2.6}$$

provided the transition function $P(s,x,t,A)$ satisfies the following not very restrictive assumption that

$$\alpha(R) = \sup_{x\in U_{\beta(R)},0<s,t<\vartheta} P(s,x,s + t,\overline{U}_R) \to 0 \text{ as } R \to \infty \tag{2.7}$$

for some function $\beta(R)$ which tends to infinity as $R \to \infty$.
Indeed, for every

$$u \in ((k - 1)\vartheta, k\vartheta)$$

it is obvious that

$$P(s,x,s + k\vartheta,\overline{U}_R)$$

$$= \left(\int_{U_{\beta(R)}} + \int_{U_{\beta(R)}}\right) P(s,x,s + u,dy)P(s + u,y,s + k\vartheta,\overline{U}_R)$$

$$\leqslant P(s,x,s + u,\overline{U}_{\beta(R)}) + \sup_{\substack{u\in((k-1)\vartheta,k\vartheta),\\ y\in U_{\beta(R)}}} P(s + u,y,s + k\vartheta,\overline{U}_R)$$

and hence, integrating both sides with respect to u from $(k - 1)\vartheta$ to $k\vartheta$, we see that

$$P(s,x,s + k\vartheta,\overline{U}_R) \leqslant \frac{1}{\vartheta}\int_{(k-1)\vartheta}^{k\vartheta} P(s,x,s + u,\overline{U}_{\beta(R)})du + \alpha(R).$$

Summation over k from 1 to n in combination with (2.7) shows that (2.6) implies (2.5). A similar argument shows that if (2.7) is satisfied, then (2.5) implies (2.6).

3. Stochastic differential equations

In Section 1 we have used the finite-difference equation (1.2) to determine a Markov chain x_n from a sequence of independent random variables ξ_n. It is natural to try to use the differential analog of (2.1) to construct continuous-time Markov processes. The formal analog of equation (1.2) for the one-dimensional case is

$$\frac{dX}{dt} = b(t,X) + \sigma(t,X)\dot{\xi}(t) \tag{3.1}$$

The random process $\dot{\xi}(t)$, by analogy with the sequence ξ_n in (1.2), must be a process with independent increments. Such a process does not exist, but equation (3.1) can nonetheless be given a rigorous meaning.

To this end, it is convenient to rewrite the equation in terms of differentials:

$$dX(t) = b(t,X)dt + \sigma(t,X)d\xi(t). \tag{3.2}$$

What properties must the process $\xi(t)$ possess? First, it must have independent increments, as the integral of a "process" $\dot{\xi}(t)$ with independent increments. Second, it must have continuous sample functions, if we wish the solution $X(t)$ of equation (3.2) to be a continuous stochastic process. As mentioned at the end of Section 1, such a process is always Gaussian. We may also assume that its mean and variance satisfy conditions (1.14), since this may always be achieved by modifying the coefficients b and σ.

Equations of type (3.1) or (3.2) were first considered by Langevin [1] as far back as 1908, shortly after Einstein and Smoluchowski had published their first papers on the theory of Brownian motion. More systematic investigations of stochastic equations began only in the thirties. The simplest and most convenient construction of the solution of the stochastic differential equation (3.2) was given by Itô [1-2]. This construction has been presented in detail in many books (Doob [3], Skorohod [1], Dynkin [3], Gihman and Skorohod [1]). We shall therefore present without proof some of Itô's theorems on the existence and properties of solutions of equation (3.2).

The basic tool for the constructions is the stochastic integral. Let $\xi(t,\omega)$ be a Wiener process on the interval $[a,b]$, defined on a probability space $(\Omega,\mathfrak{U},\mathbf{P})$. Let $\tilde{\mathcal{N}}_t$ $(t \geqslant 0)$ be a family of σ-algebras of sets in $\mathfrak{U}$, related to the Wiener process $\xi(t)$ as follows:

(1) $\tilde{\mathcal{N}}_{t_1} \subset \tilde{\mathcal{N}}_{t_2}$, if $t_1 < t_2$;

(2) $\xi(t)$ is an $\tilde{\mathcal{N}}_t$-measurable random variable for each $t \geqslant 0$;

(3) the increment $\xi(t+h)-\xi(t)$ of the process $\xi(t)$ is independent of every event $A \in \tilde{N}_t$.

For every bounded step function $f(t,\omega) = f(t)$ with jumps at points $t_1,\ldots,t_n$, such that $f(t)$ is $\tilde{N}_t$-measurable for each $t \in [a,b]$, the Itô stochastic integral is defined as the sum

$$\int_a^b f(t)d\xi(t) = \sum_{i=0}^{n-1} f(t_i)[\xi(t_{i+1}) - \xi(t_i)].$$

Using the independence of $\xi(t_{i+1}) - \xi(t_i)$ and $f(t_i)$, one easily verifies the following properties of integrals of step functions:

$$\mathbf{E}\left(\int_a^b f(t)d\xi(t)/\tilde{N}_a\right) = 0 \qquad \text{(a.s.)}, \tag{3.3}$$

$$\mathbf{E}\left(\left[\int_a^b f(t)d\xi(t)\right]^2/\tilde{N}_a\right) = \int_a^b M(f^2(t)/\tilde{N}_a)dt \qquad \text{(a.s.)}. \tag{3.4}$$

Next, the integral of an $\tilde{N}_t$-measurable function $f(t)$ such that $f^2(t)$ is in class $\mathbf{L}$ is defined by means of a passage to the limit (see Gihman and Skorohod [1], Chapter 8, Section 2). One then proves that relations (3.3) and (3.4) are valid for every function $f(t)$, provided

$$\int_a^b \mathbf{E}(f^2(t)/\tilde{N}_a)dt \qquad \text{(a.s.)}. \tag{3.5}$$

It can be proved that the stochastic integral

$$\zeta(t) = \int_a^t f(s)d\xi(s) = \int_a^b \chi_t(s)f(s)d\xi(s)$$

(where $\chi_t(s)$ is the indicator function of the set $\{s < t\}$) can be so defined that it becomes a separable almost surely continuous stochastic process. This process satisfies the Kolmogorov inequality

$$\mathbf{P}\{\sup_{t\in[a,b]} |\zeta(t)| > c/\tilde{N}_a\} \leqslant \frac{1}{c^2}\int_a^b \mathbf{E}\{f^2(s)/\tilde{N}_a\}ds. \tag{3.6}$$

Hitherto we have considered stochastic integrals in E_ℓ. It is not difficult to extend the construction to the multi-dimensional case. Let $\sigma_1(t),\ldots,\sigma_k(t)$ be vectors in E_ℓ whose components are $\tilde{N}_t$-measurable for each fixed t. Let $\xi_1(t),\ldots,\xi_k(t)$ be mutually independent $\tilde{N}_t$-measurable Wiener processes such that the random variables $\xi_i(t+h) - \xi_i(t)$ are all independent of every event in $\tilde{N}_t$ for $h > 0$. Then stochastic integrals with values in E_ℓ are defined in the following natural (coordinatewise) manner:

$$\zeta_r(t) = \int_a^t \sigma_r(s)d\xi_r(s); \qquad \zeta(t) = \sum_{r=1}^k \zeta_r(t).$$

The Itô (stochastic) differential $d\zeta(t)$ of the $\tilde{N}_t$-measurable process $\xi(t)$ is defined as

$$b(t)dt + \sum_{r=1}^k \sigma_r(t)d\xi_r(t),$$

provided $b(t)$ and $|\sigma_r|^2$ are in class **L**, are $\tilde{N}_t$-measurable for each t, and for all $a < t_1 < t_2 < b$

$$\zeta(t_2) - \zeta(t_1) = \int_{t_1}^{t_2} b(t)dt + \sum_{r=1}^k \int_{t_1}^{t_2} \sigma_r(t)d\xi_r(t). \tag{3.7}$$

Itô [2] established the following analog of the chain rule for stochastic differentials.

THEOREM 3.1. *If the function* $u(t,x)$ $(t \in [a,b],\ x \in E_\ell)$ *has continuous partial derivatives up to second order in* x *and to first order in* t, *and the process* $\zeta(t)$ *with values in* E_ℓ *has an Itô differential*

$$d\zeta(t) = b(t)dt + \sum_{r=1}^k \sigma_r(t)d\xi_r(t),$$

then the process $\eta(t) = u(t,\zeta(t))$ *also has an Itô differential*

$$d\eta(t) = \left[\frac{\partial u(t,\zeta(t))}{\partial t} + \left(b(t), \frac{\partial}{\partial x}\right)u(t,\zeta(t)) + \frac{1}{2}\sum_{r=1}^k \left(\sigma_r(t), \frac{\partial}{\partial x}\right)^2 u(t,\zeta(t))\right] + \sum_{r=1}^k \left(\sigma_r(t), \frac{\partial}{\partial x}\right)u(t,\zeta(t))d\xi_r(t) \tag{3.8}$$

(here and below $\partial/\partial x$ *denotes the vector with components* $\partial/\partial x_1, \ldots, \partial/\partial x_\ell$*).*

It is clear that the only difference between (3.8) and the

usual chain rule is the presence of the term

$$\frac{1}{2}\sum_{r=1}^{k}\left(\sigma_r(t), \frac{\partial}{\partial x}\right)^2 udt.$$

We now return to the stochastic differential equation (3.2), which will be interpreted as an equation relating the stochastic differentials of a process $X(t,\omega)$ in E_ℓ and the Wiener processes $\xi_r(t)$; written by means of integrals this becomes

$$X(t) = X(t_0) + \int_{t_0}^{t} b(s,X(s))ds$$
$$+ \sum_{r=1}^{k}\int_{t_0}^{t} \sigma_r(s,X(s))d\xi_r(s). \qquad (3.9)$$

In the sequel we shall mean by a solution of equation (3.9) on the interval $[t_0,T]$ a stochastic process $X(t)$ such that the random variable $X(t)$ is $\tilde{\mathcal{N}}_t$-measurable for each t, the integrals in (3.9) exist and equality (3.9) holds almost surely for each $t \in [t_0,T]$.

The following theorem concerns the existence, uniqueness and certain other properties of the solution of equation (3.9).

THEOREM 3.2. *Let the vectors* $b(s,x)$, $\sigma_1(s,x),\ldots,\sigma_k(s,x)$ $(s \in [t_0,T],\ x \in E_\ell)$ *be continuous functions of* (s,x), *such that for some constant* B *the following conditions hold in the entire domain of definition:*

$$|b(s,x) - b(s,y)| + \sum_{r=1}^{k} |\sigma_r(s,x) - \sigma_r(s,y)|$$
$$\leqslant B|x - y|, \qquad (3.10')$$

$$|b(s,x)| + \sum_{r=1}^{k} |\sigma_r(s,x)| \leqslant B(1 + |x|). \qquad (3.10'')$$

Then:

(1) *For every random variable* $X(t_0)$ *independent of the processes* $\xi_r(t) - \xi_r(t_0)$ *there exists a solution* $X(t)$, *unique up to equivalence*[5], *of equation* (3.9) *which is an almost surely continuous stochastic process and is unique up to equivalence*[5].

(2) *This solution is a Markov process whose transition func-*

[5] Two solutions $X_1(t)$ and $X_2(t)$ are said to be equivalent if $\mathrm{P}\{X_1(t) = X_2(t) \text{ for all } t \in [t_0,T]\} = 1$.

tion $P(s,x,t,A)$ *is a Feller function which is defined for* $t > s$ *by the relation* $P(s,x,t,A) = \mathbf{P}\{X^{s,x}(t) \in A\}$, *where* $X^{s,x}(t)$ *is a solution of the equation*

$$X^{s,x}(t) = x + \int_s^t b(u,X^{s,x}(u))du$$

$$+ \sum_{r=1}^{k} \int_s^t \sigma_r(u,X^{s,x}(u))d\xi_r(u). \tag{3.11}$$

(3) *The transition function* $P(s,x,t,A)$ *satisfies for* $h \to 0$:

$$\left.\begin{aligned} \mathbf{E}[X^{s,x}(s+h) - x] &= \int (y-x)P(s,x,s+h,dy) \\ &= b(s,x)h + O(h^{3/2}), \\ \mathbf{E}[(X_i^{s,x}(s+h) - x_i)(X_j^{s,x}(s+h) - x_j)] \\ &= a_{ij}(s,x)h + O(h^{3/2}), \\ \mathbf{P}(s,x,s+h,\overline{U}_\varepsilon(x)) &= O(h^{3/2}), \end{aligned}\right\} \tag{3.12}$$

where all estimates $O(\cdot)$ *are uniform in* s,x *in each bounded domain, and* $a_{ij}(s,x)$ *are elements of a matrix* $A(s,x)$ *such that for all* $\lambda \in E_1$

$$(A(s,x)\lambda,\lambda) = \sum_{r=1}^{k} (\sigma_r(s,x),\lambda)^2.$$

(4) *There exists a constant* k, *depending only on the dimension of the space* E_ℓ, *on the constant* B *of condition* (3.10) *and on the length* $T - t_0$ *of the interval such that for all* $s,t \in [t_0,T]$

$$\mathbf{E}|X^{s,x}(t) - x|^4 \leqslant k(t-s)^2(1 + |x|^4).$$

(5) *If the coefficients of equation* (3.9) *are independent of* s, *then the transition function of the corresponding Markov process is homogeneous; and if the coefficients are* ϑ*-periodic in* s, *then the transition function is* ϑ*-periodic.*

The proof of this theorem, except for the second part of (5), may be found in Dynkin [3], Gihman and Skorohod [1]. The proof of the second part of (5) is analogous to that of the first.

Here the matrix $A(s,x)$ is known as the diffusion matrix, and the vector $b(s,x)$ as the drift vector. Their probabilistic meaning is clear from formulas (3.12).

Let us consider the Markov process $X^{s,x}(t,\omega)$ determined by equation (3.11). We shall often have to calculate the expectation of various random variables which are measureable with respect to the evolution of this process (i.e., with respect to the σ-algebra generated by the events $\{X^{s,x}(t) \in A\}$, $A \in \mathfrak{B}$, $s < t < \infty$). Instead of writing the indices s,x in the symbols for each of these random variables, we shall sometimes attach them to the symbols **E** and **P**. For example, $\mathbf{P}_{s,x}\{X(t,\omega) \in A\} = \mathbf{P}\{X^{s,x}(t,\omega) \in A\}$. If the coefficients of equation (3.9) are independent of s, we need only consider the process $X^{0,x}(t)$[6] which will be denoted by $X^x(t)$. Accordingly, the index x for random variables involved in the process $X^x(t)$ will sometimes be attached to the symbols **E** and **P**.

Let $\mathbf{C}_2$ denote the class of functions on E which are twice continuously differentiable with respect to $x_1,\ldots,x_\ell$ and continuously differentiable with respect to t. Let $V \in \mathbf{C}_2$. Then it follows from Theorems 3.1 and 3.2 that

$$V(t,X(t)) - V(s,X(s))$$

$$= \int_s^t LV(u,X(u))du + \sum_{r=1}^{k} \int_s^t \left(\sigma_r, \frac{\partial V}{\partial x}\right) d\xi_r(u), \qquad (3.13)$$

where

$$LV(s,x) = \frac{V(s,x)}{s} + \sum_{i=1}^{\ell} b_i(s,x) \frac{\partial V(s,x)}{\partial x}$$

$$+ \frac{1}{2} \sum_{i,j=1}^{\ell} a_{ij}(s,x) \frac{\partial^2 V(s,x)}{\partial x_i \partial x_j}$$

$$\equiv \frac{\partial V}{\partial s} + \left(b, \frac{\partial}{\partial x}\right) V + \frac{1}{2} \sum_{r=1}^{k} \left(\sigma_r, \frac{\partial}{\partial x}\right)^2 V. \qquad (3.14)$$

If moreover the function V and its derivatives are bounded (or increasing no faster than a linear function of x), then, calculating the expectation in (3.13) and using the properties of stochastic integrals and Fubini's theorem, we get

$$\mathbf{E}[V(t,X(t)) - V(s,X(s))] = \int_s^t \mathbf{E}LV(u,X(u))du. \qquad (3.15)$$

---

[6] See Dynkin [3], Chapter II, Section 2.

Substituting $X(t) = X^{s,x}(t)$ into this equality, dividing both sides by $t - s$ and letting $t \to s + 0$, we readily find that

$$\lim_{h \to +0} \frac{1}{h}[\mathbf{E}_{s\ x}V(s + h, X(s + h)) - V(s,x)] = LV(s,x). \qquad (3.16)$$

Hence it follows that $\tilde{A}V(s,x) = LV(s,x)$ for all $V \in \mathbf{C}_2$ with compact support, and that for homogeneous processes $AV(x) = LV(x)$.

The operator L defined by (3.14) will be called the differential generator of the Markov process. It is clear from the definition that this is a local concept, i.e., the value of the operator at a point (s,x) is determined by the values of V in an arbitrarily small neighbourhood of the point (s,x). The left-hand side of formula (3.16) may be undefined for rapidly growing processes (for example, for the Wiener process, when $t > 0$, the function $V = \exp(x^3)$ is not in the domain of the operator T_t). The probabilistic meaning of the operator L for any function $V \in \mathbf{C}_2$ is given by the following lemma:

LEMMA 3.1[7]. *Let $X(u)$ be a process satisfying equation* (3.9) *on the time interval* $[s,T]$, $V \in \mathbf{C}_2$, τ_U *the random variable equal to the time at which the point x of the sample function of the process $X(u)$ first leaves the bounded neighbourhood U, and let* $\tau_U(t) = \min(\tau_U, t)$. *Suppose moreover that* $\mathbf{P}\{X(s) \in U\} = 1$. *Then*

$$\mathbf{E}[V(\tau_U(t), X(\tau_U(t))) - V(s,X(s))]$$

$$= \mathbf{E}\int_s^{\tau_U(t)} LV(u,X(u))du. \qquad (3.17)$$

PROOF. It is known (see Dynkin [3], Chapter 11, Section 3) that the process $Y(t) = X(\tau_U(t))$, obtained by stopping the process $X(t)$ at the instant it reaches the boundary of the domain U, has an Itô differential:

$$dY(t) = \chi_{\tau_U > t}(\omega)b(t,Y(t))dt$$

$$+ \sum_{r=1}^{k} \chi_{\tau_U > t}(\omega)\sigma_r(t,Y(t))d\xi_r(t).$$

(Since $\{\tau_U > t\} \in \tilde{\mathcal{N}}_t$, the Itô differential in the formula is defined.) Applying Theorem 3.1 to the process $Y(t)$ and the function V, we get

$$V(\tau_U(t), X(\tau_U(t)) - V(s,X(s)) = \qquad \text{(Contd)}$$

[7] Lemma 3.1 is a special case of a formula of Dynkin ([3], p. 191).

$$\text{(Contd)} \qquad = \int_s^{\tau_U(t)} LV du + \sum_{r=1}^{k} \int_s^{\tau_U(t)} \left(\sigma_r, \frac{\partial V}{\partial x}\right) d\xi_r(u). \tag{3.18}$$

(Here and below we denote

$$\int_s^{\tau_U(t)} \Phi d\xi(u) = \int_s^t \chi_{\tau_U > t}(\omega) \Phi d\xi(u).)$$

This immediately implies (3.17).

REMARK 1. Note that under the assumptions of the lemma the expectation of the random variable $V(t, X(t))$ need not exist and thus formula (3.15) may be false.

REMARK 2. Setting $X(s) = x$ in formula (3.17) and letting $t \to s + 0$, we get

$$\mathfrak{U}V = \lim_{h \to +0} \frac{\mathbf{E}_{s,x} V(\tau_U, s + h), X(\tau_U(s + h))) - V(s,x)}{\mathbf{E}_{s,x}[\tau_U(s + h) - s]}$$

$$= LV(s,x). \tag{3.19}$$

The operator on the left of (3.19) may be regarded as an extension of the differential operator L. It was first introduced in a more general setting by Dynkin in [1].

4. Conditions for regularity of the solution[8]

It follows from Theorem 3.2 that if condition (3.10) holds for all $t > t_0$ then the solution $X(t)$ of equation (3.9) is defined and continuous for all $t > t_0$. Condition (3.10) is rather restrictive. For example, it is intuitively clear that the problem $dx(t) = -x^3(t)dt + d\xi(t)$, $x(0) = x_0$, has a unique solution for all $t > 0$ (since the drift coefficient "directs" the motion to the origin), but conditions (3.10) hold for this equation only in a compact domain of the x-space. The same applies to Example 5 at the end of this section and also to an important class of stochastic equations which arise in the statistical analysis of partially observable Markov processes (see Stratonovič [3], Širjaev [1]). It is therefore of paramount importance to find other,

[8] Certain necessary and sufficient conditions for the regularity of homogeneous Markov processes of the diffusion type were obtained by Feller [1] for the temporally homogeneous case. Multidimensional temporally homogeneous processes were studied by the author in [1] where it was shown that the conditions of Theorem 4.1 are also necessary for regularity in the non-degenerate case.

broader conditions for the existence and uniqueness of the solution of equation (3.9). In this section we shall prove analogs of Theorems 1.2.2 and 1.3.3 for Itô (stochastic) equations.

If conditions (3.10) are valid in every cylinder $I \times U_R$, one can construct a sequence of functions $b_n(t,x)$ and $\sigma_r^{(n)}(t,x)$ such that for $|x| < n$

$$\sigma_r^{(n)}(t,x) = \sigma_r(t,x); \qquad b_n(t,x) = b(t,x),$$

and for each b_n, $\sigma_r^{(n)}$ satisfy conditions (3.10) everywhere in E. By Theorem 3.2, there exists a sequence of Markov processes $X_n(t)$ corresponding to the functions b_n, $\sigma_r^{(n)}$. To simplify matters, we shall limit the discussion to the case in which the distribution of $X_0(t)$ has compact support in E_ℓ. Then it is intuitively clear (and it can be proved rigorously; see Doob [2], Dynkin [3]) that the first exit times of the processes $X_m(t)$ from the set $|x| < n$ are identical for $m \geqslant n$. Let this common value be τ_n. It is also clear that the processes themselves coincide up to time τ_n, i.e.,

$$\mathbf{P}\{\sup_{t_0 \leqslant t \leqslant \tau_n} |X_n(t) - X_m(t)| > 0\} = 0, \qquad m > n.$$

Let τ denote the (finite or infinite) limit of the monotone increasing sequence τ_n as $n \to \infty$. We call the random variable τ the first exit time of the sample function from every bounded domain, or briefly the escape time to infinity. This definition is natural, since one easily shows that the values of τ are not changed if we replace the domains $U_n = \{|x| < n\}$ by any other expanding sequence of bounded domains such that the distance from the origin to the boundary tends to infinity.

We now define a new stochastic process $\tilde{X}(t)$ by setting $\tilde{X}(t) = X_n(t)$ for $t < \tau_n$. It can be shown that this is always a Markov process for $t < \tau$ (for the definition of a Markov process stopped at a random time τ, see Dynkin [2]).

We shall say that the process $X(t)$ is *regular* if

$$\mathbf{P}\{\tau = \infty\} = 1. \tag{4.1}$$

If condition (4.1) is satisfied, the process $X(t)$ is almost surely defined for all $t \geqslant t_0$. For a process satisfying the assumption of Theorem 3.2, regularity follows from continuity. The following theorem gives a more general sufficient condition for regularity[8].

THEOREM 4.1. *Suppose that conditions* (3.10) *are valid in every cylinder* $I \times U_R$ *and, moreover, that there exists a nonnegative function* $V \in \mathbf{C}_2$ *on the domain* E *such that for some constant* $c > 0$

---

[8] See p. 83 for the text of this footnote.

$$LV \leqslant cV, \tag{4.2}$$

$$V_R = \inf_{|x|>R} V(t,x) \to \infty \quad \text{as} \quad R \to \infty. \tag{4.3}$$

Then parts 1, 2 *and* 5 *of Theorem* 3.2 *hold. Part* 3 *is valid if the expectations in* (3.12) *are replaced by "truncated" expectations (i.e., for example, instead of* $\mathbf{E}[X^{s,x}(s+h) - x]$ *one considers* $\mathbf{E}\chi(\omega)[X^{s,x}(s+h) - x]$, *where* $\chi(\omega)$ *is the characteristic function of the set* $X^{s,x}(s+h,\omega) - x < k$). *Moreover, the process also satisfies the inequality*

$$\mathbf{E}V(t,X(t)) \leqslant \mathbf{E}V(t_0,X(t_0))e^{c(t-t_0)}, \tag{4.4}$$

if the expectation on the right exists.

PROOF. We first prove that under the assumptions (4.2) and (4.3) the process $\tilde{X}(t)$ constructed at the beginning of the section is regular. From (4.2) it follows that the function

$$W(t,x) = V(t,x)\exp\{-c(t-t_0)\}$$

satisfies $LW \leqslant 0$. Hence, by Lemma 3.1, for $\tau_n(t) = \min(\tau_n, t)$, we have

$$\mathbf{E}\{V(\tau_n(t),X(\tau_n(t)))\exp[-c(\tau_n(t) - t_0)]\} - \mathbf{E}V(t_0,X(t_0))$$

$$= \mathbf{E}\int_{t_0}^{\tau_n(t)} LW(u,X(u))du \leqslant 0.$$

This, together with the inequalities $\tau_n(t) \leqslant t$, $V \geqslant 0$, implies

$$\mathbf{E}V(\tau_n(t),\tilde{X}(\tau_n(t))) \leqslant e^{c(t-t_0)}\mathbf{E}V(t_0,X(t_0)). \tag{4.5}$$

From (4.5) we derive the estimate

$$\mathbf{P}\{\tau_n \leqslant t\} \leqslant \frac{e^{c(t-t_0)}\mathbf{E}V(t_0,\tilde{X}(t_0))}{\inf_{|x|\geqslant n, u>t_0} V(u,x)}.$$

Letting $n \to \infty$ and making use of (4.3), we now get (4.1); thus the process $\tilde{X}(t)$ is a solution of equation (3.9) for all $t \geqslant t_0$. This solution is unique up to equivalence. Indeed, it follows from the definition of $\tilde{X}(t)$ and from the uniqueness of the solution of (3.9) in the domain $|x| < n$ that for every pair of solutions $X(t)$ and $Y(t)$

$$\mathbf{P}\{\sup_{0<t<\tau_n} |X(t) - Y(t)| > 0\} = 0.$$

The desired result now follows by letting $n \to \infty$ and using (4.1).

Various other properties of the process just constructed can be proved in a similar manner. For example, we can prove relation (4.4) by letting $n \to \infty$ in (4.5) and using Fatou's lemma (see Halmos [1]). When constructing the process $\tilde{X}(t)$, we assumed that the distribution of $\tilde{X}(t)$ has compact support. The case of an arbitrary initial distribution may now be dealt with in the way described in Gihman and Skorohod [1], pp. 512-514.

REMARK 1. It is intuitively clear that whether a process is regular or not depends only on the behavior of the coefficients b and σ_r in a neighbourhood of the point at infinity. It is therefore natural to expect the conclusion of Theorem 4.1 to remain valid if a function V satisfying conditions (4.2), (4.3) exists merely in the domain $\{t > 0\} \times \overline{U}_R$ for some $R > 0$. One easily sees from the proof of Theorem 4.1 that in this case the sample function cannot escape to infinity before it exits from the set $|x| > R$. Using the strong Markov property of the process (see Chapter IV), one easily infers that the process is also regular in the sense of definition (4.1).

REMARK 2. In many cases (see Chapter IV) it is useful to know when the sample function of a diffusion process almost surely does not exit from a given open set D in a finite time. Sufficient conditions for the invariance of the set D in this sense can be derived from Theorem 4.1 by noting that in the proof of this theorem the assumption that the sequence U_n converges to E_ℓ is not essential. Replacing the sequence U_n by an increasing sequence of open sets D_n whose closures are contined in D, and such that $\cup D_n = D$, we get the following result.

Suppose that in every cylinder $I \times D_n$ the coefficients b and σ_r satisfy conditions (3.10) *and there exists a function $V(t,x)$, twice continuously differentiable in x and continuously differentiable in t in the domain $I \times D$, which satisfies condition* (4.2) *and the condition*

$$\inf_{t>0,\, x \in D \setminus D_n} V(t,x) \to \infty \quad \textit{as} \quad n \to \infty.$$

Then the conclusion of Theorem 4.1 *holds provided that also* $\mathbf{P}\{X(t_0) \in D\} = 1$. *Moreover the solution satisfies the relation*

$$\mathbf{P}\{X(t) \in D\} = 1 \quad \textit{for all} \quad t \geqslant t_0.$$

As the following theorem shows, conditions (4.2), (4.3) are in a certain sense "the best possible".

THEOREM 4.2. *Suppose that conditions* (3.10) *hold in every cylinder* $(t > 0) \times U_R$ *and that, moreover, there exists in the domain* $E = (t > 0) \times E_\ell$ *a nonnegative bounded function* $V(t,x) \in \mathbf{C}_2$ *such that for some* $c > 0$

$$LV \geqslant cV. \tag{4.6}$$

Then for all points s,x such that $V(s,x) > 0$ the process $X^{s,x}(t)$ defined by equation (3.11) *up to the time $\tau^{s,x}$ of first exit from every bounded domain, is not regular. More precisely, for each $\varepsilon > 0$ we have*

$$\mathrm{P}\left\{\tau^{s,x} - s < \frac{1}{c}\ln\frac{\sup_E V}{V(s,x)} + \varepsilon\right\} > 0.$$

PROOF. Exactly as in the proof of Theorem 4.1, we apply Lemma 3.1 and condition (4.6) to get the relation

$$\mathbf{E}V(\tau_n^{s,x}(t), X(\tau_n^{s,x}(t)))\exp\{-c(\tau_n^{s,x}(t) - s)\} \geqslant V(s,x).$$

Since V is bounded, this implies that

$$\mathbf{E}\exp\{-c(\tau_n^{s,x}(t) - s)\}\sup_E V \geqslant V(s,x).$$

Letting $n \to \infty$ and putting $\tau^{s,x}(t) = \min(\tau^{s,x}, t)$, we get

$$\mathbf{E}\exp\{-c(\tau^{s,x}(t) - s)\}\sup_E V \geqslant V(s,x). \tag{4.7}$$

Assume that $\mathbf{P}\{\tau^{s,x} > t\} = 1$ for some $t > s$. Then (4.7) implies the inequality

$$\exp\{-c(t - s)\} > V(s,x)/\sup_E V,$$

which is in contradiction with $t - s > c^{-1}\ln(\sup_E V/V(s,x))$. This completes the proof.

We append a very brief discussion of the situation arising under the assumptions of Theorem 4.2 when the process $X^{s,x}(t)$ is not regular. In this case equation (3.11) determines the process only up to the random time $\tau^{s,x}$. How does the process continue for $t > \tau^{s,x}$? There exists an infinite set of possible continuations. For example, we can set $X^{s,x}(\tau^{s,x} + 0) = y \in E$ (a jump to the point y after escape to infinity), and we can stipulate other additional conditions determining the evolution of the process after $\tau^{s,x}$. There is an extensive literature on the continuation of Markov processes after the time at which the sample function has hit the boundary. The problem is closely connected with the description of all possible ways in which this process can reach the boundary (assuming that it is a Martin boundary). The case of a one-dimensional temporally homogeneous process defined by equation (3.11) (and a somewhat more general one) has been thoroughly studied by Feller [3] and Ventcel' [1]. For the multi-dimensional case extremely interesting results have been obtained by Ventcel' [2], Ueno [2] and others.

EXAMPLE 1. Let conditions (3.10) hold everywhere in E. Then the function $V = (|x|^2 + 1)^{n/2}$ satisfies the assumptions of Theorem 4.1 for every $n > 0$. Hence the solution $X(t)$ of equation (3.9) exists and is almost surely bounded on every finite time interval (this result follows from Theorem 3.2). Moreover, for some constant c_n we have the estimate

$$\mathbf{E}|X(t)|^n < e^{c_n(t-t_0)}\mathbf{E}|X(t_0)|^n.$$

EXAMPLE 2. Suppose that condition (3.10') holds in every bounded x-domain and that for $x \in E$

$$|b| < B(1 + |x|\ln|x|), \tag{4.8}$$

$$\sum_{r=1}^{k} |\sigma_r|^2 < B(1 + |x|^2\ln|x|). \tag{4.9}$$

Then, using the auxiliary function $V = \ln(|x|^2 + 1)$ and applying Theorem 4.1, we conclude that the solution of equation (3.9) is regular.

It follows from (4.8) that the scalar product $(b(s,x),x)$ increases no faster than $|x|^2\ln(|x|^2 + 1)$ as $x \to \infty$. The next example shows that if the rate of increase of $(b(s,x),x)$ is slightly higher, the process fails to be regular.

EXAMPLE 3. Suppose that condition (4.9) holds and that for some $\varepsilon > 0$

$$(b(s,x),x) > |x|^2[\ln(|x|^2 + 1)]^{1+\varepsilon}.$$

Then the process $X(t)$ is not-regular for every initial condition. To prove this it suffices to apply Theorem 4.2 with the auxiliary function

$$V(x) = \exp\{-[\ln(|x|^2 + 1)]^{\varepsilon}\}.$$

In this example, the sample function escapes to infinity because of large drift (the phenomenon also occurs for $\sigma \equiv 0$). It is hard to find examples in which the lack of regularity is due to large diffusion.

EXAMPLE 4. Consider a process $X(t)$ in E_ℓ defined by equation (3.9). Then b and σ are scalar functions and the operator L has the form

$$L = \frac{\partial}{\partial t} + b(s,x)\frac{\partial}{\partial x} + \frac{1}{2}\sigma^2(s,x)\frac{\partial^2}{\partial x^2}.$$

Using the same auxiliary function as in Example 3, we see that if

$$\sigma^2 > B(x^2 + 1)[\ln(x^2 + 1)]^{1+\varepsilon},$$

then the process $X(t)$ is not regular for every function $b(s,x)$ which, with respect to the second variable, increases at most linearly. However, when the function $|b(s,x)|$ increases faster than linearly, the process may nonetheless be regular. For example, when $b(s,x) = -x^5$ and $\sigma = x^3$, regularity follows from Theorem 4.1 where we should take the auxiliary function $(x^2 + 1)^\alpha$.

EXAMPLE 5. The assumptions of Theorem 3.1 fail to hold even for the familiar radio-engineering system described by the Van der Pol equation driven by white noise of constant intensity σ^2. This system can be described by the Itô equations:

$$dx_1 = x_2 dt; \qquad dx_2 = [-x_1 + \varepsilon(1 - x_1^2)x_2]dt + \sigma d\xi(t).$$

The auxiliary fuction

$$V = \frac{1}{2}\left[x_2 + \varepsilon\left(\frac{x_1^3}{3} - x_1\right)\right]^2 + \frac{x_1^2}{2} + \frac{\sigma^2}{4\varepsilon}$$

satisfies the assumptions of Theorem 4.1 since

$$\begin{aligned} LV &= x_2\frac{\partial V}{\partial x_1} + [\varepsilon(1 - x_1^2)x_2 - x_1]\frac{\partial V}{\partial x_2} + \frac{\sigma^2}{2}\frac{\partial^2 V}{\partial x_2^2} \\ &= -\varepsilon\frac{x_1^4}{3} + \varepsilon x_1^2 + \frac{\sigma^2}{2} \leqslant 2\varepsilon V. \end{aligned}$$

Thus the process is regular.

5. Stationary and periodic solutions of stochastic differential equations

In Section 2 we gave conditions implying the existence of periodic solutions which were stationary Markov processes, stated in terms of the properties of the transition functions. These conditions are of little use for stochastic differential equations, since the transition functions of such processes are usually not expressible in terms of the coefficients of the equation. Fortunately, however, one can state simple conditions in terms of Ljapunov functions for the required properties of the transition function. The following theorems are analogous to Theorem 2.3.1.

THEOREM 5.1. *Suppose that the coefficients of equation* (3.9) *are independent of* t *and satisfy conditions* (3.10) *in* U_R *for every* R, *and that there exists a function* $V(x) \in \mathbf{C}_2$ *in* E_ℓ *with the properties*

$$V(x) \geqslant 0, \tag{5.1}$$

$$\sup_{|x|>R} LV(x) = -A_R \to -\infty \quad as \quad R \to \infty. \tag{5.2}$$

Suppose moreover that the process $X^x(t)$ *is regular for at least one* $x \in E_\ell$. *Then there exists a solution of equation* (3.9) *which is a stationary Markov process.*

THEOREM 5.2. *Suppose that the coefficients of equation* (3.9) *are* ϑ*-periodic in* t *and satisfy conditions* (3.10) *in every cylinder* $I \times U$, *and suppose further that there exists a function* $V(t,x) \in \mathbf{C}_2$ *in* E *which is* ϑ*-periodic in* t, *and satisfies condition* (5.2) *and the condition*

$$\inf_{|x|>R} V(t,x) \to \infty \quad as \quad R \to \infty. \tag{5.3}$$

Then there exists a solution of equation (3.9) *which is a* ϑ*-periodic Markov process*[9].

PROOF OF THEOREM 5.1. Let $X^x(t)$ be a regular solution of equation (3.9) and let $V(x)$ satisfy conditions (5.1), (5.2). Lemma 3.1 implies that

$$\mathbf{E}V(X^x(\tau_n(t))) - V(x) = \mathbf{E}\int_0^{\tau_n(t)} LV(X^x(u))du.$$

Estimating the right-hand side of this equality by means of the obvious inequality

$$LV(X^x(u)) \leqslant -\chi_{\{|X^x(u)|>R\}}(\omega)A_R + \sup_{x \in E_\ell} LV(x),$$

we get

$$A_R\mathbf{E}\int_0^{\tau_n(t)} \chi_{\{|X^x(u)|>R\}}(\omega)du \leqslant c_1 t + c_2.$$

---

[9] Condition (5.2) of Theorems 5.1 and 5.2 may often be replaced by the weaker condition that $LV \leqslant -1$ outside some compact set (see Chapter IV).

Since the process $X^x(t)$ is regular, it follows that almost surely $\tau_n(t) \to t$ as $n \to \infty$. Moreover, the sequence of integrals in the last inequality is bounded by t. Letting $n \to \infty$ and then changing the order of integration, we obtain the inequality

$$\frac{1}{t}\int_0^t P(x,u,\overline{U}_R)du < \frac{c_3}{A_R} . \tag{5.4}$$

It follows from (5.4), (5.2) and Theorem 2.1 that there exists a stationary initial distribution. The solution of equation (3.9) with this initial distribution is obviously stationary. Q.E.D.

PROOF OF THEOREM 5.2. Arguments similar to those used in proving (5.4) yield the inequality

$$\frac{1}{T}\int_0^T P(s,x,s + u,\overline{U}_R)du < \frac{c_3}{A_R} .$$

Therefore, to prove the theorem we need only to show that condition (2.7) is satisfied and then use the remark following Theorem 2.2.

To prove (2.7), we can again use the method of auxiliary functions. We may assume without loss of generality that the function V satisfying (5.2) and (5.3) also satisfies (5.1) (otherwise we may add to it a constant). Further, it follows from the assumptions of the theorem that $L(V(t,x) - kt) \leqslant 0$ for a sufficiently large constant k. Using this inequality, the regularity of the process (which follows from Theorem 4.1), and Lemma 3.1, we easily obtain that

$$\mathbf{E}_{s,x}V(t,X(t)) \leqslant k(t - s) + V(s,x).$$

Together with Čebyšev's inequality, this implies

$$P(s,x,t,\overline{U}_R) \leqslant \frac{k(t - s) + V(s,x)}{\inf\limits_{|x|>R} V(t,x)} .$$

Thus condition (2.7) will hold if $\beta(R)$ is chosen so that

$$\frac{\sup\limits_{|x|<\beta(R)} V}{\inf\limits_{|x|>R} V} \to 0 \quad \text{as} \quad R \to \infty.$$

This is possible because (5.3) holds. The proof is complete.

To demonstrate the range of application of Theorems 5.1 and 5.2, we present a few examples.

EXAMPLE 1. Consider the auxiliary function

$$V_1 = \sum_{i,j=1}^{\ell} c_{ij}x_i x_j = (Cx,x),$$

where C is a positive definite matrix. We have

$$LV_1 = 2[(Cx,b(t,x)) + \mathrm{tr}(A(t,x)C)].$$

Thus a sufficient condition for the existence of a stationary (periodic) solution of equation (3.9) in the case when the coefficients are independent of t (ϑ-periodic in t) is that for some positive definite matrix C

$$(Cx,b(t,x)) + \mathrm{tr}(A(t,x)C) \to -\infty \quad \text{as } |x| \to \infty,$$

and that the coefficients b and σ_r satisfy conditions (3.10) in every cylinder $I \times U_R$.

EXAMPLE 2[10]. Let us investigate conditions for the existence of a stationary process at the output of a system described by a Lienard equation driven by Gaussian white noise $\dot{\xi}(t)$.

The system is described by the equation

$$x'' + f(x)x' + g(x) = \sigma\dot{\xi}(t), \tag{5.5}$$

where σ^2 is the intensity of the white noise at the input (assumed constant for simplicity's sake). Setting $y(t) = x'(t)$, one easily sees that the pair $(x(t),y(t))$ is a Markov process satisfying the following system of Itô equations

$$dx = y\,dt; \qquad dy = [-yf(x) - g(x)]dt + \sigma d\xi(t)$$

with the differential generator

$$L = y\frac{\partial}{\partial x} - [yf(x) + g(x)]\frac{\partial}{\partial y} + \frac{\sigma^2}{2}\frac{\partial^2}{\partial y^2} + \frac{\partial}{\partial t}.$$

As in Example 2 of Section 2.3, we set

$$F(x) = \int_0^x f(u)du; \qquad G(x) = \int_0^x g(u)du; \qquad p(x) = \gamma \,\mathrm{arctg}\, x$$

and consider the function

$$V_1(x,y) = \frac{y^2}{2} + [F(x) - p(x)]y + G(x) + \int_0^x f(u)[F(u) - p(u)]du + k,$$

---

[10] Due to Nevel'son.

which is analogous to the auxiliary function of the above-mentioned example. It is easy to show that conditions (5.1) - (5.3) hold for $V_1(x,y)$ for a suitable choice of γ and k, if the conditions

$$\begin{aligned} &\operatorname{sign} g(x) = \operatorname{sign} x && \text{for } |x| > x_0, \\ &g(x)F(x) - \delta|g(x)| \to \infty && \text{as } |x| \to \infty, \\ &G(x) + \delta\int_0^x \frac{F(u)}{1+u^2}\,du \to \infty && \text{as } |x| \to \infty \end{aligned} \tag{5.6}$$

hold for some $\delta > 0$, $x_0 > 0$.

Thus, if conditions (5.6) are satisfied, the system (5.5) has a stationary output. Conditions (5.6) will obviously hold for the process at the output of a system described by the Van der Pol equation with $f = \varepsilon(x^2 - 1)$, $g = x$. One can give similar conditions which imply the existence of a periodic solution in the case when the driving process contains a periodic component.

EXAMPLE 3. The following fact is well-known in the theory of systems of ordinary differential equations: If x_0 is an asymptotically stable equilibrium point of an autonomous system $dx/dt = F(x)$ and $f(t)$ is ϑ-periodic, then for sufficiently small ε the system $dx/dt = F(x) + \varepsilon f(t)$ has a ϑ-periodic solution in a neighbourhood of the equilibrium point. The possibility of extending this result to systems of differential equations describing Markov processes was suggested by Krasovskiĭ. The method developed in the section can be employed to this end.

Let $X(t)$ be a homogeneous stochastic process described by the system (3.9). We consider another process which differs from (3.9) by the presence of a "force" $f(x,t)$ which is ϑ-periodic in t (the dependence on x means that the value of f may depend on the state of the system):

$$dY = b(Y)dt + \sum_{r=1}^{k} \sigma_r(Y)\xi_r(t) + f(t,Y)dt. \tag{5.7}$$

Assume that the unperturbed system ($f(t,Y) \equiv 0$) has a stationary solution, and that there exists a function V satisfying conditions (5.1) - (5.3). Then the system (5.7) will have a ϑ-periodic solution, provided the additional condition $(\operatorname{grad} V, f) < c$ holds for some constant c. The proof follows directly from the fact that then the function V satisfies the assumptions of Theorem 5.2 for the process $Y(t)$. It is not difficult to apply this type of reasoning also when the perturbing force itself is a periodic stochastic process. We shall not dwell on this here.

6. Stochastic equations and partial differential equations

The method of investigation of Markov processes based on studying the properties of the solution of some generalized differential equation for the sample functions of the process was, historically speaking, a later development. An earlier method originating in the work of Einstein and Smoluchowski, was subsequently perfected by Kolmogorov, Feller, Dynkin and others. The main idea of the earlier method is to investigate the properties of the solutions of differential equations whose unkowns are expectations of various functions of the processes in question. It turns out that in many cases these expectations are solutions of boundary-value problems for linear parabolic and elliptic equations. The converse is also true. The solution of the first boundary-value problem for the geneal linear elliptic or parabolic equation admits a probabilistic representation in terms of the expectations of certain functionals of the process $X(t)$.

We shall consider this representation in a few special cases. In so doing we shall confine ourselves to domains in E which are Cartesian products of a certain domain $U \subset E_\ell$ having a sufficiently smooth boundary and of a closed interval on the real axis[11].

Throughout this section we shall consider classical solutions of equations of the type

$$Lu \equiv \frac{\partial u}{\partial s} + \sum_{i=1}^{\ell} b_i(s,x) \frac{\partial u}{\partial x} + \frac{1}{2} \sum_{i,j=1}^{\ell} a_{ij}(s,x) \frac{\partial^2 u}{\partial x_i \partial x_j} = -g(s,x).$$

By a solution we shall mean a function $u(s,x)$ which is twice continuously differentiable in x and continuously differentiable in t, and such that $Lu = -g$ at each point of the given domain.

In case the function u does not depend on s, we shall retain the notation Lu for the "elliptic" operator

$$\sum_{i=1}^{\ell} b_i(x)\frac{\partial u}{\partial x_i} + \frac{1}{2} \sum_{i,j=1}^{\ell} a_{ij}(x)\frac{\partial^2 u}{\partial x_i \partial x_j}$$

(this operator is not necessarily elliptic, since the matrix A may be singular).

LEMMA 6.1. *Let $f(s,x)$ be a function, bounded and continuous on the closure of the domain $(t_0,t) \times U$, and satisfying the equation $Lf = 0$ on that domain. Assume that there exists a unique and, if the domain U is unbounded, regular Markov process $X(t)$*

[11] There are numerous papers on the probabilistic representation of solutions of the first boundary-value problem for elliptic and parabolic equations. See e.g., Dynkin [3] and the bibliography given there.

associated with the operator L up to the time τ_U at which the process reaches the boundary of U. Then

$$f(s,x) = \mathbf{E}_{s,x} f(\tau_U(t), X(\tau_U(t))). \tag{6.1}$$

PROOF. The proof for a bounded domain U follows easily from Lemma 3.1. For $n \to \infty$ we consider an increasing sequence of domains $U^{(n)} \uparrow U$ such that for every n there exists a function $f_n(s,x)$ which is equal to $f(s,x)$ for $x \in U^{(n)}$ and has two continuous derivatives with respect to x and one with respect to s for all $x \in E_\ell$, $s \in (t_0,t)$. Applying Lemma 3.1 to the function $f_n(s,x)$ and then letting $n \to \infty$, we get (6.1). The proof for an unbounded domain is similar, except that one considers a sequence of "truncated" domains $U_n = U \cap \{|x| < n\}$ and, using the regularity of the process, one lets $n \to \infty$.

REMARK 1. The stochastic process $X(\tau_U(t))$ is almost surely confined either to the "lateral" surface of the cylinder $[t_0,t] \times T$ or to its "upper" base $\{s = t\} \times U$. Hence it follows that under the assumptions of Lemma 6.1 there exists at most one solution of the equation $Lu = 0$ which takes on given values on the set $\Gamma_1 = [t_0,t] \times \Gamma \cup \{s = t\} \times U$. In particular, if $U = E_\ell$, we get from (6.1) the following formula for the solution of the Cauchy problem for a parabolic equation:

$$f(s,x) = \mathbf{E}_{s,x} f(t,X(t)) = \int_{E_\ell} f(t,y) P(s,x,t,dy). \tag{6.2}$$

REMARK 2. If the domain U is bounded, then the assertion of Lemma 6.1 will hold, for instance, if we require that the coefficients b, σ_r satisfy conditions (3.10) in the domain $(t_0,t) \times U$. For an unbounded domain U, it is sufficient to require that conditions (3.10) hold in every compact set and that there exists a function V satisfying conditions (4.2), (4.3) (see Theorem 4.1).

REMARK 3. It is a well known fact in the theory of differential equations that a sufficient condition for the existence of solutions both to the mixed problem and the Cauchy problem for a parabolic equation is that, in addition to (3.10), the following nondegeneracy condition

$$\sum_{i,j=1}^{\ell} a_{ij}(s,x)\lambda_i\lambda_j > m(s,x) \sum_{i=1}^{\ell} \lambda_i^2 \tag{6.3}$$

is satisfied (here $m(s,x)$ is a positive continuous function on E). Thus, if (6.3) holds, there exists a function f taking on given values on Γ_1 and such that $Lf = 0$. Consequently, formulas (6.1), (6.2) furnish a classical solution of these boundary-value problems for the equation $Lu = 0$. If the function $f(s,x)$ defined by

(6.1) or (6.2) is not differentiable, we may regard it as a generalized solution of the equation $Lu = 0$. (It is readily shown that it satisfies the equation $\mathfrak{A}u = 0$; see (3.19).) However, it can be shown that condition (6.3) need not hold in order for the function $f(t,x)$ defined by (6.2) to be differentiable provided that the "initial" function $f(t,x)$ is sufficiently smooth (see, e.g., Gihman and Skorohod [1]). The situation is different for the solution of the mixed problem.

REMARK 4. Comparing formula (6.2) with the formula

$$f(s,x) = \int f(t,y)p(s,x,t,y)dy,$$

which expresses the solution of the Cauchy problem for a parabolic equation in terms of a fundamental solution $p(s,x,t,y)$, one can verify that if the parabolic equation $Lu = 0$ has a fundamental solution, then the transition function $P(s,x,t,A)$ has a density with respect to Lebesgue measure in E_ℓ and this density is $p(s,x,t,y)$.

Let us now consider a homogeneous process $X(t)$.

LEMMA 6.2. *Suppose that the coefficients* b *and* σ_r *are independent of* t *and that they satisfy conditions* (3.10) *in every bounded domain. Let* U *be a bounded domain in* E_ℓ. *Let* $f(x)$ *be twice continuously differentiable with respect to* x_i *in* U, *continuous on the closure of* U, *and such that* $Lf = 0$. *Assume moreover that for every point* $x \in U$

$$\mathbf{P}_x\{\tau_U < \infty\} = 1. \tag{6.4}$$

Then

$$f(x) = \mathbf{E}_x f(X(\tau_U)) = \int_\Gamma \mathbf{P}_x\{X(\tau_U) \in dy\}f(y) \tag{6.5}$$

holds. Formula (6.5) *remains valid for an unbounded domain* U *if the process* $X(t)$ *is regular and the function* $f(x)$ *is bounded in* U.

PROOF. It will suffice to prove the lemma for a function $f(x) \in \mathbf{C}_2$; the general case can then be treated by means of a suitable limiting process, as in the proof of Lemma 6.1. If $f(x) \in \mathbf{C}_2$, we infer from Lemma 3.1 that

$$f(x) = \mathbf{E}_x\{f(X(\tau_U))\chi_{\tau_U \leq t}(\omega)\} + \mathbf{E}_x\{f(X(t))\chi_{\tau_U > t}(\omega)\}$$

holds for each t. Letting $t \to \infty$ and using (6.4), we get (6.5).

REMARK 5. It follows from Lemma 6.2 that in particular the Dirichlet problem for the "elliptic" equation

$$\sum_{i=1}^{\ell} b_i(x)\frac{\partial u}{\partial x_i} + \frac{1}{2}\sum_{i,j=1}^{\ell} a_{ij}(x)\frac{\partial^2 u}{\partial x_i \partial x_j} = 0$$

in the domain U has at most one solution if condition (6.4) holds. It is known from the theory of differential equations that the Dirichlet problem has a solution if the coefficients satisfy conditions (3.10) and (6.3). If condition (6.3) fails to hold, the generalized solution (6.5) may turn out to be discontinuous. In the next section we shall present a sufficient condition for (6.4) to hold; this will be related to the existence of an auxiliary (Ljapunov) function. Comparing formula (6.5) with the formula (see Miranda [1])

$$f(x) = \int_{\Gamma} K(x,y)f(y)\sigma_{\Gamma}(dy)$$

(where $K(x,y)$ is the normal derivative of the Green function), we see that the measure $\mathbf{P}_x\{X(\tau_U) \in \gamma\}$ has density $K(x,y)$ with respect to the "surface area" $\sigma_\Gamma(dy)$ on Γ, and this density is bounded above and below uniformly in x on every compact subset of U.

Let us now discuss the probabilistic representation of the solution of the inhomogeneous equation

$$Lf = -g. \tag{6.6}$$

We need only consider zero boundary conditions.

LEMMA 6.3. *Let U be a bounded domain in E_ℓ. Let $g(s,x)$ be a continuous bounded function. Let $f(s,x)$ be a function twice continuously differentiable with respect to x_i and continuously differentiable with respect to t in the domain $(t_0,t) \times U$, which is continuous on the closure of this domain, vanishes on the set $\Gamma_1 = [t_0,t] \times \Gamma \cup \{s = t\} \times U$, and satisfies equation* (6.6). *Assume moreover that in every domain which is bounded with respect to x, conditions* (3.10) *are valid. Then*

$$f(s,x) = \mathbf{E}_{s,x}\int_s^{\tau_U(t)} g(u,X(u))du. \tag{6.7}$$

This representation remains valid for an unbounded domain U if the process $X(t)$ is regular and the functions f and g are bounded in $(t_0,t) \times U$.

The proof, analogous to the proofs of Lemmas 6.1 and 6.2, is left to the reader.

LEMMA 6.4. *Let the coefficients b and σ_r be independent of t and suppose they satisfy conditions* (3.10) *in every bounded domain. Let $g(x)$ be continuous and bounded in $U \cup \Gamma$ and let $f(x)$ be twice continuously differentiable and bounded in U, vanishing on Γ, and such that*

$$Lf = \sum_{i=1}^{\ell} b_i(x)\frac{\partial f}{\partial x_i} + \frac{1}{2}\sum_{i,j=1}^{\ell} a_{ij}(x)\frac{\partial^2 f}{\partial x_i \partial x_j} = -g.$$

Assume moreover that for $x \in U$

$$\mathbf{E}_x \tau_U < c. \tag{6.8}$$

Then the function f *can be written as*

$$f(x) = \mathbf{E}_x \int_0^{\tau_U} g(X(t))dt.$$

The proof is analogous to that of Lemma 6.2.

REMARK 6. Condition (6.8) is satisfied, e.g., if U is a bounded domain in which the nondegeneracy condition (6.3) holds with a function $m(s,x) = m(x)$ which is positive on the closure of U (see Corollary 2 in the next section). For an unbounded domain, condition (6.8) does not hold, as a rule. Then Lemma 6.4 is of little use. In the next section we shall prove a better result for this case.

7. Conditions for recurrence and finiteness of mean recurrence time

Let U_1 be some (bounded or unbounded) domain, and denote its complement $\overline{U}_1$ by U. A process $X(t)$ is said to be *recurrent relative to the domain* U_1 (or U*-recurrent*) if it is regular, and for every $s,x \in U$

$$\mathbf{P}_{s,x}\{\tau_U < \infty\} = 1. \tag{7.1}$$

Recurrence is an extremely important concept for the investigation of the properties of sample functions for large time values. A simple condition for recurrence is given by

THEOREM 7.1. *A process* $X(t)$ *is recurrent relative to the domain* U_1 *if it is regular and there exists in* $I \times U$ *a nonnegative function* $V(s,x)$*, twice continuously differentiable with respect to* x *and continuously differentiable with respect to* s*, such that*

$$LV(s,x) \leqslant -\alpha(s), \tag{7.2}$$

where $\alpha(s) \geqslant 0$ *is a function for which*

$$\beta(t) = \int_0^t \alpha(s)ds \to \infty \quad as \quad t \to \infty. \tag{7.3}$$

Moreover, the expectation of the random variable $\beta(\tau_U)$ *exists and satisfies the inequality*

$$\mathbf{E}_{s,x}\beta(\tau_U) \leqslant \beta(s) + V(s,x). \tag{7.4}$$

PROOF. We define a random variable $\tau_U^{(n)}(t)$ by $\tau_U^{(n)}(t) = \min\{\tau_U, t, \tau_n\}$. (As before, $\tau_n = \inf\{t: |X(t)| = n\}$.) By Lemma 3.1,

$$\mathbf{E}_{s,x}V(\tau_U^{(n)}(t), X(\tau_U^{(n)}(t))) - V(s,x) = \mathbf{E}_{s,x}\int_s^{\tau_U^{(n)}(t)} LV(u,X(u))du$$

and therefore, by (7.2),

$$\mathbf{E}_{s,x}\beta(\tau_U^{(n)}(t)) \leqslant \beta(s) + V(s,x).$$

Since $\tau_U^{(n)}(t) \to \tau_U(t)$ almost surely as $n \to \infty$ (since the process is regular), it follows via Fatou's lemma that

$$\mathbf{E}_{s,x}\beta(\tau_U(t)) \leqslant \beta(s) + V(s,x). \tag{7.5}$$

Hence it follows that

$$\mathbf{P}_{s,x}\{\tau_U \geqslant t\} \leqslant \frac{\beta(s) + V(s,x)}{\beta(t)}.$$

Letting $t \to \infty$ and using (7.3), we see that (7.1) holds. The other part of the theorem follows from (7.5) if we let $t \to \infty$ and again use Fatou's lemma.

COROLLARY 1. *Suppose that the assumptions of Theorem* 7.1 *hold, with a function* $\alpha(s) = cs^{n-1}$ $(c > 0)$. *Then the random variable* τ_U *has an n-th moment, and*

$$\mathbf{E}_{s,x}\tau_U^n - s^n \leqslant \frac{n}{c} V(s,x).$$

In particular we see that the recurrence time for the bounded domain U *has finite expectation, if* $\inf_{|x|>R} V \to \infty$ *as* $R \to \infty$, $V \geqslant 0$, *and* $LV \leqslant -c$ *in the domain* $I \times U$ *for some positive constant* c.

COROLLARY 2. *Let the domain* $U = \overline{U}_1$ *be bounded with respect to one of the coordinates, i.e.,* $U \subset \{x_i^{(0)} < x_i < x_i^{(1)}\}$. *Suppose that the diffusion is nonsingular in the same coordinate uniformly in s and that the function* $b_i(s,x)$ *is bounded above or below, so that*

$$0 < a_0 < a_{ii}(s,x), \quad b_i(s,x) < b_0 \quad (\text{or } \; b_i(s,x) > b_0). \tag{7.6}$$

Consider the auxiliary function

$$V(s,x) = e^{\gamma s}[k - (x_i + \beta)^{2n}],$$

where the constants γ, k, β *and* n *will be specified later. Obviously,*

$$LV(s,x) = e^{\gamma s}[\gamma k - \gamma(x_i + \beta)^{2n} - 2nb_i(s,x)(x_i + \beta)^{2n-1}$$

$$- 2n(2n - 1)a_{ii}(s,x)(x_i + \beta)^{2n-2}].$$

Assume for definiteness that $b_i(s,x) < b_0$. *We now choose* $\beta < 0$ *to be large enough in absolute value so that* $x_i + \beta < 1$, *we determine the number* n *by the condition* $(2n - 1)a_0 > b_0 \sup_U |x_i + \beta| + 1$, *and finally we set* $k = \sup_U (x_i + \beta)^{2n}$, $\gamma = k^{-1}$.

Then there is a positive constant c *such that the inequality* $LV(s,x) < -ce^{\gamma s}$ *holds.*

We may now draw the following conclusion from Theorem 7.1:

If $U \subset \{x_i^{(0)} < x_i < x_i^{(1)}\}$ *and conditions* (7.6) *hold in* U, *then the process* $X(t)$ *is recurrent relative to* U_1. *Moreover the random variable* τ_U *then has moments of arbitrary order, actually* $\mathbf{E}_{s,x}e^{\gamma\tau_U}$ *exists for sufficiently small* γ.

A similar conclusion is valid if instead of (7.6) we assume that

$$b_i(s,x) > b_0 > 0 \qquad (\text{or } b_i(s,x) < - b_0 < 0) \tag{7.7}$$

holds in U. In this case, we must set $V(s,x) = e^{\gamma s}(k - x_i)$. Note that conditions (7.6) are satisfied if U is bounded, the coefficients of the operator L are bounded in U, and the matrix A is nonsingular in this domain.

Let us now devote some further attention to temporally homogeneous processes, confining the discussion for the sake of simplicity to the case in which the nondegeneracy condition (6.3) holds in every bounded domain.

THEOREM 7.2. *Let* U_1 *be a bounded domain whose boundary* Γ *is regular*[12] *relative to* $U = \overline{U}_1$ *and suppose that the coefficients* b, σ_r *are independent of* t *and satisfy conditions* (3.10) *in every compact set. Assume further* (6.3), *and that the corresponding Markov process* $X(t)$ *is regular.*

Then the process $X(t)$ *is recurrent relative to* U_1 *if and only if the Dirichlet problem*

---

[12] The term "regular" is used here in the sense customary in the theory of elliptic equations (see Petrovskiĭ [1]).

$$Lu = \sum_{i=1}^{\ell} b_i(x) \frac{\partial u}{\partial x} + \frac{1}{2} \sum_{i,j=1}^{\ell} a_{ij}(x) \frac{\partial^2 u}{\partial x_i \partial x_j} = 0;$$
$$u|_\Gamma = f(s) \tag{7.8}$$

has a unique bounded solution in U.

PROOF. Let $\tau_U^{(n)} = \min(\tau_U, \tau_n)$ denote the time at which the sample function of the process first exits from the domain $U \cap U_n$ ($U_n = \{|x| < n\}$). It follows from the assumptions of the theorem and from Corollary 2 above that $P_x\{\tau_U^{(n)} < \infty\} = 1$. Applying Lemma 6.2, we see that the function

$$\mathbf{P}_x\{X(\tau_U^{(n)}) \in \Gamma\} = u_n(x)$$

is the unique solution in $U \cap U_n$ of the problem

$$Lu = 0; \qquad u|_\Gamma = 1; \qquad u|_{|x|=n} = 0.$$

The sequence $u_n(x)$, is a monotone sequence of bounded solutions of the equation $Lu = 0$. Thus it is obvious that it converges, as $n \to \infty$, to a limit which is also a solution in U of this equation and satisfies the condition $u|_\Gamma = 1$. (This follows from Harnack's second theorem for the equation $Lu = 0$; see Serrin [1].) If the Dirichlet problem in U has a unique bounded solution, then evidently

$$\lim_{n\to\infty} u_n(x) = 1. \tag{7.9}$$

On the other hand, it is obvious that

$$\{\tau_U < \infty\} \subset \bigcup_{n=1} \{X(\tau_U^{(n)}) \in \Gamma\},$$

whence it follows in view of (7.9) that

$$\mathbf{P}_x\{\tau_U < \infty\} = \lim_{n\to\infty} \mathbf{P}\ \{X(\tau_U^{(n)}) \in \Gamma\} = 1. \tag{7.10}$$

This proves the sufficiency.

Now assume that the process $X(t)$ is U_1-recurrent, and suppose that the Dirichlet problem in U has two bounded solutions. Suppose that their difference $v(x)$, which satisfies the condition $v|_\Gamma = 0$ on Γ, is absolutely bounded by some constant k. It follows from Lemma 6.2 that for every n

$$v(x) = \mathbf{E}_x v(X(\tau_U^{(n)})) \leqslant k(1 - \mathbf{P}_x\{X(\tau_U^{(n)}) \in \Gamma\}).$$

Letting $n \to \infty$ and using the fact that $\mathbf{P}_x\{\tau_U^{(n)} \in \Gamma\} \to \mathbf{P}_x\{\tau_U < \infty\}$ by virtue of the regularity of the process, we get $v\ x \equiv 0$. This contradiction concludes the proof.

REMARK. The regularity of the process $X(t)$ is not needed for the sufficiency part of the proof.

THEOREM 7.3. *Under the assumptions of Theorem 7.2, a necessary and sufficient condition for the existence of the expectation of the random variable* τ_U *is that there exists in U a function* $V(x)$, *twice continuously differentiable and nonnegative, such that*

$$LV(x) = -1.$$

The function $u(x) = \mathbf{E}_x\tau_U$ *is then the smallest positive solution in U of the problem*

$$Lu = -1, \qquad u|_\Gamma = 0. \tag{7.11}$$

PROOF. We first assume that there exists a function $V(x)$ satisfying the conditions of the theorem. Then $\mathbf{E}_x\tau_U$ exists and by Theorem 7.1, we have $\mathbf{E}_x\tau_U \leqslant V(x)$. Since $\tau_U^{(n)} \uparrow \tau_U$ almost surely as $n \to \infty$, it follows that $\mathbf{E}_x\tau_U^{(n)} \to \mathbf{E}_x\tau_U$. By Lemma 6.4, the function $u_n(x) = \mathbf{E}_x\tau_U^{(n)}$ is a solution in U of the problem

$$Lu_n(x) = -1; \qquad u_n|_\Gamma = 0. \tag{7.12}$$

Hence the function $v_n(x) = u_{n+1}(x) - u_n(x)$ is L-harmonic (i.e., satisfies the equation $Lv_n = 0$) in the domain $U \cap \{|x| < n\}$. It is also clear that $v_n(x) > 0$. The function $u(x)$ is the sum of the series

$$u(x) = u_{n_0}(x) + \sum_{k=n_0}^{\infty} v_k(x). \tag{7.13}$$

It is known that the sum of a convergent series of positive L-harmonic functions is also an L-harmonic function. Thus (7.13) implies that the function $u(x)$ is twice continuously differentiable and satisfies the equation $Lu = -1$. Because of the assumed regularity properties of the boundary Γ, we have $u|_\Gamma = 0$. We claim that u is the smallest positive solution of the problem (7.11). To prove this we consider another positive solution $W(x)$ of (7.11). Since the function $u_n(x)$ satisfies the boundary conditions $u_n|_\Gamma = 0$; $u_n|_{|x|=n} = 0$, it follows from the maximum principle for elliptic equations (see Petrovskiĭ [1]) that $u_n(x) \leqslant W(x)$ in $U \cap \{|x| < n\}$ for all n. Letting $n \to \infty$, we get $u(x) \leqslant W(x)$, as required.

The necessity part of the proof is even simpler. If $u(x_0) = \mathbf{E}_{x_0}\tau_U < \infty$ for at least one $x_0 \in U$, then, using the representation (7.13), we verify that the sequence $u_n(x_0)$ converges to $u(x_0)$. Hence it follows from Harnack's second theorem for L-harmonic functions that the series (7.13) converges uniformly on every compact set. Its limit $u(x)$ also satisfies the equation $Lu = -1$. Thus there exists a function $V(x)$ satisfying the required conditions (for example, we can set $V(x) = u(x)$).

8. Further conditions for recurrence and finiteness of mean recurrence time[13]

In this section we derive a few corollaries from the results of Section 7. We also consider some examples.

EXAMPLE 1. Considering the auxiliary function $V(x) = \sum_{i,j} b_{ij}x_i x_j = (Bx,x)$, we infer from Theorems 4.1 and 7.1 that the process $X(t)$ is recurrent relative to the domain U if for some function $\alpha(s) \geqslant 0$ satisfying condition (7.3) and a nonsingular positive definite matrix B the condition

$$(Bx,b(s,x)) + \operatorname{tr}(A(s,x)B) \leqslant -\alpha(s) \tag{8.1}$$

holds in $\{t > 0\} \times U$. It follows from Corollary 1 to Theorem 7.1 that the process $X(t)$ has finite mean recurrence time for U_1 if condition (8.1) holds with a function $\alpha(s) \leqslant -1$.

By imposing on the process an additional, not too stringent restriction, one can derive a more convenient condition for recurrence. In the following lemma the domain $U_1 \in E_\ell$ is assumed bounded.

LEMMA 8.1. *Suppose that the process $X(t)$ almost surely exits from each bounded domain in a finite time. Then a sufficient condition for U_1-recurrence is that there exist a nonnegative function $V(t,x)$ in the domain $\{t > 0\} \times U$ such that*

$$V_R = \inf_{t>0,|x|\geqslant R} V(t,x) \to \infty \quad as \quad R \to \infty,$$

$$LV \leqslant 0.$$

PROOF. Let $\tau_U, \tau_R, \tau_{U,R}$ denote the time of first exit of the sample function of the process from the domains U, U_R and $U \cap U_R$ respectively. Denote $\min(\tau_{U,R},t)$ by $\tau_{U,R}(t)$. For each $s \geqslant 0$,

---

[13] For the nondegenerate case, results resembling those of this section have been established by Wonham [1]. His methods were different from ours.

$x \in U$, it follows from the assumptions of the lemma and from Lemma 3.1 that

$$E_{s,x} V(\tau_{U,R}(t), X(\tau_{U,R}(t)) \leqslant V(s,x).$$

Hence, as in the proof of Theorem 7.1, we get

$$E_{s,x} V(\tau_{U,R}, X(\tau_{U,R})) \leqslant V(s,x).$$

This inequality and Čebyšev's inequality imply that

$$\mathbf{P}_{s,x}\{\tau_R < \tau_U\} \leqslant \frac{V(s,x)}{V_R} \to 0 \quad \text{as} \quad R \to \infty.$$

Since moreover the process is continuous, we have $\mathbf{P}_{s,x}\{\tau_R = \tau_U\} = 0$. These relations yield the assertion of the lemma, since

$$\mathbf{P}_{s,x}\{\tau_U < \infty\} \geqslant \mathbf{P}_{s,x}\{\tau_U < \tau_R\} \to 1 \quad \text{as} \quad R \to \infty.$$

REMARK. The first assumption of the lemma (recurrence of the process $X(t)$ relative to the exterior of every bounded domain) holds for instance when either of the conditions (7.6) or (7.7) is satisfied in every bounded domain.

LEMMA 8.2. *A sufficient condition for recurrence of the process* $X^{s_0,x_0}(t)$ *defined by the equation*

$$X^{s_0,x_0}(t) = x_0 + \int_{s_0}^{t} b(u, X^{s_0,x_0}(u))du + \int_{s_0}^{t} \sum_{r=1}^{k} \sigma_r(u, X^{s_0,x_0}(u))d\xi_r(u), \tag{8.2}$$

relative to the domain $U_1 = \overline{U}$ *with boundary* Γ *is that there exists in* $\{t > 0\} \times U$ *a function* $V(t,x)$ *such that*

$$\sup_{x\in\Gamma} V(t,x) \leqslant 0; \qquad LV \geqslant 0; \qquad \sup_{(t>0)\times U} V(t,x) < k < \infty. \qquad V(s_0,x_0) > 0. \tag{8.3}$$

PROOF. It follows easily from $LV \geqslant 0$ and from Lemma 3.1 that

$$E_{s_0,x_0} (\tau_U^{(n)}(t), X(\tau_U^{(n)}(t))) \geqslant V(s_0,x_0),$$

where $\tau_U^{(n)}(t) = \min(\tau_n, \tau_U, t)$. Hence in view of (8.3), we have

$$k\mathbf{P}_{s_0,x_0}\{\tau_U > \min(\tau_n,t)\} \geq V(s_0,x_0). \tag{8.4}$$

It will now be convenient to distinguish two cases: (a) the process $X^{s_0,x_0}(t)$ is regular in U, so that $\lim \tau_n = \tau = \infty$ almost surely; (b) the process is nonregular, i.e., $\tau < \infty$ with positive probability. In case (a) we let $n \to \infty$ in (8.4) and we conclude that $\mathbf{P}_{s_0,x_0}\{\tau_U = \infty\} > 0$. In case (b) the process is nonrecurrent, since it is nonregular. This proves the lemma.

Lemmas 8.1 and 8.2 easily yield necessary and sufficient conditions for recurrence in the one-dimensional case.

EXAMPLE 2. Let the process $X(t)$ be defined by the equation

$$dX(t) = b(X(t))dt + \sigma(X(t))d\xi(t),$$

where b and σ are continuously differentiable in E_ℓ and $\sigma(x) \neq 0$. The generating operator for this process is

$$L = \frac{\partial}{\partial t} + b(x)\frac{\partial}{\partial x} + \frac{1}{2}\sigma^2(x)\frac{\partial^2}{\partial x^2}.$$

Consider the functions

$$Q(x) = \exp\left\{-2\int_0^x \frac{b(z)}{\sigma^2(z)}dz\right\}; \qquad W(x) = \int_0^x Q(y)dy.$$

It is easily seen that $LW = 0$. If moreover $W(x) \to \pm\infty$ as $x \to \pm\infty$, then the function $W(x)\operatorname{sign}x$ satisfies the assumptions of Lemma 8.1, and therefore the process $X(t)$ is recurrent relative to every segment containing the origin. But if the function $W(x)$ is bounded in the domain $x > 0$ $(x < 0)$, then the function $W(x) - \varepsilon$ $(-W(x) - \varepsilon)$ satisfies the assumptions of Lemma 8.2.

Thus, the condition

$$\int_0^x \exp\left\{-2\int_0^y \frac{b(z)}{\sigma^2(z)}dz\right\}dy \to \pm\infty \quad \text{as} \quad x \to \pm\infty$$

is necessary and sufficient for a temporally homogeneous one-dimensional process to be recurrent.

LEMMA 8.3. *Let U_1 be a bounded domain with boundary Γ. Assume that there exist in the domain $\{t > 0\} \times U$ functions $V(t,x)$ and $W(t,x)$ such that*

(1) $V > 0$; $W|_\Gamma \leq 0$; $LV \geq 1$; $LW \geq 0$;

(2) *for some increasing system of bounded domains $U_n \supset U_1$ with boundaries Γ_n,*

$$\frac{\inf_{\Gamma_n} V}{\sup_{\Gamma_n} W} = A_n \to \infty \quad as \quad n \to \infty;$$

(3) the process $X(t)$, defined in the domain $U \cap U_n$ up to the time $\tau^{(n)}$ at which it first reaches the boundary of this domain, satisfies the condition

$$\mathbf{P}_{s,x}\{\tau^{(n)} < \infty\} = 1$$

for all n. Then $\mathbf{E}_{s,x}\tau_U = \infty$ for all points s,x such that $W(s,x) > 0$.

PROOF. By assumption, the function $V - A_nW$ satisfies in $U \cap U_n$ the conditions

$$L(V - A_nW) \leqslant 1; \qquad (V - A_nW)|_{\Gamma_n} \geqslant 0; \qquad (V - A_nW)|_{\Gamma} \geqslant 0.$$

This together with Lemma 3.1 and $X(\tau^{(n)}) \in \Gamma \cup \Gamma_n$, implies that

$$\begin{aligned} A_nW(s,x) - V(s,x) & \\ &\leqslant \mathbf{E}_{s,x}\, V(\tau^{(n)}, X(\tau^{(n)})) - A_nW(\tau^{(n)}, X(\tau^{(n)})) \\ &\qquad + A_nW(s,x) - V(s,x) \\ &\leqslant \mathbf{E}_{s,x}(\tau^{(n)} - s). \end{aligned}$$

Since $A_n \to \infty$ as $n \to \infty$, we see from the above that

$$\lim_{n\to\infty} \mathbf{E}_{s,x}\tau^{(n)} = \infty$$

for all points s,x such that $W(s,x) > 0$. The assertion of the lemma is now obvious if we observe that $\tau^{(n)} < \tau_U$ almost surely.

REMARK. It is not hard to find conditions, in terms of auxiliary functions implying the validity of assumption (3) of Lemma 8.3 (see Section 7). In particular, this assumption holds if (3.10) and the nondegenerary condition (6.3) are valid in every domain bounded with respect to x.

EXAMPLE 3. We conclude from Lemma 8.1 that a process $X(t)$ in E_ℓ satisfying the nondegeneracy condition (6.3) is recurrent relative to every bounded domain U_1 containing the origin when there exists a positive definite symmetric matrix B such that,

$$(Bx, b(t,x)) + \mathrm{tr}(A(t,x)B) \leqslant \frac{2(BA(t,x)Bx, x)}{(Bx, x)} \tag{8.5}$$

holds in $U = \overline{U}_1$. The condition for recurrence given by inequality (8.5) is more general than that given by inequality (8.1), since the right-hand side of (8.5) is always positive. The sufficiency of condition (8.5) can be proved using the auxiliary function $V(x) = \ln(Bx,x) + k$.

We now consider some particular consequences of (8.5). First let $B = A = J$, where J is the $\ell \times \ell$ identity matrix. Then (8.5) becomes $(x,b(t,x)) \leqslant 2 - \ell$. Hence, in particular, if $\ell = 1$ or $\ell = 2$, then the process $X(t)$ associated with the operator

$$L = \frac{\partial}{\partial t} + \sum_{i=1}^{\ell} b\ (t,x) \frac{\partial}{\partial x_i} + \frac{1}{2} \sum_{i=1}^{\ell} \frac{\partial^2}{\partial x_i^2}, \tag{8.6}$$

is recurrent, provided $(x,b(t,x)) \leqslant 0$ outside some bounded domain. For example, the Wiener process on the line and the Wiener process in the plane ($b \equiv 0$) are recurrent.

Lemma 8.2 yields various sufficient conditions for recurrence relative to U_1 (and hence also relative to every domain contained in U_1).

EXAMPLE 4. Consider the auxiliary function

$$V(x) = 1 - k(Bx,x)^{-\alpha}.$$

It is readily seen that

$$LV(x) = 2k\alpha(Bx,x)^{-\alpha-1}$$
$$\times \left[(Bx,b(t,x) + \operatorname{tr}(A(t,x)B) - \frac{2(1+\alpha)(BA(t,x)Bx,x)}{(Bx,x)} \right].$$

If for some α

$$(Bx,b(t,x)) + \operatorname{tr}(A(t,x)B) \geqslant \frac{2(1+\alpha)(BA(t,x)Bx,x)}{(Bx,x)}, \tag{8.7}$$

then for a suitable choice of k and U_1 the function V satisfies the assumptions of Lemma 8.2. Hence we may conclude that the process $X(t)$ is nonrecurrent if there exist a positive definite symmetric matrix B and a constant $\alpha > 0$ for which inequality (8.7) holds outside some bounded domain. Setting $B = A = J$, we infer from (8.4) that if

$$(Bx,b(x)) \geqslant 2 - \ell + \varepsilon$$

in a neighbourhood of the point at infinity, then the process described by the operator (8.6) is nonrecurrent. In particular, it follows that the Wiener process is nonrecurrent in every space of dimension higher than 2.

It is readily seen that the condition that B be positive definite can be considerably weakened; if suffices to assume that B

is positive semi-definite and that at least one of its eigenvalues is nonzero. In this case the neighborhood of the point at infinity must be replaced by a (connected) component of the set $(Bx,x) > y > 0$.

For example, let B be a matrix all of whose elements except b_{ii} are zeros, and $b_{ii} = 1$. It then follows from (8.7) that the process $X(t)$ is nonrecurrent if for some i and $\varepsilon > 0$ the condition

$$x_i b_i(t,x) \geqslant (1 + \varepsilon) a_{ii}(t,x)$$

holds in the domain $x_i > \gamma$ or $x_i < -\gamma$ (where γ is sufficiently large).

EXAMPLE 5. We set $V = (B_1 x,x)$, $W = (B_2 x,x)^\alpha - k$, where B_1 and B_2 are positive definite symmetric matrices, $0 < \alpha < 1$, and the constant k is sufficiently large. Obviously,

$$LV = 2(B_1 x,b(t,x)) + 2\mathrm{tr}(A(t,x)B_1),$$

$$LW = 2\alpha V^{\alpha-1}\left[(B_2 x,b(t,x)) + \mathrm{tr}(A(t,x)B_2) - 2(1 - \alpha)\frac{(B_2 A(t,x)B_2 x,x)}{(B_2 x,x)}\right].$$

Applying Lemma 8.3, we see that the mean recurrence time in U_1 is infinite if there are positive definite matrices B_1, B_2 and a number $\varepsilon > 0$ such that in U

$$(B_1 x,b(t,x)) + \mathrm{tr}(A(t,x)B_1) < \text{const.},$$

$$(B_2 x,b(t,x)) + \mathrm{tr}(A(t,x)B_2) > \varepsilon\frac{(B_2 A(t,x)B_2 x,x)}{(B_2 x,x)}.$$

We again consider the process associated with the operator (8.6). Setting $B_1 = B_2 = J$, we see that the mean recurrence time in a bounded domain for this process does not exist, if for some $\varepsilon > 0$ the scalar product $(x,b(x))$ satisfies

$$-\ell + \varepsilon < (x,b(x)) < \text{const}$$

for $|x| > R$. In particular, these inequalities are valid for the Wiener process in E_ℓ for every ℓ.

CHAPTER IV

ERGODIC PROPERTIES OF SOLUTIONS OF STOCHASTIC EQUATIONS[1]

1. Kolmogorov classification of Markov chains with countably many states

In the preceding chapters we found some sufficient conditions for the existence of a stationary Markov process defined by a stochastic equation. The following two questions are also of great interest: (a) When is the stationary Markov process associated with a given stochastic equation unique? (b) When can it be said that a Markov process with arbitrary initial distribution from a given class converges in some sense to a stationary one?

In this chapter we shall consider these and some related questions in terms of the properties of the coefficients of the operator L. Using the results of Chapter III, one can reformulate the results in terms of Ljapunov functions.

The material in this chapter is organized similarly as in the well-known paper of Kolmogorov [2] on Markov chains with countably many states E_i and discrete time. In that paper Kolmogorov divides the classes of communicating essential[2] states of the chain into the two types of recurrent and nonrecurrent (or transient) ones.

---

[1] The main results of this chapter are an extended and improved version of Sections 2 and 3 of the author's paper [1]. Under different assumptions, results similar to those given below in Sections 4 and 5 have been derived by Maruyama and Tanaka [2]. The existence proof given in Sections 6 and 7 for the limit of the transition function is similar to that of Theorem 3 in Il'in and Has'minskiĭ. For the discrete time case, the existence of the limit of the transition function has been proved under similar assumptions by Nagaev [1].

[2] A state E_i is said to be unessential if the transition probability from E_i to some other state E_j is positive, but the reverse transition from E_i to E_j is impossible.

A class of states E_i is said to be *recurrent* if the probability L_{ii} of reaching E_i at least once from E_i is equal to 1. (It can be proved that in a single class either $L_{ii} < 1$ for all i or all L_{ij} are equal to 1.) A nonrecurrent class is said to be *transient*. It is easy to prove (see below in Section 2 the analog of this result for continuous time) that the n-step transition probability $p_{ij}^{(n)}$ from E_i to E_j for a transient class tends to zero as $n \to \infty$. Hence it follows that no stationary distribution exists for a transient class of states.

Kolmogorov now divides recurrent classes into two types. A class of states is said to be *positive* if the mean recurrence time M_{ii} for a state E_i, starting from E_i, is finite. (It can be proved that in a single class either $M_{ii} = \infty$ for all i or all M_{ij} are finite.) Otherwise the class is said to be a *null class*. It can be shown that in a null recurrent class $p_{ij}^{(n)} \to 0$ as $n \to \infty$. For a positive class it can be shown under a certain additional assumption (that the class consists of a single subclass) that

$$\lim_{n\to\infty} p_{ij}^{(n)} = \pi_j \neq 0.$$

Thus within a positive class of states consisting of a single subclass a limit distribution π_j will ultimately be established, irrespective of the initial distribution. It is easily shown that this limit distribution is stationary.

Kolmogorov has proved that a necessary and sufficient condition for the existence of a stationary distribution within a given class A is that $M_{ii} < \infty$ holds for at least one state $E_i \in A$.

In this chapter we shall derive sufficient conditions for the existence of a stationary distribution and we shall prove theorems on the limit behavior of the transition probability for continuous Markov processes described by stochastic differential equations in terms of functions analogous to L_{ij} and M_{ij}. These results will be derived in terms of functions analogous to L_{ij} and M_{ij}. The only significant modfication is that recurrence for a single state must be replaced by recurrence for a compact set.

As before, we shall use the symbols $\mathbf{E}_x$ and $\mathbf{P}_x$ to designate the expectation of a random variable and the probability of an event when these are determined by the evolution of a temporally homogeneous Markov process $X^{0,x}(t)$. In this and the following chapters we shall make constant use of the strong Markov property, introduced by Dynkin and Juškevič [1]. In Chapter III we considered the numerical random variable τ_U equal to the first exit time of the sample path from the domain U, $\tau_U(t) = \min(\tau_U, t)$, and other variables, all possessing the property that the event "the random variable assumes a value smaller than s" depends only on the evolution of the process up to time s, i.e.

$$\{\tau \leqslant s\} \in \mathcal{N}_s. \tag{1.1}$$

A random variable τ satisfying condition (1.1) will be called a *Markov time* or a *random variable independent of the future*. Roughly speaking, a Markov process is said to be *strongly Markov* if the future is independent of the past not only for a fixed instant of time but also for any Markov time τ.

A Markov process such that the transition function $P(s,x,t,A)$ is Borel-measurable as a function of (s,x) is said to be *strongly Markov* if for any Markov time τ, $\tilde{\mathcal{N}}_\tau$-measurable random variable $\eta \geqslant \tau$, $x \in E_\ell$ and $A \in \mathfrak{B}$,

$$\mathbf{P}\{X^{s,x}(\eta) \in A/\tilde{\mathcal{N}}_\tau\} = P(\tau, X^{s,x}(\tau), \eta, A) \qquad \text{(a.s.)}. \tag{1.2}$$

It can be proved that any right-continuous process with Feller transition function is strongly Markov (Dynkin and Juškevič [1]). Therefore, the solution of an Itô equation (3.3.9) is also a strong Markov process. This fact will be repeatedly used in the sequel.

2. Recurrence and transience

In Sections 3.7 and 3.8 we studied conditions which imply the recurrence solutions of stochastic equations relative to a domain U, i.e., conditions under which the paths issuing from any point $x \in E_\ell \setminus U$ almost usrely reach the set U. In this section we shall show that for temporally homogeneous processes with nonsingular diffusion matrix the recurrence property does not depend on the choice of the bounded domain U.

Let $X(t)$ be a regular temporally homogeneous Markov process in E_ℓ described by the stochastic equation

$$dX(t) = b(X)dt + \sum_{r=1}^{k} \sigma_r(X) d\xi_r(t). \tag{2.1}$$

Here and in the next section we shall assume that the diffusion matrix

$$A(x) = ((a_{ij}(x))), \qquad a_{ij}(x) = \sum_{r=1}^{k} \sigma_r^i(x)\sigma_r^j(x)$$

of the process $X(t)$ is nonsingular, i.e., the smallest eigenvalue of the matrix $A(x)$ is bounded away from zero in every bounded domain.

LEMMA 2.1. *If $X(t)$ is recurrent relative to some bounded domain U, then it is recurrent relative to any nonempty domain in E_ℓ.*

PROOF. It will suffice to prove that the process is recurrent relative to a domain U_0 with regular boundary Γ_0 such that $U_0 \cup \Gamma_0 \subset U$. We may also assume that the boundary Γ of U is regular (otherwise we replace U by a suitable domain containing U).

Let U_1 be a domain with regular boundary Γ_1 such that $U \cup \Gamma \subset U_1$. Let $x \in \overline{U}_0$ be arbitrary. We claim the $\mathbf{P}_x\{\tau_{\overline{U}_0} < \infty\} = 1$.

Assuming for definiteness that $x \in \overline{U}_1$, we consider the following random variables: τ_1', the time at which the path of the process first reaches the set Γ, and τ_1, the first time after τ_1' at which the path reaches Γ_1. We now define two random variables inductively: τ_n', the first time after τ_{n-1} at which the path reaches Γ, and τ_n, the first time after τ_n' at which it reaches Γ_1 (Figure 1). By the assumption that the process is recurrent relative to U and by Corollary 2 to Theorem 3.7.1, it follows that $\tau_n < \infty$ almost surely.

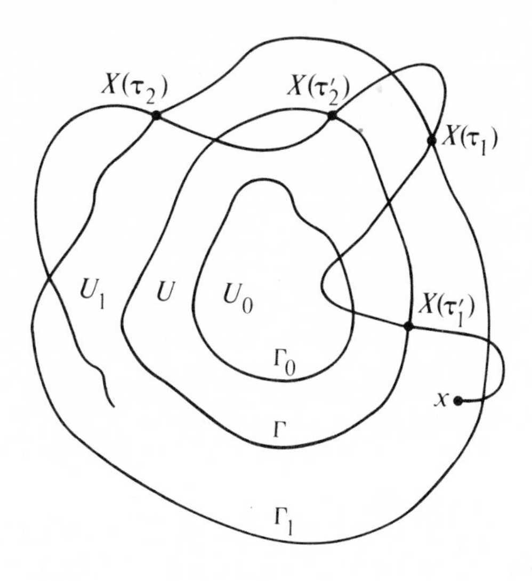

Figure 1

Set $U_3 = U_1 \setminus U_0$. By Lemma 3.6.2, the function

$$u(x) = \mathbf{P}_x\{X(\tau_{U_0}) \in \Gamma_0\}$$

is a solution of the elliptic equation

$$Lu = 0$$

in the domain U_3 satisfying the boundary conditions

$$u|_{\Gamma_0} = 1; \qquad u|_{\Gamma_1} = 0.$$

By the strong maximum principle for solutions of elliptic equations,

$$\max_{x\in\Gamma} \mathbf{P}_x\{X(\tau_{U_3}) \in \Gamma_1\} = p_0 < 1. \tag{2.2}$$

Using the fact that the random variables $\tau_1', \tau_1, \tau_2', \tau_2, \ldots$ are Markov times and applying the strong Markov property of the process $X(t)$, and also (2.2), we get

$$\begin{aligned}
\mathbf{P}_x\{\tau_{\bar{U}_0} = \infty\} &\leqslant \sup_{z\in\Gamma_1} \mathbf{P}_z\left\{\bigcap_{i=1}^{\infty} (X(\tau_{U_3}^{(i)}) \in \Gamma_1)\right\} \\
&= \sup_{z\in\Gamma_1} \int_{\Gamma} \mathbf{P}_z\{X(\tau_1') \in dy\} \int_{\Gamma_1} \mathbf{P}_y\{X(\tau_{U_3}) \in dz_1\} \\
&\qquad \times \mathbf{P}_{z_1}\left\{\bigcap_{i=1}^{\infty} (X(\tau_{U_3}^{i}) \in \Gamma_1)\right\} \\
&\leqslant p_0 \sup_{z\in\Gamma_1} \mathbf{P}_z\left\{\bigcap_{i=1}^{\infty} (X(\tau_{U_3}^{(i)}) \in \Gamma_1)\right\},
\end{aligned}$$

where $\tau_{U_3}^{i}$ denotes the first time after τ_i' at which the path exits from the set U_3.

These inequalities imply that $\mathbf{P}_x\{\tau_{\bar{U}_0} = \infty\} = 0$, since $p_0 < 1$. This completes the proof of the lemma.

In view of this lemma the following definitions are natural. A regular process $X(t)$ described by an equation (2.1) with non-singular diffusion matrix is said to be *recurrent* if there exists a bounded domain U such that for all $x \in \overline{U}$,

$$\mathbf{P}_x\{\tau_{\bar{U}} < \infty\} = 1.$$

If there exist a non-empty domain U and a point $x \in \overline{U}$ such that $\mathbf{P}_x\{\tau_U < \infty\} < 1$, the process is said to be *transient*.

It follows from the definition of recurrence and from Lemma 2.1 that the sample path of a recurrent process is almost surely dense in E_ℓ and it prevails an infinite time in every fixed neighbourhood of any point. The situation is different for transient processes.

LEMMA 2.2. *If the process $X(t)$ is transient, then for any compact set K and any $x \in E_\ell$ the probability of the event "the sample path of the process $X^x(t)$ never passes through the set K after some random but finite instant of time $t_0(\omega)$" equals* 1.

PROOF. Let K be a compact set in E_ℓ, and let $U_R \subset K$ a ball of radius R containing K. Since the process is transient, the value of the function

$$u(x) = \mathbf{P}_x\{\tau_{\bar{U}_R} < \infty\}$$

is less than 1 at some point $x \in \overline{U}_R$. This function satisfies the equation $Lu = 0$ (as the limit of a monotone sequence of L-harmonic functions; see Section 3.7). By the strong maximum principle,

$$\max_{|x|=R+1} \mathbf{P}_x\{\tau_{\bar{U}_R} < \infty\} = q < 1. \tag{2.3}$$

Let $x \in \overline{U}_R$. We now define the random variables $\tau_1', \tau_1, \tau_2', \ldots$ as follows: $\tau_1' = \tau_{U_{R+1}}$, τ_1 is the first time after τ_1' at which the set $|x| = R$ is reached, τ_1' is the first time after τ_1 at which the set $|x| = R + 1$ is reached, and so on. For example, let us estimate the probability $\mathbf{P}_x\{\tau_2 < \infty\}$. Using the strong Markov property of the process and also (2.3), we get

$$\begin{aligned}
&\mathbf{P}_x\{\tau_2 < \infty\} \\
&= \int_{|y_1|=R+1} \mathbf{P}_x\{X(\tau_1') \in dy_1\} \int_{|z|=R} \mathbf{P}_{y_1}\{\tau_{U_R} < \infty; X(\tau_{\bar{U}_R}) \in dz\} \\
&\quad \times \int_{|y_2|=R+1} \mathbf{P}_z\{X(\tau_2') \in dy_2\} \mathbf{P}_{y_2}\{\tau_{\bar{U}_R} < \infty\} \\
&\leqslant q^2.
\end{aligned}$$

Similarly, one shows that $\mathbf{P}_x\{\tau_n < \infty\} \leqslant q^n$.

Thus $\{\tau_n < \infty\}$ is a sequence of events whose probabilities decrease in geometric progression. By the Borel-Cantelli lemma, this implies that almost surely only finitely many of these events will occur. This proves the lemma.

COROLLARY. Since $\{X(t) \in K\} \subset \{t_0(\omega) > t\}$, it follows that a transient process $X(t)$ satisfies

$$\mathbf{P}_x\{X(t) \in K\} = P(x,t,K) \to 0 \quad \text{as} \quad t \to \infty \tag{2.4}$$

for every compact set $K \subset E_\ell$.

LEMMA 2.3. *If the process $X(t)$ is transient, then the random variable ζ_K equal to the total time which the sample path of the process spends in the set K satisfies*

$$\mathbf{E}_x|\zeta_K|^n < c_n < \infty. \tag{2.5}$$

for any $n < 0$.

PROOF. It is sufficient to prove this when K is the ball $|x| \leqslant R$. Let $q < 1$ be the number defined by (2.3). The random variable τ_{R+1} defined as the time to the first exit of the path of $X(t)$ from the set $\{|x| < R+1\}$ has bounded expectation, due to Corollary 2 to Theorem 3.7.1. By Čebyšev's inequality, there is a constant $t > 0$ such that

$$\mathbf{P}_x\{\tau_{R+1} \geqslant t\} < \frac{1-q}{2}. \tag{2.6}$$

Since for any $x \in K$

$$\mathbf{P}_x\{\zeta_K \geqslant t\} \leqslant \mathbf{P}_x\{\tau_{R+1} \geqslant t\} + \sup_{|y|=R+1} P_y\{\tau_{\bar{U}_R} < \infty\}, \tag{2.7}$$

it follows from (2.3) and (2.6) that

$$\mathbf{P}_x\{\zeta_K \geqslant t\} \leqslant \frac{1+q}{2} = q_1 < 1$$

for $x \in K$. It is clear that (2.7) is valid for all $x \in E_\ell$. A necessary condition for the occurrence of the event $\{\zeta_K \geqslant 2t\}$ is obviously that of the event $\{\zeta_K \geqslant t\}$ occurs and that beginning from the time at which ζ_K is first equal to t, the path remains in K for a time not less than t. Hence (the rigorous argument involves the strong Markov property)

$$\mathbf{P}_x\{\zeta_K \geqslant 2t\} \leqslant q_1^2, \quad \ldots, \quad \mathbf{P}_x\{\zeta_K \geqslant nt\} \leqslant q_1^n. \tag{2.8}$$

Inequalities (2.8) guarantee the existence of bound moments for all powers of the random variable ζ_K, since

$$\mathbf{E}_x\zeta_K^n \leqslant \sum_{m=0}^{\infty} \mathbf{P}_x\{\zeta_K > mt\}[(m+1)t]^n$$

for $x \in K$, and hence also for all $x \in E_\ell$.

COROLLARY. By virtue of the equality

$$\mathbf{E}_x\zeta_K = \mathbf{E}\int_0^\infty \chi_K(X(t))dt = \int_0^\infty P(x,t,K)dt$$

it follows from Lemma 2.3 that the function $\int_0^\infty P(x,t,K)dt$ is bounded in E_ℓ.

REMARK. It is readily seen that the proofs of the lemmas in this section essentially use far weaker properties of the process $X(t)$ than the nonsingularity of the diffusion matrix. For example, the following generalization of Lemma 2.1 is valid.

LEMMA 2.1'. *The solution $X(t)$ of the stochastic equation* (2.1) *is recurrent relative to the domain U_2 when it is recurrent relative to U_1 and*

$$\inf_{x \in U_1} P(x,T,U_2) > 0 \tag{2.9}$$

for some $T > 0$.

It is easy to see that condition (2.9) is satisfied if the domain U_1 is bounded and the transition probability of $X(t)$ has an everywhere positive density.

3. Positive and null processes

Suppose that the conditions formulated at the beginning of Section 2 are satisfied, and let the process $X(t)$ be recurrent. As in the case of a process with countably many states, the asymptotic behavior of the transition probability depends essentially on whether the mean recurrence time for a bounded domain is finite or infinite.

LEMMA 3.1. *Under the above assumptions, if $\mathbf{E}_{x_0}\tau_{\bar{U}}$ is finite for some bounded domain U and $x_0 \in \bar{U}$, then $\mathbf{E}_x\tau_{\bar{U}_0}$ is finite for all nonempty domains U_0 and all $x \in \bar{U}_0$.*

PROOF. Let $\mathbf{E}_{x_0}\tau_{\bar{U}} < \infty$ for some bounded domain U and $x_0 \in \bar{U}$. It was shown in the proof of Theorem 3.7.3 that then $\mathbf{E}_x\tau_{\bar{U}} < \infty$ for *all* $x \in \bar{U}$.

We must prove that $\mathbf{E}_x\tau_{\bar{U}_0} < \infty$ for any other non-empty domain U_0. As in the proof of Lemma 2.1, we need only deal with the case in which U and U_0 are domains with regular boundaries Γ and Γ_0 respectively and $U_0 \cup \Gamma_0 \subset U$.

As in the proof of Lemma 2.1, we construct an auxiliary domain U_1 and consider the corresponding times $\tau_1', \tau_1, \ldots, \tau_n', \tau_n, \ldots$ (see Figure 1). We shall call the portion of the sample path of the process from τ_{n-1} to τ_n the n-th cycle, and we set $\tau_0 = 0$. The event A: "the path of $X(t)$ reaches the set U_0" can first occur either during the first cycle, or during the second, etc.. If it occurs during the n-th cycle, then obviously,

$$\tau_{\bar{U}_0} < \tau_n.$$

On the other hand, it follows by (2.2) and the strong Markov process that the probability that A will not occur during the $n - 1$

preceding cycles is majorized by p_0^{n-1}. Hence we have the estimate

$$\mathbf{E}_x\tau_{\bar{U}_0} < \mathbf{E}_x\tau_1 + p_0\mathbf{E}_x\tau_2 + \ldots + p_0^{n-1}\mathbf{E}_x\tau_n + \ldots .$$

Without loss of generality, we may assume that $x \in \Gamma_1$. Obviously,

$$\sup_{x\in\Gamma_1} \mathbf{E}_x\tau_1 \leqslant \sup_{x\in\Gamma_1} \mathbf{E}_x\tau_{\bar{U}} + \sup_{x\in\Gamma} \mathbf{E}_x\tau_{U_1} = B < \infty.$$

Therefore

$$\mathbf{E}_x\tau_{\bar{U}_0} \leqslant B + 2Bp_0 + \ldots + nBp_0^{n-1} + \ldots < \infty$$

as required.

In accordance with Kolmogorov's terminology for Markov chains with countably many states, a recurrent process such that the mean recurrence time for some (hence for each) bounded set is finite will be called a *positive process*. Otherwise we have a *null process*.

Sufficient conditions for a process to be positive, null, recurrent, or transient, in terms of the coefficients b and σ_r, were given above (Sections 3.7, 3.8).

4. Existence of a stationary distribution

In Sections 4 through 7 we shall study ergodic properties of positive Markov processes. First, in Section 4, we shall establish the existence of a stationary distribution for such processes. This will enable us to apply the ergodic theorem for stationary processes and thus, in Section 5, to establish the strong law of large numbers for functions of diffusion-type Markov processes.

Next, in Section 6 and 7, we shall prove a theorem which states that, under certain restrictions, the transition probability of a temporally homogeneous process from a point x to a set A in time t tends to a limit as $t \to \infty$. This limit is independent of the "point of departure" x. It equals the stationary distribution $\mu(A)$. In Section 8 we discuss certain generalizations of previous results. In Section 9 we shall prove corollaries concerning the behavior of the solutions of parabolic differential equations for large time values. Sections 10 and 11 are devoted to a more detailed consideration of a null process on the real line.

The main assumptions adopted in Section 4 through 7 may be described as follows. We shall stipulate that the process $X(t)$ has finite mean recurrence time for some bounded domain U, and within this domain all sample paths "mix sufficiently well" (while outside U the diffusion matrix may be as strongly singular as desired; for example, the process may be deterministic outside U).

More precisely, we shall adopt the following assumption (B):

There exists a bounded domain $U \subset E_\ell$ *with regular boundary* Γ, *having the following properties:*

(B.1) *In the domain* U *and some neighbourhood thereof, the smallest eigenvalue of the diffusion matrix* $A(x)$ *is bounded away from zero.*

(B.2) *If* $x \in E_\ell \setminus U$, *the mean time* τ *at which a path issuing from* x *reaches the set* U *is finite, and* $\sup_{x \in K} \mathbf{E}_x\tau < \infty$ *for every compact subset* $K \subset E_\ell$.

We consider a domain U_1 with sufficiently smooth boundary such that condition (B.1) holds in U_1 and $U \cup \Gamma \subset U_1$.

The construction used above in the proofs of Lemmas 2.1 and 3.1 is basic for what follows. This construction divides an arbitrary sample path of the process into "cycles":

$$[\tau_0,\tau_1); \qquad [\tau_1,\tau_2); \qquad \ldots; \qquad [\tau_n,\tau_{n+1}); \qquad \ldots$$

Here $\tau_0 = 0$, and the times $\tau_1', \tau_1, \tau_2', \tau_2, \ldots$ are defined inductively: τ_{n+1}' is the first time after τ_n at which the set Γ is reached, and τ_{n+1} is the first time after τ_{n+1}' at which the path reaches Γ_1 (see Figure 1 on p. 112).

The process $X(t)$ is U-recurrent by condition (B.2) and U_1-recurrent by condition (B.1) and Corollary 2 to Theorem 3.7.1. Hence, all the random variables $\tau_1' < \tau_1 < \ldots < \tau_n' < \tau_n < \ldots$ are almost surely finite.

Suppose that $X(0) = x \in \Gamma_1$, and consider the sequence $X(\tau_i) = \tilde{X}_i$. It follows from the strong Markov property of $X(t)$ that this sequence is a Markov chain on Γ_1. Let $\tilde{P}(x,\gamma)$ denote the one-step transition probability of this chain, and set

$$\mathbf{E}_x f(\tilde{X}_1) = \int_{\Gamma_1} \tilde{P}(x,dy) f(y).$$

Let us first establish some properties of the process $\tilde{X}_n$. We denote by $\tilde{P}^{(n)}(x,\gamma)$ the n-step transition probability of this process.

LEMMA 4.1. *The Markov chain* $X_1,\ldots,X_n,\ldots$ *has a unique stationary distribution* $\mu(\gamma)$, *which satisfies*

$$|\tilde{P}^{(n)}(x,\gamma) - \mu(\gamma)| < k^n \tag{4.1}$$

uniformly in $\gamma \in \Gamma_1$ *for some constant* $k < 1$.

PROOF. By Remark 5 in Section 3.6, for all x the measure $P_x(X(\tau_{U_1}) \in \gamma)$ has a density $\sigma_{\Gamma_1}(\gamma)$ on Γ_1, relative to surface area, which is bounded away from zero. Hence the obvious equality

$$\tilde{P}(x,\gamma) = \int \mathbf{P}_x\{X(\tau_1') \in dz\}\mathbf{P}_z\{X(\tau_{U_1}) \in \gamma\}$$

implies that $\tilde{P}(x,\gamma)$ has the same property. Now it is well-known (see Doob [3], p. 197) that this condition is sufficient for the existence of a unique stationary distribution $\tilde{\mu}(\gamma)$ of the Markov chain $\tilde{X}_n$ and for (4.1) to hold.

LEMMA 4.2. *Let* τ *be a Markov time,* $\mathbf{E}_x\tau < \infty$ *and* $f(x)$ *a Borel-measurable function. Then*

$$\mathbf{E}_x\int_0^\tau f(X(t + s))ds = \mathbf{E}_x\int_0^\tau \mathbf{E}_{X(s)}f(X(t))ds. \tag{4.2}$$

PROOF. Since τ is a Markov time, the characteristic function $\chi_{s<\tau}(\omega)$ of the set $\{s < \tau\}$ is $\tilde{\mathcal{N}}_s$-measurable. Therefore

$$\mathbf{E}_x\int_0^\tau f(X(t + s))ds = \int_0^\infty \mathbf{E}_s\mathbf{E}_s\{\chi_{s<\tau}f(X(t + s))/\tilde{\mathcal{N}}_s\}ds$$

$$= \mathbf{E}_x\int_0^\infty \chi_{s<\tau}\mathbf{E}_x\{f(X(t + s))/\tilde{\mathcal{N}}_s\}ds$$

$$= \mathbf{E}_x\int_0^\tau \mathbf{E}_x\{f(X(t + s))/\mathcal{N}_s\}ds.$$

This, together with (3.1.7), (3.1.8), implies the assertion.

THEOREM 4.1. *If* (B) *holds, then the Markov process* $X(t)$ *has a stationary distribution* $\mu(A)$.

PROOF. Let $A \in \mathfrak{B}$. Let τ^A denote the time spent by the path of $X(t)$ in the set A during the first cycle. We define a measure $\nu(A)$ by

$$\nu(A) = \int_{\Gamma_1} \tilde{\mu}(dx)\mathbf{E}_x\tau^A. \tag{4.3}$$

Then, for any continuous bounded function $f(x)$,

$$\int_{\Gamma_\ell} f(X)\nu(dx) = \int_{\Gamma_1} \tilde{\mu}(dx)\mathbf{E}_x\int_0^{\tau_1} f(X(t))dt. \tag{4.4}$$

Recall that $\tilde{\mu}(\gamma)$ is the stationary distribution of the Markov chain $\tilde{X}_n$, i.e., for any bounded Borel-measurable function $g(x)$ on Γ_1 we have

$$\int_{\Gamma_1} \mathbf{E}_x g(\tilde{X}_1)\tilde{\mu}(dx) = \int_{\Gamma_1} g(x)\tilde{\mu}(dx). \tag{4.5}$$

From (4.4) and (4.2) we see by performing the substitution $t + s = u$ that

$$\begin{aligned}
\int_{E_\ell} & \mathbf{E}_x f(X(t))\nu(dx) \\
&= \int_{\Gamma_1} \tilde{\mu}(dx)\mathbf{E}_x \int_0^{\tau_1} \mathbf{E}_{X(s)} f(X(t))ds \\
&= \int_{\Gamma_1} \tilde{\mu}(dx)\mathbf{E}_x \int_0^{\tau_1} f(X(t + s))ds \\
&= \int_{\Gamma_1} \tilde{\mu}(dx)\mathbf{E}_x \int_t^{t+\tau_1} f(X(u))du \\
&= \int_{\Gamma_1} \tilde{\mu}(dx)\mathbf{E}_x \int_0^{\tau_1} f(X(u))du + \int_{\Gamma_1} \tilde{\mu}(dx)\mathbf{E}_x \int_{\tau}^{\tau_1+t} f(X(u))du \\
&\qquad - \int_{\Gamma_1} \mu(dx)\mathbf{E}_x \int_{\tau_1}^{t} f(X(u))du.
\end{aligned} \tag{4.6}$$

It follows from (4.5) that

$$\begin{aligned}
\int_{\Gamma_1} \tilde{\mu}(dx)\mathbf{E}_x \int_{\tau_1}^{\tau_1+t} f(X(u))du &= \int_{\Gamma_1} \tilde{\mu}(dx)\mathbf{E}_x \mathbf{E}_{\tilde{X}_1} \int_0^t f(X(u))du \\
&= \int_{\Gamma_1} \tilde{\mu}(dx)\mathbf{E}_x \int_0^t f(X(u))du.
\end{aligned} \tag{4.7}$$

By (4.6) and (4.7), we have

$$\int_{E_\ell} \mathbf{E}_x f(X(t))\nu(dx) = \int_{E_\ell} f(x)\nu(dx). \tag{4.8}$$

We now see from (4.8) that the measure given by $\mu(A) = \nu(A)/\nu(E_\ell)$ defines the required stationary distribution.

REMARK. The measure $\nu(A)$ is invariant (i.e. it satisfies condition (4.8)) even if no assumption is made concerning the finiteness of $\mathbf{E}_x\tau_1$. It is sufficient to require that the process $X(t)$ is U-recurrent. In this case the measure ν is merely σ-finite and $\nu(E_\ell) = \infty$. For details, see Maruyama and Tanaka [2], Has'minskiĭ [1].

5. Strong law of large numbers

Chung's proof [1] of the law of large numbers for random variables $\zeta_n = \sum_{i=1}^{n} f(X_i)$, where X_i is a recurrent Markov chain with countably many states, employs the following method. Each random variable ζ_n is split into components. The k-th component contains the terms $\sum f(X_i)$ for $\tau_k \leqslant i < \tau_{k+1}$ (τ_k is the time at which a fixed state of the chain is reached for the k-th time). According to this approach, ζ_n may be described as a sum of independent random variables and a certain (generally small) remainder term. Therefore the law of large numbers ζ_n follows from the corresponding law for sums of independent random variables. Maruyama and Tanaka [1] adopt a similar approach to prove the law of large numbers for one-dimensional diffusion processes. This reduction is impossible for many-dimensional processes, since a many-dimensional process does not generally return to the initial point. However, the law of large numbers can be used for stationary sequences.

THEOREM 5.1. *Suppose that condition* (B) *holds, and let* μ *be the stationary distribution of the process* $X(t)$, *constructed in the proof of Theorem* 4.1. *Let* $f(x)$ *be a function integrable with respect to the measure* μ. *Then*

$$\mathbf{P}\left\{\frac{1}{T}\int_0^T f(X(t))\,dt \xrightarrow[T\to\infty]{} \int_{E_\ell} f(x)\mu(dx)\right\} = 1 \tag{5.1}$$

for all $x \in E_\ell$.

PROOF. We first prove (5.1) for the initial distribution

$$\mathbf{P}\{X(0) \in \gamma\} = \mu(\gamma), \qquad \gamma \in \Gamma_1, \tag{5.2}$$

where $\tilde{\mu}(\gamma)$ is the stationary distribution of the Markov chain $\tilde{X}_n$, existing according to Lemma 4.1. Under this assumption the sequence of random variables

$$\eta_n = \int_{\tau_n}^{\tau_{n+1}} f(X(t))\,dt$$

is a random sequence which is stationary in the narrow sense, and if follows from (4.3) that

$$\mathbf{E}\eta_n = \int_{E_\ell} f(x)\nu(dx).$$

We easily see from (4.1) that the sequence η_n is metrically transitive.

Let $\nu_1(T)$ denote the number of cycles completed up to time T. It is obvious that

$$\int_0^T f(X(t))dt = \sum_{n=0}^{\nu_1(T)} \eta_n + \int_{\tau_{\nu_1(T)}}^T f(X(t))dt. \tag{5.3}$$

We may assume without loss of generality that $f(x) > 0$. It then follows from (5.3) that

$$\sum_{n=0}^{\nu_1(T)} \eta_n \leqslant \int_0^T f(X(t))dt \leqslant \sum_{n=0}^{\nu_1(T)+1} \eta_n. \tag{5.4}$$

Since the sequence $\eta_1,\ldots,\eta_n,\ldots$ is stationary and metrically transitive, the law of large numbers for such sequences (see Gihman and Skorohod [1]) implies the relation

$$\mathbf{P}\left\{\frac{1}{k}\sum_{n=0}^{k} \eta_n \xrightarrow[k\to\infty]{} \int_{E_\ell} f(x)\nu(dx)\right\} = 1. \tag{5.5}$$

In particular for $f(x) \equiv 1$ we get from the above that

$$\mathbf{P}\left\{\frac{\tau_k}{k} \xrightarrow[k\to\infty]{} \int_{E_\ell} \tilde{\mu}(dx)\mathbf{E}_x\tau = \nu(E_\ell)\right\} = 1. \tag{5.6}$$

We claim that almost surely,

$$\frac{\tau_{\nu_1(T)}}{T} \to 1 \quad \text{as} \quad T \to \infty. \tag{5.7}$$

Using (5.6) and the fact that $\nu_1(T) \to \infty$ as $T \to \infty$, we see that almost surely,

$$\lim_{T\to\infty} \frac{\tau_{\nu_1(T)}}{\tau_{\nu_1(T)}} = \lim_{T\to\infty} \frac{\tau_{\nu_1(T)}}{\nu_1(\ \ ) + 1} \lim_{T\to\infty} \frac{\nu_1(T)}{\tau_{\nu_1(T)}} = 1.$$

Hence, using the inequalities $\tau_{\nu_1(T)} \leqslant T \leqslant \tau_{\nu_1(T)+1}$ we get (5.7). From (5.4) - (5.7) we readily infer that

$$\mathbf{P}\left\{\frac{1}{T}\int_0^T f(X(t))dt = \frac{\int_0^T f(X(t))dt}{\nu_1(T)}\,\frac{\nu_1(T)}{\tau_{\nu_1(T)}}\,\frac{\tau_{\nu_1(T)}}{T}\xrightarrow[T\to\infty]{}\int_{E_\ell} f(x)\mu(dx)\right\}$$

$$= 1.$$

This proves (5.1) for the initial distribution (5.2). It follows that (5.1) is valid for almost all points $x \in \Gamma_1$ with respect to the measure $\tilde{\mu}(\gamma)$ (hence also with respect to the measure $\sigma_{\Gamma_1}(\gamma)$).

Let x be any point of E_ℓ. It is clear from the proof of Lemma 4.1 that the measure $\mathbf{P}_x\{X(\tau_1) \in \gamma\}$ is absolutely continuous with respect to the measure $\sigma_{\Gamma_1}(\gamma)$ for any $x \in E_\ell$. Hence, using the equalities

$$\mathbf{P}_x\left\{\lim_{T\to\infty}\frac{1}{T}\int_0^T f(X(t))dt = a\right\}$$

$$= \mathbf{P}_x\left\{\lim_{T\to\infty}\frac{1}{T}\int_{\tau_1}^T f(X(t))dt = a\right\}$$

$$= \int_{y\in\Gamma_1}\mathbf{P}_x\{X(\tau_1)\in dy\}\mathbf{P}_y\left\{\lim_{T\to\infty}\frac{1}{T}\int_0^T f(X(t))dt = a\right\},$$

we get the assertion of the theorem for all $x \in E_\ell$.

COROLLARY 1. *If $f(x)$ is bounded, then*

$$\lim_{T\to\infty}\frac{1}{T}\int_0^T \mathbf{E}_x f(X(t))dt = \int_{E_\ell} f(x)\mu(dx). \tag{5.8}$$

In particular, if $A \in \mathfrak{B}$, then

$$\frac{1}{T}\int_0^T P(x,t,A)dt \to \mu(A) \quad \text{as} \quad T\to\infty. \tag{5.9}$$

These relations follow from (5.1) by Lebesgue's dominated convergence theorem.

COROLLARY 2. *If* (B) *holds, then the stationary distribution of the process $X(t)$ is unique.*

Indeed, let $\mu_1(A)$ be another stationary distribution. Then

$$\int_{E_\ell}\mu_1(dx)P(x,t,A) = \mu_1(A). \tag{5.10}$$

Integrating (5.10) with respect to t from 0 to T and using (5.9), we get $\mu(A) = \mu_1(A)$.

The results (5.8) and (5.9) can be consdierably strengthened. To be precise, one can prove that the transition function itself (not merely its Cesàro mean) tends to $\mu(A)$ as $t \to \infty$, provided certain additional assumptions are made. We present a general outline of one of the proofs of this theorem (Doob [1]). A Markov process $X(t)$ with stationary initial distribution (the existence of the latter was proved in Section 4) generates a dynamical system in the space of sample paths of the process. This system has a finite invariant measure. Employing the ergodicity property of this system (which follows from the results of this section), we conclude that there is no angular variable and that the μ-singular component of the transition probability tends to 0 as $t \to \infty$. (It can be shown that condition (B) is sufficient to guarantee these properties). Applying the Von Neumann-Koopman mixing theorem (see Hopf [1], pp. 36-37). one can prove that the transition probabilities tend to a limit. However, this method of proof, for all its generality, is probably difficult to extend to the temporally non-homogeneous case. Therefore in Section 6 and 7 we shall employ a different method which does not depend on the general theory of dynamical systems and can be generalized to nonhomogeneous processes.

6. Some auxiliary results

In this section we shall derive some further simplifications of condition (B), concerning the properties of the stationary distribution μ and the transition function $P(x,t,A)$ of the process $X(t)$. These properties will be used in our proof (Section 7) of the theorem on the limiting behavior of the transition probabilities as $t \to \infty$.

LEMMA 6.1. *If condition* (B) *holds for* U, *then the stationary distribution* $\mu(A)$ *of the process* $X(t)$, *constructed in Section* 4, *possesses the property*

$$\inf_{x \in U} \mu(U_\delta(x)) = \beta(\delta) > 0 \quad \text{for} \quad \delta > 0. \tag{6.1}$$

PROOF. It is clear from the construction of $\mu(A)$ in Section 4 that this measure is proportional to the mean time spent by the path of the process in the set A during one cycle, provided the initial distribution on Γ_1 coincides with $\tilde{\mu}(\gamma)$. This time is at least equal to the time $\tau^A(\Gamma_1)$ spent in the set A during a half-cycle (from the time the set Γ is reached to the time Γ_1 is reached).

The function $u(x) = \mathbf{E}_x \tau^{U_\delta(x_0)}(\Gamma_1)$ can be expressed as

$$u(x) = \mathbf{E}_x \int_0^{\tau_{U_1}} \chi_{U_\delta(x_0)}(X(t))dt.$$

Hence, applying Lemma 3.6.4[3], we see that the function $u(x)$ is a solution in U_1 of the problem

$$Lu + \chi_{U_\delta(x_0)}(x) = 0, \qquad u|_{\Gamma_1} = 0.$$

Since the Green function of this problem is bounded away from zero in every domain which, together with its boundary, lies in U_1, this implies the inequality

$$\inf_{x \in \Gamma, x_0 \in U} \mathbf{E}_x \tau^{U_\delta(x_0)}(\Gamma_1) > 0.$$

This inequality, together with the above-mentioned estimate

$$\mu(U_\delta(x_0)) \geqslant \inf_{x \in \Gamma, x_0 \in U} \mathbf{E}_x \tau^{U_\delta(x_0)}(\Gamma_1)$$

yields the assertion of the lemma.

LEMMA 6.2. *If condition* (B) *holds, then the function* $P(x,t,A)$ *is a uniformly continuous function of* x *for* $x \in U$, $t > t_0$, $A \in \mathfrak{B}$ *(where* $t_0 > 0$ *is arbitrary).*

PROOF. We have the following identity for events:

$$\{X(t) \in A\} = P\{X(t) \in A, \tau_{U_1}^1 > t\} \cup \{X(t) \in A, \tau_{U_1} \leqslant t\}.$$

Therefore, by the strong Markov property of the process $X(t)$, we get

$$P(x,t,A) = P\{X(t) \in A, \tau_{U_1} > t\} \tag{6.2}$$

$$+ \int_{\Gamma_1} \int_{u=0}^{t} P\{\tau_{U_1} \in du, X(\tau_{U_1}) \in dz\} P(z, t-u, A).$$

---

[3] In Lemma 3.6.4, the integrand is assumed to be continuous. However, by "sandwiching" the function $\chi_{U_\delta(x_0)}(x)$ between two continuous functions and using the properties of solutions of elliptic equations (recall that L is a nonsingular elliptic operator in U_1), one easily shows that Lemma 3.6.4 is valid also for this function. Analogous remarks apply to the other boundary-value problems considered in this and the following section.

By Lemma 3.6.1, this formula implies that the function $u(t,x) = P(x,t,A)$ satisfies in the cylinder $(t > 0) \times U_1$ the following non-degenerate parabolic equation

$$Lu(t,x)$$

$$= -\frac{\partial u}{\partial t} + \sum_{i=1}^{\ell} b_i(x)\frac{\partial u}{\partial x_i} + \frac{1}{2}\sum_{i,j=1}^{\ell} a_{ij}(x)\frac{\partial^2 u}{\partial x_i \partial x_j} = 0. \qquad (6.3)$$

The boundary conditions for this equation are[4]

$$u(0,x) = \chi_A(x); \qquad u(t,x)|_{x\in\Gamma_1} = P(x,t,A)|_{x\in\Gamma_1}. \qquad (6.4)$$

Let us use now certain known estimates for the solutions of equation (6.3) with bounded measurable boundary conditions (6.4) (see Friedman (1), Theorem 3). According to these, the function $u(t,x)$ has bounded derivatives with respect to x in every cylinder $\{t > t_0\} \times K$, where $K \subset U_1$ is compact and $t_0 > 0$. The proof is complete.

LEMMA 6.3. *If condition (B) holds, then for any $x \in E_\ell$ and $\varepsilon > 0$ there exist numbers $R > 0$ and $t_0(x) > 0$ such that*

$$P(x,t,\bar{U}_R) = P_x\{|X(t)| > R\} < \varepsilon \qquad (6.5)$$

for $t > t_0(x)$.

PROOF. Given an arbitrary $\varepsilon > 0$, it follows from Lemma 6.2 that there exists a $\delta > 0$ such that for $x_0 \in U \cup \Gamma$, $|x - x_0| < \delta$, $A \in \mathfrak{B}$, $t > t_0$,

$$|P(x,t,A) - P(x_0,t,A)| < \frac{\varepsilon}{4}. \qquad (6.6)$$

Let $\beta(\delta)$ be the number defined by (6.1). Choose $R > 0$ large enough so that the stationary measure of the set $\bar{U}_R = \{|x| \geq R\}$ satisfies

$$\mu(\bar{U}_R) < \frac{\varepsilon\beta(\delta)}{4}. \qquad (6.7)$$

We shall now make use of the equality (see (4.8))

$$\mu(\bar{U}_R) = \int \mu(dx)P(x,t,\bar{U}_R). \qquad (6.8)$$

By this equality, we get from (6.6) and (6.7) the estimate

[4] See footnote on p. 125.

$$\frac{\beta(\delta)\varepsilon}{4} > \mu(\bar{U}_R) = \int \mu(dx)P(x,t,\bar{U}_R)$$

$$\geqslant \int_{U_\delta(x_0)} \mu(dx)P(x,t,\bar{U}_R) \geqslant \left[P(x_0,t,\bar{U}_R) - \frac{\varepsilon}{4}\right]\mu(U_\delta(x_0))$$

for $x_0 \in U$. Thus, in view of (6.1) we have the inequality

$$P(x_0,t,\bar{U}_R) \leqslant \frac{\varepsilon}{2}, \tag{6.9}$$

valid for $x_0 \in U \cup \Gamma$, $t > t_0$. Since $t_0 > 0$ is arbitrary, inequality (6.9) is valid for all $t > 0$.

Thus the lemma is proved for $x_0 \in U \cup \Gamma$. Now let $x_0 \in \overline{U \cup \Gamma}$. Since the process $X(t)$ is recurrent relative to the set U, there exists a number $t_0(x_0) > 0$ such that

$$\mathbf{P}_{x_0}\{\tau_{\bar{U}} > t_0(x)\} < \frac{\varepsilon}{2} . \tag{6.10}$$

By arguments similar to those used for showing (6.2), we deduce that

$$P(x_0,t,\bar{U}_R) = \int_\Gamma \int_{s=0}^{t} \mathbf{P}_{x_0}\{\tau_{\bar{U}} \in ds, X(\tau_{\bar{U}}) \in dy\}P(y,t - s,\bar{U}_R)$$

$$+ \mathbf{P}_{x_0}\{\tau_{\bar{U}} > t, |X(t)| > R\}$$

for $t > t_0(x)$. Inequalities (6.9) and (6.10) imply (6.5). The proof is complete.

LEMMA 6.4. *If condition (B) holds, then there exists a constant $T > 0$ such that for any $\delta > 0$*

$$\inf_{\substack{t>0, x \in U \cup \Gamma \\ x_0 \in U \cup \Gamma}} \mathbf{E}_x \int_t^{t+T} \chi_{U_\delta(x_0)}(X(s))ds > 0 \tag{6.11}$$

(i.e., the mean time spent by the path, starting from x, in the δ-neighborhood of x_0 during the time interval $[t,t + T]$ is bounded away from zero uniformly in t, $x_0 \in U$, $x \in U$).

PROOF. Let K be a compact set such that

$$P(x,t,K) > \frac{1}{2} \tag{6.12}$$

for all $x \in U$, $t > 0$. Such a set K exists by (6.9). Now, since

$E_x\tau_{\bar{U}}$ is bounded in every compact set, we can choose a constant $T \geqslant 0$ such that

$$P_x\{\tau_{\bar{U}} > T - 1\} < \frac{1}{2} \tag{6.13}$$

holds for all $x \in K$.

We now consider the function

$$\mathbf{E}_x \int_0^{\tau_{\bar{U}}(t)} \chi_{U_\delta(x_0)}(X(s))ds = v^{(\delta)}(t,x).$$

By Lemma 3.6.3, this function is a solution in $\{t > 0\} \times \bar{U}_1$ of the parabolic equation[5]

$$Lv^{(\delta)}(t,x) + \chi_{U_\delta(x_0)}(x) = 0,$$

satisfying the boundary conditions

$$v^{(\delta)}(0,x) = 0; \qquad v^{(\delta)}(t,x)|_{x\in\Gamma_1} = 0.$$

Since the operator is nondegenerate in the above domain, it follows from the general properties of solutions of parabolic equations that

$$\inf_{x\in U\cup\Gamma, x_0\in U\cup\Gamma} v^{(\delta)}(1,x) > 0. \tag{6.14}$$

Next, we see that for $x \in U$

$$\mathbf{E}_x \int_t^{t+T} \chi_{U_\delta(x_0)}(X(s))ds$$

$$\geqslant \int_{y\in K} P(x,t,dy)\mathbf{E}_y \int_0^T \chi_{U_\delta(x_0)}(X(s))ds$$

$$\geqslant \int_{y\in K} P(x,t,dy) \int_{z\in U\cup\Gamma} \int_{u=0}^{T-1} \mathbf{P}_y\{\tau_{\bar{U}} \in du, X(\tau_{\bar{U}}) \in dz\} v^{(\delta)}(1,z).$$

Using (6.12), (6.13) and (6.14), we infer from the last inequality that (6.11) holds.

LEMMA 6.5. *If condition (B) is satisfied, then for any $\delta > 0$ there exists a constant $\alpha_\delta > 0$ such that for all $x \in E_\ell$, $x_0 \in U$ and $t > t_0(x)$*

---

[5] See footnote on p. 125.

$$P(x,t,U_\delta(x_0)) > \alpha_\delta.$$

PROOF. (1) Since

$$\mathbf{E}\int_t^{t+T} \chi_{U_\delta(x_0)}(X(s))ds = \int_t^{t+T} P(x,s,U_\delta(x_0))ds,$$

it follows from Lemma 6.4 that for any $x \in U$ there exists a sequence $s_1, s_2, \ldots, s_n, \ldots$ ($s_n \in [(n-1)T, nT]$) such that

$$P(x,s_n,U_\delta(x_0)) > \gamma_\delta > 0.$$

Now let $t \in [(n-1)T, nT]$. Then

$$P(x,t,U_\delta(x_0)) \geqslant \int_{y \in U_{\delta/2}(x_0)} P(x,s_{n-1},dy)P(y,t-s_{n-1},U_\delta(x_0)). \tag{6.15}$$

It is clear that $0 < t - s_{n-1} < 2T$ and

$$P(y,t-s_{n-1},U_\delta(x_0)) \geqslant \mathbf{P}_y\{X(t-s_{n-1}) \in U_\delta(x_0), \tau_{U_1} > t - s_{n-1}\}. \tag{6.16}$$

By Lemma 6.1, the function $w(t,y) = \mathbf{P}_y\{X(t) \in U_\delta(x_0), \tau_{U_1} > t\}$ is a solution in $\{t > 0\} \times U_1$ of the equation $Lw(t,y) = 0$ satisfying the boundary conditions

$$w(0,y) = \chi_{U_\delta(x_0)}(y), \qquad w(t,y)|_{y\ \Gamma_1} = 0.$$

From general properties of solutions of parabolic equations we readily conclude that

$$\inf_{\substack{0 \leqslant t \leqslant 2T, y \in U_{\delta/2}(x_0) \\ x_0 \in U}} w(t,y) > 0.$$

Using this estimate in combination with (6.15) and (6.16) we deduce that for all $x \in U$, $x_0 \in U$, $t > 0$, $\delta > 0$

$$P(x,t,U_\delta(x_0)) \geqslant 2\alpha_\delta > 0.$$

(2) Now let $x \in \overline{U}$. Take $t_0(x)$ such that $P_x\{\tau_{\overline{U}} > t_0(x)\} < 1/2$. Then for $t > t_0(x)$

$$P(x,t,U_\delta(x_0)) \geqslant$$ (Contd)

$$\text{(Contd)} \geq \int_{z\in\Gamma}\int_{u=0} P\,\{\tau_{\bar{U}} \in du, X(\tau_{\bar{U}}) \in dz\}P(x,t-u,U_\delta(x_0))$$

$$\geq \frac{1}{2}\,2\alpha\delta = \alpha\delta.$$

This complete the proof of the lemma.

7. Existence of the limit of the transition function

Lemma 6.3 guarantees that the process $X^x(t)$ is bounded in probability for all $x \in E_\ell$. As we observed in Section 2.1, this implies that the family of distributions $P(x,t_n,A)$ is weakly compact for any sequence $t_n \to \infty$. In this section we shall show that the limit does not depend on which sequence $t_n \to \infty$ is chosen. The proof uses essentially Lemmas 6.2 and 6.5.

THEOREM 7.1. *If condition* (B) *holds, then for any continuous bounded function* $f(x)$ *and any Borel subset* A *with boundary* Γ *such that* $\mu(\Gamma) = 0$, *we have*

$$T_t f(x) = \int_{E_\ell} P(x,t,dy)f(y) = \int_{E_\ell} f(y)\mu(dy), \tag{7.1}$$

$$P(x,t,A) \to \mu(A) \tag{7.2}$$

as $t \to \infty$.

PROOF. Let $f(x)$ be an arbitrary continuous bounded function in E_ℓ. Then the function $T_t f(x)$ is also bounded. It is also continuous in x uniformly in $t > t_0$ ($t_0 > 0$) on any compact subset K of U_1 (see proof of Lemma 6.2). Hence there exists a sequence t_n tending to infinity as $n \to \infty$ such that $\lim_{n\to\infty} T_{t_n} f(x) = c(x)$, where $c(x)$ is a function continuous on U_1, and the convergence is uniform in U.

We set

$$c_- = \min_{U\cup\Gamma} c(x) = c(x_1); \qquad c_+ = \max_{U\cup\Gamma} c(x) = c(x_2),$$
$$a = \inf_{x\in E_\ell} \varliminf_{t\to\infty} T_t f(x); \qquad b = \sup_{x\in E_\ell} \varlimsup_{t\to\infty} T_t f(x). \tag{7.3}$$

To prove (7.1), it will suffice to show that $a = b$. For if $a = b$, then the function $T_t f(x)$ has a limit independent of x as $t \to \infty$. This limit is then equal to $\int f(y)\mu(dy)$, by Theorem 5.1 (see Corollary 1 to that theorem).

By (7.3), there exist for any $\varepsilon > 0$ a point $x_0 \in E_\ell$ and a sequence $t_n' \to \infty$ such that

$$T_{t_n'} f(x_0) < a + \varepsilon. \tag{7.4}$$

Also, for any $x \in E_\ell$

$$T_t f(x) > a - \varepsilon \tag{7.5}$$

if it is sufficiently large. Moreover, for any compact subset $K \subset E_\ell$ there exists a $t_0 > 0$ such that inequality (7.5) holds for $t > t_0$ and all $x \in K$. (Indeed, if A_{t_0} consists of all points x such that (7.5) is true for $t > t_0$, then the sets A_t obviously possess the property that $A_s \subset A_t$ for $s < t$ and the union $\bigcup\limits_{t>0} A_t$ coincides with E_ℓ).

Our next task is to prove that $c_+ \leqslant a$. To do this, we shall assume that

$$c_+ > a \tag{7.6}$$

and we shall derive a contradiction.

Set $\gamma = (c_+ - a)/2$ and choose a number $\delta > 0$ such that $c(x)$ $c_+ - \gamma$ for $x \in U_\delta(x_2)$. Since the sequence $T_{t_n} f(x)$ converges to $c(x)$ uniformly in U, we can find a number n_0 such that for $x \in U_\delta(x_2)$ and $n > n_0$, we also have

$$T_{t_n} f(x) > c_+ - \gamma. \tag{7.7}$$

Now, using Lemma 6.3, we choose a compact set $K_1 \supset U$ such that

$$P(x_0, t, \overline{K}_1) < \varepsilon \tag{7.8}$$

for all $t > t_1$.

We next take $t_2 > 0$ large enough so that inequality (7.5) holds for all $x \in K_1$ if $t > t_2$. Finally, using Lemma 6.5, we choose t_3 such that

$$P(x_0, t, U_\delta(x_2)) > \alpha_\delta \tag{7.9}$$

for $t > t_3$.

We now consider some fixed number t_n $(n > n_0)$, and then a number t_k' such that $t_k' - t_n \geqslant \max(t_1, t_2, t_3)$; it follows from (7.4), (7.5) and (7.8) that

$$a + \varepsilon > T_{t_n'} f(x_0)$$

$$= \int T_{t_n'} f(y) P(x_0, t_k' - t_n, dy) \geqslant \qquad \text{(Contd)}$$

$$\text{(Contd)} \quad \geq \int_{U_\delta(x_2)} (c_+ - \gamma)P(x_0, t_k' - t_n, dy)$$

$$+ \int_{K_1 \setminus U_\delta(x_2)} (a - \varepsilon)P(x_0, t_k' - t_n, dy) - \varepsilon \max_{E_\ell} |f(x)|.$$

Since $\gamma = (c_+ - a)/2$, this inequality can be rewritten as

$$a + \varepsilon \geq (a - \varepsilon)P(x_0, t_k' - t_n, K_1) \tag{7.10}$$

$$+ \gamma P(x_0, t_k' - t_n, U_\delta(x_2)) - \varepsilon \max_{E_\ell} |f(x)|$$

$$\geq (a - \varepsilon)(1 - \varepsilon) + \gamma P(x_0, t_k' - t_n, U_\delta(x_2)) - \varepsilon \max_{E_\ell} |f(x)|.$$

From (7.9) and (7.10) follows now the estimate

$$\gamma < \frac{\varepsilon(2 + a + \max_{E_\ell} |f(x)|)}{a_\delta}.$$

Since ε is arbitrarily small, it follows that $\gamma \leq 0$ and this contradicts our assumption that $c_+ > a$.

Consequently, $c_+ < a$. Similarly, we show that $c_- \geq b$. Thus $c_+ \leq a \leq b \leq c_- \leq c_+$, and so $a = b$. As we have mentioned above, this implies (7.1).

The validity of (7.1) means that the measure $P(x,t,A)$ converges weakly to $\mu(A)$ as $t \to \infty$ for every fixed x. It is known (see Prohorov [1]) that this implies the second assertion of the theorem.

8. Some generalizations

In this section we shall show how to generalize the results of Sections 4 through 7 to a wider class of Markov processes and to diffusion processes in other phase spaces. At the end of the section we shall generalize the theorems of Sections 5 and 7 to the nonergodic case.

1. We assumed above that the diffusion matrix is nonsingular on the domain U_1. An analysis of the proof shows that in fact a weaker condition is sufficient; for example, we need only require that there exist a sufficiently smooth positive Green function for the first boundary-value problem for the parabolic equation $Lu = 0$ in $(t > 0) \times U$.

2. Many of the results of Sections 4 through 7 can be proved for Markov processes in a general Banach space. Neither is there

any need to confine the discussion to processes with continuous sample paths. Indeed, the decomposition of the path into cycles, which was the basic construction in the previous sections, carries over to processes whose paths may have discontinuities of the first kind (jumps). For details, see Maruyama and Tanaka [2] and Has'minskiĭ [1].

All the results of Sections 4 through 7, without exception, carry over to diffusion processes on smooth manifolds. In particular, a process with nonsingular diffusion matrix on a compact smooth manifold is always positive. This follows from the fact that the mean exit time of the sample path of a nonsingular diffusion process from a bounded domain is bounded (see Corollary to Theorem 3.7.1).

3. As was already mentioned above, the construction used in Section 4 to yield the stationary distribution carries over to U-recurrent processes with infinite mean recurrence time for U. This measure is invariant in the sense of (4.8). By a slight modification of the arguments in Section 5, one can then prove the following generalization of Theorem 5.1.

THEOREM 8.1. *Suppose that condition* (B.1) *holds, and let the process* $X(t)$ *be* U*-recurrent. Let* $\nu(A)$ *be the measure defined by* (4.3) *and let* $f(x)$ *and* $g(x)$ *be functions integrable with respect to this measure such that* $\int g(x)\nu(dx) \neq 0$.

Then for all $x \in E_\ell$

$$\mathbf{P}_x\left\{\lim_{T\to\infty} \frac{\int_0^T f(X(t))dt}{\int_0^T g(X(t))dt} = \frac{\int f(x)\nu(dx)}{\int g(x)\nu(dx)}\right\} = 1. \tag{8.1}$$

This result, deduced from somewhat different assumptions, may be found in Maruyama and Tanaka [2], Has'minskiĭ [1].

Now, if the expected time at which U is reached is infinite, we have

$$\nu(E_\ell) = \infty, \tag{8.2}$$

and thus, using a monotone increasing sequence of functions $g_n(x)$ such that $g_n(x) \uparrow 1$ as $n \to \infty$, we get from (8.1) that

$$\mathbf{P}_x\left\{\frac{1}{T}\int_0^T f(X(t))dt \xrightarrow[T\to\infty]{} 0\right\} = 1$$

for any ν-integrable function $f(x)$. Taking expectations and applying the Lebesgue theorem, as in (5.8) and (5.9), we conclude that if $f(x)$ is ν-measurable and K is a compact set,

$$\frac{1}{T}\int_0^T \mathbf{E}_x f(X(t))dt \to 0, \qquad \frac{1}{T}\int_0^T P(x,t,K)dt \to 0 \tag{8.3}$$

as $T \to \infty$.

By methods similar to those used in Section 6,7, one can show that for a null process the assumptions of Theorem 8.1 imply

$$\lim_{t\to\infty} \mathbf{E}_x f(X(t)) = 0, \qquad \lim_{t\to\infty} P(x,t,K) = 0. \tag{8.4}$$

We shall not go into the proof of these formulas.

4. It was proved in Sections 4 through 7 that if condition (B) holds, then the Markov process $X(t)$ has a unique stationary distribution, whose "domain of attraction" is the entire space. However, one frequently encounters the situation in which there exist domains of attraction with different stationary distributions. To bring such situations into the range of our discussion, we need only modify condition (B.2)

Let us say that a domain D is *invariant* for the process $X(t)$ if $P_x\{X(t) \in D\} = 1$ for all $t > 0$, $x \in D$.

Suppose that, besides condition (B.1) *as formulated in Section 4, the following condition* (B.2') *is satisfied:*

There exists a set $D \supset U$, *invariant for the process* $X(t)$, *such that* $D = \bigcup_{n=1}^{\infty} K_n$ *(*K_n *are compact subsets) and* $\sup_K \mathbf{E}_x\tau < \infty$ *for all* n.

Analysing the previous proofs, we see that under conditions (B.1), (B.2') there exists a unique stationary distribution μ for the process $X(t)$, such that $\mu(E_\ell \setminus D) = 0$. Moreover, for all $x \in U$, the strong law of large numbers (5.1) and the theorem on the limiting behavior (7.1) and (7.2) of the transition probability are valid.

Call F a *set of inessential states* of the process $X(t)$ if

$$P(x,t,F) \to 0 \tag{8.5}$$

for all $x \in E_\ell$, $t \to \infty$.

Now suppose that the entire phase space E_ℓ can be decomposed as the union of a finite or countable family of invariant sets D_i and a set of inessential states $F = E_\ell \setminus \cup D_i$. Assume further that each of the sets D_i contains a "mixing region" U_i, i.e., U_i and D_i satisfy conditions (B.1), (B.2').

Each set D_i is called an *ergodic* set of the process $X(t)$. Suppose a specific stationary distribution $\mu_i(A)$ is established in each ergodic set. Since the sets D_i are mutually disjoint, the measures $\mu_i(A)$ are singular with respect to each other. It is easily shown that for any positive constants $k_1, k_2, \ldots$ such that $\sum_i k_i = 1$, the measure

$$\mu(A) = \sum_i k_i \mu_i(A) \tag{8.6}$$

is also stationary.

The converse is also valid: Any distribution which is stationary for the process $X(t)$ is expressible in the form (8.6).

In fact, let $\mu(A)$ be some stationary distribution. Then

$$\mu(A) = \int \mu(dx) P(x,t,A). \tag{8.7}$$

Setting $A = F$, letting $t \to \infty$ and using (8.5), we conclude that $\mu(F) = 0$. Now let $B \subset D_i$. Then

$$\mu(B) = \sum_i \int_{D_i} \mu(dx) P(x,t,B) = \int_{D_i} \mu(dx) P(x,t,B) \tag{8.8}$$

by the invariance of the set D_i. The invariance of D_i and (8.8) imply that the measure ν defined by $\nu(A) = \mu(A \cap D_i)/\mu(D_i)$, is a stationary distribution for $X(t)$ such that $\nu(E_\ell \setminus D_i) = 0$. Since there is exactly one such measure, we get

$$\mu(A \cap D_i) = \mu(D_i) \mu_i(A).$$

Using this and the condition $\mu(F) = 0$, we finally get

$$\mu(A) = \sum_i \mu(D_i) \mu_i(A).$$

Suppose the point $X(0) = x$ is an ergodic set D_i. Then, as we already have seen above, the strong law of large numbers (5.1) and the theorem on the limiting behavior of the transition probability (7.12), (7.13) are valid, provided we set $\mu = \mu_i$ in these formulas.

We now consider the case $X(0) = x \in F$. First, since the event $B = \{X(t) \in F \text{ for all } t > 0\}$ implies the event $\{X(t) \in F\}$, it follows by (8.5) that $\mathbf{P}_x(B) = 0$. Using this fact and the strong law of large numbers (which is valid in each ergodic set), we get the equality

$$\mathbf{P}\left\{\lim_{T\to\infty} \frac{\int^T f(X(t))dt}{T} = \xi\right\} = 1.$$

Here ξ is the random variable defined by

$$f_i = \int f(x) \mu_i(dx),$$

when the path $X(t)$ leads from F to the set D_i. Let $\pi_i(x)$ denote the probability that a path issuing from x "settles" in the set D_i. Since $P_x(B) = 0$, it follows that $\sum_i \pi_i(x) = 1$. By the above, we have $P_x\{\xi = f_i\} = \pi_i(x)$.

Similar arguments yield the conclusion that

$$\lim_{t\to\infty} P(x,t,A) = \sum_i \pi_i(x)\mu_i(A)$$

for $x \in F$.

9. Stabilization of the solution of the Cauchy problem for a parabolic equation

In previous sections we applied properties of solutions of parabolic and elliptic differential equations in order to study diffusion processes. Recent years have seen the publication of numerous papers in which, conversely, probabilistic methods are employed to study the properties of solutions of second-order parabolic and elliptic equations. In this section we shall see that the results proved in Sections 4 through 8 yield information about the behavior of the solutions of the Cauchy problem for a parabolic equation for large time values.

Let $X(t)$ be a temporally homogeneous Markov process, regular in E_ℓ, with differential generator

$$L = \frac{\partial}{\partial s} + \sum_{i=1}^{\ell} b_i(x)\frac{\partial}{\partial x} + \frac{1}{2}\sum_{i,j=1}^{\ell} a_{ij}(x)\frac{\partial^2}{\partial x_i \partial x_j}$$

$$\equiv \frac{\partial}{\partial s} + \left(b(x), \frac{\partial}{\partial x}\right) + \frac{1}{2}\sum_{r=1}^{k}\left(\sigma_r(x), \frac{\partial}{\partial x}\right)^2 .$$

Henceforth we shall assume in this book that the coefficients $b(x)$, $\sigma_r(x)$ satisfy a Lipschitz condition

$$\sum_{r=1}^{k} |\sigma_r(x) - \sigma_r(y)| + |b(x) - b(y)| < B|x - y|$$

in every compact subset $K \subset E_\ell$, where the Lipschitz constant B may depend on K. Let us assume now also that the nondegeneracy condition (B.1) holds in every compact set.

It is well known (see Eidel'man [1]) that the above conditions are sufficient for the existence of a solution of the problem

$$Lu = 0; \qquad u(0,x) = f(x) \tag{9.1}$$

in the domain $(s < 0) \times E_\ell$ for any bounded continuous function $f(x)$. It follows from Lemma 3.6.1 and the remarks following its proof that this solution is unique and that it can be written as

$$u(s,x) = \mathbf{E}_{s,x} f(X(0)).$$

The process $X(t)$ is temporally homogeneous. Therefore, making the substitution $s = -t$, we see that the function

$$u(-t,x) = u_1(t,x) = \mathbf{E}_x f(X(t)) \tag{9.2}$$

is a solution of the following boundary-value problem, equivalent to problem (9.1):

$$\frac{\partial u_1}{\partial t} = \left(b(x), \frac{\partial}{\partial x}\right) u_1 + \frac{1}{2} \sum_{r=1}^{k} \left(\sigma_r(x), \frac{\partial}{\partial x}\right)^2 u_1, \qquad u_1(0,x) = f(x). \tag{9.3}$$

LEMMA 9.1. *If the process $X(t)$ is transient, then for any continuous initial function $f(x)$ with compact support the solution $u_1(t,x)$ of problem* (9.3) *tends to zero as $t \to \infty$. Moreover, the function*

$$v(x) = \int_0^\infty u_1(t,x)\,dt$$

is bounded in E_ℓ.

PROOF. Let $f(x) = 0$ for $x \in E_\ell \setminus K$. Then (9.2) implies

$$|u_1(t,x)| \leqslant \max_{E_\ell} |f(x)| P(x,t,K). \tag{9.4}$$

Hence the first assertion of the theorem follows by (2.4). The second assertion follows from (9.4) and the Corollary to Lemma 2.3.

REMARK 1. It can be shown that the function $v(x)$ satisfies the elliptic equation

$$\sum_{i=1} b_i(x) \frac{\partial v}{\partial x_i} + \frac{1}{2} \sum_{i,j=1} a_{ij}(x) \frac{\partial^2 v}{\partial x_i \partial x_j} = -f(x) \tag{9.5}$$

and

$$v(x) \to 0 \text{ as } |x| \to \infty. \tag{9.6}$$

Thus the conditions of the theorem are sufficient for the existence of a solution of problem (9.5), (9.6) in E_ℓ.

Consider the sufficient conditions for transience proved in Chapter III. Using these and Lemma 9.1, one can derive sufficient conditions for the solution of problem (9.3) to tend to zero as $t \to \infty$ and for the existence of a solution of problem (9.5), (9.6). These conditions will then be given in terms of the coefficients of the equation.

LEMMA 9.2. *Let $X(t)$ be a null recurrent process. Then, for any continuous initial function $f(x)$ with compact support, the solution $u_1(t,x)$ of problem* (9.2) *tends to zero as $t \to \infty$. If $f(x) \geqslant 0$ and $f(x) \not\equiv 0$, then*

$$\int_0^T u_1(t,x)dt \to \infty \quad \text{as} \quad T \to \infty.$$

PROOF. The first assertion follows from (9.4) and (8.4), and the second from the equality

$$\int_0^T u_1(t,x)dt = \mathbf{E}_x \int_0^T f(X(t))dt$$

and the fact that a recurrent process spends an infinite time in the neighbourhood of any point

Stronger results can be obtained if $X(t)$ is a positive process.

LEMMA 9.3. *Let $X(t)$ be a positive process. Then, for any continuous bounded function $f(x)$, the solution $u_1(t,x)$ of problem* (9.3) *tends to a constant as $t \to \infty$. This constant is given by*

$$\int_{E_\ell} f(x)\mu(dx),$$

where $\mu(A)$ is the stationary distribution of the process $X(t)$.

The proof follows from the representation (9.2) of the solution of problem (9.3) and from Theorem 7.1.

The following Lemma may be useful for the actual computation of the limit of $u_1(t,x)$ as $t \to \infty$ [6].

LEMMA 9.4. *The distribution $\mu(A)$ has a density $p_0(x)$ with respect to Lebesgue measure in E_ℓ. This density is the unique bounded solution of the equation*

$$L^*p_0 \equiv \frac{1}{2} \sum_{i,j=1}^{\ell} \frac{\partial^2}{\partial x_i \partial x_j} (a_{ij}(x)p_0) - \sum_{i=1}^{\ell} \frac{\partial}{\partial x_i} (b_i(x)p_0) = 0, \tag{9.7}$$

---

[6] Formula (9.7) for the function $p_0(x)$ is due to Kolmogorov [1]. Lemmas 9.4 and 9.5 were proved in Section 4 of Il'in and Has'minskiĭ [1] under more general assumptions.

satisfying the additional condition

$$\int_{E_\ell} p_0(x)dx = 1. \tag{9.8}$$

PROOF. The stationary distribution $\mu(A)$ satisfies the condition

$$\mu(A) = \int_{E_\ell} \mu(dx)P(x,t,A). \tag{9.9}$$

Since the measure $P(x,t,A)$ has a density $p(x,t,y)$ with respect to Lebesgue measure, which is simply Green's function[7] of equation (9.3), it follows from (9.9) that the measure $\mu(A)$ also has a density $p_0(x)$; i.e.

$$\mu(A) = \int_A p_0(x)dx. \tag{9.10}$$

From (9.9) and (9.10) we get the formula

$$p_0(y) = \int_{E_\ell} p_0(x)p(x,t,y)dx. \tag{9.11}$$

It is well known (see Il'in, Kalašnikov and Oleĭnik [1]) that the solution $v(t,x)$ of the problem

$$\frac{\partial v}{\partial t} = L^*v; \qquad v(0,x) = p_0(x),$$

can be written as

$$v(t,x) = \int_{E_\ell} p_0(y)p(y,t,x)dy.$$

This, toegether with (9.11) shows immediately that the function $p_0(x)$ satisfies equation (9.7). Before proving uniqueness, we need the following lemma.

LEMMA 9.5. *If $X(t)$ is a positive recurrent process, then the function $p(x,t,y)$ satisfies the relation*

$$\lim_{t\to\infty} p(x,t,y) = p_0(y). \tag{9.12}$$

PROOF. By the Chapman-Kolomogorov equation

---

[7] The existence of Green's function under these assumptions follows, e.g., from results of Eĭdel'man [1] and Mihaĭlov [1].

$$p(x,t,y) = \int_{E_\ell} p(x,t-1,z)p(z,1,y)dz,$$

the function $p(x,t,y)$ satisfies

$$p(x,t,y) = \mathbf{E}_x p(X(t-1),1,y).$$

Hence, by Theorem 7.1, we have

$$\lim_{t\to\infty} p(x,t,y) = \int_{E_\ell} p_0(z)p(z,1,y)dz$$

for every y. This equality and (9.11) imply the desired assertion.

We can now complete the proof of Lemma 9.4. Let $q_0(x)$ be another solution of problem (9.7), (9.8). Then this is also a solution of the problem

$$\frac{\partial v}{\partial t} = L^* v; \qquad v(0,x) = q_0(x).$$

Therefore

$$q_0(x) = \int q_0(z)p(z,t,x)dz$$

and if we now let $t \to \infty$ and recall (9.12), we see that $q_0(x) = p_0(x)$. This completes the proof of Lemma 9.4.

REMARK 1. Recall that sufficient conditions for recurrence and positivity were established in Chapter III. Using the results of Sections 2 and 3, we can derive necessary and sufficient conditions for a process $X(t)$ to be recurrent, transient, null or positive. Such conditions will then be given in terms of the properties of the operator

$$L_1 = \sum_{i=1}^{\ell} b_i(x)\frac{\partial}{\partial x_i} + \frac{1}{2}\sum_{i,j=1}^{\ell} a_{ij}(x)\frac{\partial^2}{\partial x_i \partial x_j}.$$

For example, a regular process $X(t)$ is recurrent if and only if there exists a unique solution of the exterior Dirichlet problem for the equation $L_1 u = 0$ in some (hence every, see Section 2) bounded domain D. The process is positive if and only if the exterior Dirichlet problem for the equation $L_1 u = -1$ has a positive solution. These statements follow from Theorems 3.7.2, 3.7.3 and the results of Sections 2 and 3.

REMARK 2. Many results of this chapter can be generalized to the temporally non-homogeneous case. For example, in Il'in and Has'minskiĭ [1] and Has'minskiĭ [7] sufficient conditions are established in terms of Ljapunov functions for a nonhomogeneous Markov process to be bounded in probability. It has been shown that, under certain assumptions on the coefficients, a temporally nonhomogeneous process possesses a certain analog of a stationary distribution, i.e., a distribution to which all others ultimately converge (see Il'in and Has'minskiĭ [1], Theorem 5). It is not difficult to prove that if the transition function is periodic, this distribution is periodic as a function of time and determines a periodic Markov process.

10. Limit relations for null processes

In Section 5 we studied the behavior of the functional

$$\zeta(T) = \int_0^T f(X(t))dt$$

as $T \to \infty$, assuming condition (B) to hold. We showed that $\zeta(T)/T$ converges to a constant a.s., and $\mathbf{E}\zeta(T)/T$ converges to the same constant. If $X(t)$ is a null-process, the random variable $T^{-1}\zeta(T)$ converges in probability to zero. This leads one to expect the existence of a non-trivial limit distribution of $\zeta(T)$ for normalizing factors other than T^{-1}.

The analogous problem for Markov chains with countably many states was studied by Feller in his well known paper [1]. He proved there that the limit distribution of the number of hits in each state of the chain depends essentially on the distribution of the random variable τ equal to the length of one cycle (see Section 5). Since in the case at hand we have $\mathbf{E}\tau = \infty$, it is natural to conjecture that τ belongs to the domain of attraction of a stable law with exponent $\alpha < 1$. It turns out that in this case the limit distribution of $T^{-\alpha}\zeta(T)$ for a one-dimensional diffusion process coincides with that established by Feller in [1]. What is new here is that one can establish conditions for convergence to various laws, and also the values of the normalizing factors in terms of the coefficients of the differential generator of the Markov process $X(t)$.

Our exposition will be based on certain known facts. For the reader's convenience we summarize these in the following two lemmas.

The reader can find the proofs in Sections 13.5 and 13.6 of Feller [4].

LEMMA 10.1. *Let $\alpha > 0$ and let $G(x)$ be a function, monotone on the half-line* $[0,\infty)$, *whose Laplace transform*

$$g(s) = \int_0^\infty e^{-sx} dG(x)$$

exists for $s > 0$. *Then either of the relations*

$$G(x) \sim \frac{cx^{\alpha}}{\Gamma(1+\alpha)}, \qquad x \to \infty,$$

$$g(s) \sim cs^{-\alpha}, \qquad s \to 0$$

implies the other. (As usual, $\Gamma(\alpha)$ *denotes the gamma-function.)*

LEMMA 10.2. *Let* $0 < \alpha < 1$, *and let* $\xi_1, \ldots, \xi_n, \ldots$ *be independent identically distributed random variables,* $F(x) = P\{\xi_1 \geqslant x\}$, *and* $F(0) = 1$. *Let* $G_\alpha(x)$ *be the distribution function of a stable law with exponent* α, *whose Laplace transform is* $\exp\{-s^\alpha\}$. *Then any one of the relations*

$$F(x) \sim \frac{cx^{-\alpha}}{\Gamma(1-\alpha)}, \qquad x \to \infty, \tag{10.1}$$

$$\varphi(s) = \mathbf{E}e^{-s\xi_i} \sim 1 - cs^{\alpha}, \qquad s \to 0, \tag{10.2}$$

$$P\left\{\frac{\xi_1 + \ldots + \xi_n}{(nc)^{1/\alpha}} < x\right\} \to G_\alpha(x), \qquad n \to \infty \tag{10.3}$$

implies the two others.

Let $\nu(T)$ denote the integer-valued random variable defined by

$$\xi_1 + \ldots + \xi_{\nu(T)} \leqslant T, \qquad \xi_1 + \ldots + \xi_{\nu(T)} + \xi_{\nu(T)+1} > T.$$

The following lemma is a very special case of the above-mentioned results.

LEMMA 10.3. *Let the random variables* ξ_i *satisfy the assumptions of Lemma* 10.2 *and suppose* (10.1) *holds. Then*

$$\mathbf{P}\left\{\frac{\nu(T)}{T^\alpha} > \frac{x}{c}\right\} \to G_\alpha(x^{-1/\alpha}),$$

$$\mathbf{E}\nu(T) \sim \frac{T^\alpha}{c\Gamma(1+\alpha)} \tag{10.4}$$

also as $T \to \infty$.

PROOF. The definition of $\nu(T)$ implies that

$$\mathbf{P}\{\nu(T) \geqslant k\} = \mathbf{P}\{\xi_1 + \ldots + \xi_k < T\}.$$

Using (10.3) and setting $k = \left[\frac{xT^\alpha}{c}\right]$, we get the first assertion of the lemma. Next, we have

$$\mathbf{E}\nu(T) = \sum_{k=1}^{\infty} \mathbf{P}\{\nu(T) \geqslant k\} = \sum_{k=1}^{\infty} \mathbf{P}\{\xi_1 + \ldots + \xi_k < T\}.$$

Applying Lemma 10.2 again, we see that

$$A(s) = \int_0^{\infty} e^{-sT} d\mathbf{E}\nu(T) = \sum_{k=1}^{\infty} [\varphi(s)]^k$$

$$= \frac{\varphi(s)}{1 - \varphi(s)} \sim \frac{1}{c} s^{-\alpha} \qquad (10.5)$$

as $s \to 0$. The relations (10.4) now follow from (10.5) and Lemma 10.1.

We now proceed to investigate the limiting distribution of the functional $\zeta(T)$ for a temporally homogeneous recurrent diffusion process $X(T)$ on the line.

We may assume without loss of generality that the drift coefficient of this process vanishes (otherwise we can perform a transformation of coordinates $x_1 = W(x)$; see Example 2 in Section 3.8). Thus the process $X(t)$ is described by the differential operator

$$\frac{\partial}{\partial t} + \frac{1}{2} \sigma^2(x) \frac{\partial^2}{\partial x^2} \qquad (\sigma^2(x) > 0). \qquad (10.6)$$

The uniqueness and recurrence of the process $X(t)$ associated with the operator (10.6) follow from the results of Chapter III, provided the function $\sigma(x)$ satisfies a Lipschitz condition in every compact set and moreover $0 < \sigma^2(x) < k(x^2 + 1)$, where k is a constant, for $x \in E$. Throughout Sections 10 and 11 we shall assume that these conditions are satisfied.

As in Section 5, we shall start from the expansion (compare (5.3))

$$\zeta(T) = \sum_{n=1}^{\nu(T)} \eta_n + \rho(T), \qquad \rho(T) = \int_{\tau_{\nu(T)}}^{T} f(X(t))dt. \qquad (10.7)$$

Here η_k is the increment of the functional $\zeta(T)$ during the k-th cycle. It will be convenient to define now a cycle as the portion of the path of $X(t)$ beginning at $x = 0$ and extending up to the first time of return to zero after the point $x = 1$ has been reached. With this definition of cycles consider the random variables τ_n equal to the time at which the n-th cycle ends ($\tau_0 = 0$). They satisfy

$$\eta_n = \int_{\tau_{n-1}}^{\tau_n} f(X(t))dt$$

and they are independent and identically distributed, if $x(0) = 0$. The lengths of the cycles, $\xi_n = \tau_n - \tau_{n-1}$, are also obviously independent and identically distributed. To study their distribution we shall need the following lemma.

LEMMA 10.4. *Let $\tau^{(0)}$ ($\tau^{(1)}$) be the time needed by the path of the process $X(t)$ to reach the point $x = 0$ ($x = 1$). Then the function*

$$u_0(s,x) = \mathbf{E}_x \exp\{-s\tau^{(0)}\} \qquad (u_1(s,x) = \mathbf{E}_x \exp\{-s\tau^{(1)}\})$$

is the unique bounded solution of the equation

$$\frac{1}{2}\sigma^2(x)\frac{d^2u}{dx^2} - su = 0 \tag{10.8}$$

in the domain $x > 0$ ($x < 1$) satisfying the condition $u_0(s,0) = 1$ ($u_1(s,1) = 1$).

PROOF. We set $\tau_n^{(0)} = \min\{\tau^{(0)}, \tau^{(n)}\}$, where $\tau_n^{(n)}$ is the time at which the path reaches the point $x = n$. We also set $\chi_n = 1$ if $X(\tau_n^{(0)}) = 0$ and $\chi_n = 0$ if $X(\tau_n^{(0)}) = n$. Inequalities for the random variables imply corresponding inequalities for their expectations:

$$v_n(s,x) = \mathbf{E}_x[\exp\{-s\tau_n^{(0)}\}\chi_n]$$
$$\leqslant u_0(s,x) \leqslant \mathbf{E}_x \exp\{-s\tau_n^{(0)}\} = w_n(s,x).$$

It is known (Dynkin [3]) that the functions $v_n(s,x)$ and $w_n(s,x)$ are solutions of equation (10.8) in the domain $0 < x < n$, satisfying the boundary conditions $v_n(s,0) = 1$, $v_n(s,n) = 0$ and $w_n(s,0) = 1$, $w_n(s,n) = 1$. Since the process $X(t)$ is regular and recurrent, we have

$$\lim_{n\to\infty} w_n(s,x) = \lim_{n\to\infty} v_n(s,x) = u_n(s,x).$$

On the other hand, it is clear that any solution of equation (10.8) which is bounded in $x > 0$ and such that $u(s,0) = 1$, lies between $v_n(x)$ and $w_n(s,x)$ for $0 \leqslant x \leqslant n$. This completes the proof of the lemma.

REMARK. The length of a single cycle ξ_n is equal to the sum of two independent random variables: the lengths of the half-cycles. Hence the functions $u_0(s,x)$, $u_1(s,x)$ and the Laplace transform of the distribution of ξ_n satisfy the equality

$$\mathbf{E}e^{-s\xi_n} = \mathbf{E}_0 e^{-s\tau^{(1)}} \mathbf{E}_1 e^{-s\tau^{(0)}} = u_1(s,0)u_0(s,1). \tag{10.9}$$

LEMMA 10.5. *If the integral*

$$f = 2\int_{-\infty}^{\infty} \frac{f(x)}{\sigma^2(x)}\, dx \tag{10.10}$$

is absolutely convergent, then the expectation of the random variable η_k *exists and is equal to* f.

PROOF. We have

$$\mathbf{E}\eta_k = \mathbf{E}_0\int_0^{\tau^{(1)}} f(X(t))dt + \mathbf{E}_1\int_0^{\tau^{(0)}} f(X(t))dt. \tag{10.11}$$

The random variable $\int_0^{\tau^{(0)}} f(X(t))dt$ has an expectation if the sequence $\mathbf{E}_1\int_0^{\tau_n^{(0)}} |f(X(t))|dt$ has a finite limit as $n\to\infty$. By Lemma 3.6.4, the function

$$u_n(x) = \mathbf{E}_x\int_0^{\tau_n^{(0)}} |f(X(t))|dt$$

is a solution in the domain $0 < x < n$ of the problem

$$\frac{\sigma^2(x)}{2}\frac{d^2u_n}{dx^2} + |f(x)| = 0; \qquad u_n(0) = u_n(x) = 0.$$

This implies easily that

$$\mathbf{E}_x\int_0^{\tau} f(X(t))\, dt = \lim u_n(x) = 2\int_0^x dz\int_z^{\infty} \frac{|f(y)|}{\sigma^2(y)}\, dy,$$

and similarly

$$\mathbf{E}_x\int_0^{\tau^{(0)}} f(X(t))dt = 2\int_0^x dz\int_z^{\infty} \frac{f(y)}{\sigma^2(y)}\, dy.$$

In an analogous way one proves the equality

$$\mathbf{E}_x\int_0^{\tau^{(1)}} f(X(t))dt = 2\int_x^1 dz\int_{-\infty}^{z} \frac{f(y)}{\sigma^2(y)}\, dy$$

valid for $x < 1$. Together with (10.11), these equalities imply the desired assertion.

11. Limit relations for null processes (continued)[8]

THEOREM 11.1. *Suppose that the distribution of the length of a single cycle* ξ_n *belongs to the domain of normal attraction of a stable law with exponent* $\alpha < 1$, *i.e.*,

$$\mathbf{P}\{\xi_n \geqslant T\} \sim cT^{-\alpha} \qquad (T \to \infty). \tag{11.1}$$

Assume that the integral (10.10) is absolutely convergent and non-zero.

Then the process $X(t)$ defined by the operator (10.6) satisfies the relations

$$\mathbf{P}\left\{\frac{\int_0^T f(X(t))dt}{\bar{f}T^\alpha} < \frac{x}{c\Gamma(1 - \sigma)}\right\} \to 1 - G_\alpha(x^{-1/\alpha}), \tag{11.2}$$

$$\mathbf{E}_x \int_0^T f(X(t))dt \sim \frac{\bar{f}T^\alpha}{c\Gamma(1 + \alpha)\Gamma(1 - \alpha)} \,. \tag{11.3}$$

PROOF. (1) Since the process $X(t)$ is recurrent, we may assume without loss of generality that $X(0) = 0$. We now observe that the random variables $\nu(T)$ in the expansion (10.7) and the variables ξ_i satisfy the assumptions of Lemma 10.3. Therefore, as $T \to \infty$,

$$\mathbf{P}\left\{\frac{\nu(T)}{T^\alpha} > \frac{x}{c\Gamma(1 - \alpha)}\right\} \sim G_\alpha(x^{-1/\alpha}), \tag{11.4}$$

$$\mathbf{E}\nu(T) \sim \frac{T^\alpha}{\Gamma(1 - \alpha)\Gamma(1 + \alpha)} \,. \tag{11.5}$$

It follows from the expansion (10.7) that

$$\frac{\zeta(T)}{T^\alpha} = \frac{\eta_1 + \dots + \eta_{\nu(T)}}{\nu(T)} \frac{\nu(T)}{T^\alpha} + \frac{\rho(T)}{T^\alpha} \,. \tag{11.6}$$

It is clear that $\nu(T) \to \infty$ almost surely as $T \to \infty$. Hence, by the strong law of large numbers, we get

$$\mathbf{P}\left\{\frac{\eta_1 + \dots + \eta_{\nu(T)}}{\nu(T)} \to \bar{f} \quad \text{as} \quad T \to \infty\right\} = 1.$$

---

[8] The theorems of this section are proved in Has'minskiĭ [13]; see also Eskin [1,2].

This equality, together with (11.4) and (11.6), imply that (11.2) holds, provided that $\rho(T)T^{-\alpha} \to 0$ in probability as $T \to \infty$.

(2) If the function $f(x)$ is non-negative (the general case can be reduced to this by expressing $f(x)$ as the difference of two non-negative functions), then obviously $0 \leqslant \rho(T) \leqslant \eta_{\nu(T)+1}$. It will therefore suffice to prove that $T^{-\alpha}\eta_{\nu(T)+1}$ converges in mean to zero, i.e., that

$$\lim_{T\to\infty} T^{-\alpha}\eta_{\nu(T)+1} = 0. \tag{11.7}$$

For any n, the random variables η_{n+1}, ξ_{n+1} are independent of the history of the process up to time τ_n. Hence

$$\begin{aligned}\mathbf{E}\eta_{\nu(T)+1} &= \mathbf{P}\{\tau_1 \geqslant T\}\mathbf{E}\{\eta_1 | \tau_1 \geqslant T\} \\ &\quad + \sum_{n=1}^{\infty} \int_{y=0}^{T} \mathbf{P}\{\tau_n \in dy, \xi_{n+1} > T - y\} \\ &\qquad\qquad \times \mathbf{E}\{\eta_{n+1} | \xi_{n+1} > T - y\} \\ &= \mathbf{P}\{\tau_1 \geqslant T\}\mathbf{E}\{\eta_1 | \tau_1 \geqslant T\} \\ &\quad + \int_{y=0}^{T} \mathbf{E}(\eta_{n+1}\chi_{n+1}) d\left(\sum_{n=1} \mathbf{P}\{\tau_n < y\}\right),\end{aligned}$$

where $\chi_{n+1} = 1$ if $\xi_{n+1} \geqslant T - y$, and $\chi_{n+1} = 0$ if $\xi_{n+1} < T - y$.

The function $\varphi(T - y) = \mathbf{E}(\eta_{n+1}\chi_{n+1})$ is obviously independent of n, positive, bounded by $\mathbf{E}\eta_i = \bar{f}$, and convergent to zero as $T - y \to \infty$. Moreover, $\sum_{n=1}^{\infty} \mathbf{P}\{\tau_n < y\} = \mathbf{E}\nu(y)$ (see Section 10). Hence, by (11.5), it follows

$$\begin{aligned}&T^{-\alpha}\int_0^T \varphi(T - y)d\mathbf{E}\nu(y) \\ &\leqslant \frac{\mathbf{E}\nu(T - \sqrt{T})}{T^\alpha} \sup_{s>\sqrt{T}} \varphi(s) + \left[\frac{\mathbf{E}\nu(T)}{T^\alpha} - \frac{\mathbf{E}\nu(T - \sqrt{T})}{T^\alpha}\right]\bar{f} \to 0\end{aligned}$$

as $T \to \infty$. This proves (11.7), and hence also (11.2).

(3) As before, we may assume without loss of generality that $f(x) \geqslant 0$. It then follows from (10.7) and (11.7) that

$$\lim_{T\to\infty} \frac{\mathbf{E}_0\int_0^T f(X(t))dt}{T^\alpha} = \lim_{T\to\infty} \frac{\mathbf{E}_0(\eta_1 + \ldots + \eta_{\nu(T)+1})}{T^\alpha}.$$

We now observe that the event $\{\nu(T) + 1 = n\}$ depends only on the history of the process $X(t)$ up to time τ_n. Thus, applying Wald's identity (see Kolmogorov and Prohorov [1]), we see that

$$\mathbf{E}[\eta_1 + \dots + \eta_{\nu(T)+1}] = (\mathbf{E}\nu(T) + 1)f.$$

Together with (11.5), the last two equalities imply (11.3) for $x = 0$. Since the integral of the function $f(X(t))$ up to the time at which the point $x = 0$ is reached has finite expectation, it follows now that (11.3) is valid for any x. This completes the proof of the theorem.

We now present conditions for the relation (11.1) to hold. By (10.9) and Lemmas 10.2 and 10.4, a sufficient condition is that the solutions $u_0(s,x)$ and $u_1(s,x)$ of equation (10.8) satisfy

$$u_i(s,x) \sim 1 - c_i(x)s^\alpha, \qquad i = 0,1$$

as $s \to +0$. We are thus led to consider the asymptotic behavior as $s \to +0$ of the unique bounded solution of the problem

$$\tilde{y}_s'' - sp(x)\tilde{y}_s = 0, \qquad \tilde{y}_s(0) = 1. \tag{11.8}$$

To simplify the notation, we have set here $p(x) = 2/\sigma^2(x)$. We shall compare the solution of problem (11.8) with the solution of the "standard" problem

$$y_s'' - Asx^\beta y_s = 0, \qquad y_s(0) = 1, \tag{11.9}$$

with $\beta > -1$, $A > 0$. Determination of the required asymptotic behavior for equation (11.9) is trivial. In fact, by the substitution $\xi = (As)^\alpha x$, $\alpha = (\beta + 2)^{-1}$ we reduce (11.9) to the problem

$$\frac{d^2 Z_\beta}{d\xi^2} - \xi^\beta Z_\beta = 0, \qquad Z_\beta(0) = 1. \tag{11.10}$$

Integrating equation (11.10), we easily see that if $\beta > -1$, the function $Z_\beta(\xi)$ has a derivative at zero. which must be bounded because of the boundedness of the function Z_β and the structure of equation (11.10). Therefore $Z_\beta(\xi) \sim 1 - c_\alpha\xi$ $(\xi \to 0)$. Hence we conclude that

$$y_s(x) = Z_\beta(A^\alpha s^\alpha x) \sim 1 - c_\alpha A^\alpha x s^\alpha \tag{11.11}$$

as $s \to 0$. Following Eskin [1], we shall express the solution $y_s(x)$ of equation (11.8) in the tentative form

$$\tilde{y}_s(x) = y_s(x)\exp\left\{\int_0^x Q(s,y)\,dy\right\}. \tag{11.12}$$

A simple argument shows that the function Q must satisfy the Riccati equation

$$\frac{dQ}{dx} + \frac{2y_s'(x)}{y_s(x)} Q = s[p(x) - Ax^\beta] - Q^2,$$

or the equivalent integral equation

$$Q(s,x) = - sZ_\beta^{-2}(A^\alpha s^\alpha x)\int_x^\infty p(y) - Ay^\beta \; Z_\beta^2(A^\alpha s^\alpha y)dy$$

$$+ Z_\beta^{-2}(A^\alpha s^\alpha x)\int_x^\infty Q^2(s,y)Z_\beta^2(A^\alpha s^\alpha y)dy. \tag{11.13}$$

We set

$$W(s,x) = - sZ_\beta^{-2}(A^\alpha s^\alpha x)\int_x^\infty [p(y) - Ay^\beta]Z_\beta^2(A^\alpha s^\alpha y)dy.$$

THEOREM 11.2. *Let $p(x) \geqslant 0$ be a function such that the function W is defined for $x \geqslant 0$, and for some function $\varphi(s)$ which tends to zero as $s \to 0$, let*

$$|W(s,x)| < s^\alpha(1 + s^\alpha x)^{\beta/2}\varphi(s). \tag{11.14}$$

Then for any $x > 0$ the solution $\tilde{y}_s(x)$ of problem (11.8) *admits the asymptotic representation*

$$\tilde{y}_s(x) \sim 1 - c_\alpha A^\alpha x s^\alpha \tag{11.15}$$

as $s \to +0$. (Here $-c_\alpha = Z_\beta'(0)$.)

Henceforth we shall assume that $A = 1$, since the general case reduces to this by a simple substitution. To prove the theorem we need the following

LEMMA 11.1. *Let $Z(x)$ be a solution of the problem* (11.10) *which is bounded for $x > 0$. Then*

$$Z_\beta^{-2}(x)\int_x^\infty (1 + y)^\nu Z_\beta^2(y)dy < c(1 + x)^{\nu-\beta/2} \tag{11.16}$$

for all real γ.

(Throughout this section the letter c will designate any positive constant, possibly dependent on β and γ.)

PROOF. The proof of the lemma follows immediately from the well known representation

$$Z_\beta(x) = cx^{1/2}K_\alpha(2\alpha x^{1/2\alpha}) \tag{11.17}$$

of $Z_\alpha(x)$ in terms of the Bessel function $K_\alpha(x)$ (see Kamke [1]) and the asymptotic representation

$$\sqrt{x}K_\alpha(x) \sim \sqrt{\frac{\pi}{2}}\, e^{-x} \qquad (x \to \infty) \tag{11.18}$$

(see Watson [1]).

We note also that formula (11.17) provides an easy way of computing the constant c_α in (11.15):

$$c_\alpha = \frac{\alpha^{2\alpha}\Gamma(1 - \alpha)}{\Gamma(1 + \alpha)} . \tag{11.19}$$

PROOF OF THEOREM 11.2. We set

$$Q_0(s,x) = W(s,x),$$

$$Q_{n+1}(s,x) = W(s,x) + Z_\beta^{-2}(s^\alpha x)\int_x^\infty Q_n^2(s,y)Z_\beta^2(s^\alpha y)dy. \tag{11.20}$$

We shall prove by induction that

$$|Q_n(s,x)| \leqslant 2B(s,x) \tag{11.21}$$

holds for sufficiently small $s > 0$, where

$$B(s,x) = s^\alpha(1 + s^\alpha x)^{\beta/2}\varphi(s).$$

For $n = 0$ this inequality follows from the assumptions of the theorem. Assuming it to be true for some n, we conclude from (11.14), (11.19) and (11.16) that

$$\begin{aligned} &|Q_{n+1}(s,x)| \\ &\qquad \leqslant B(s,x) + 4\varphi^2(s)s^{2\alpha}Z_\beta^{-2}(s^\alpha x)\int_x^\infty (1 + s^\alpha y)^\beta Z_\beta^2(s^\alpha y)dy \\ &\qquad \leqslant B(s,x) + 4\varphi^2(s)s^\alpha Z_\beta^{-2}(s^\alpha x)\int_{s^\alpha x}^\infty (1 + y)^\beta Z_\beta^2(y)dy \\ &\qquad \leqslant B(s,x) + 4\varphi^2(s)s^\alpha c(1 + s^\alpha x)^{\beta/2}. \end{aligned}$$

Now let $s_0 > 0$ be small enough so that $4\varphi(s)c < 1$ for all $s < s_0$. The desired inequality $|Q_{n+1}(s,x)| \leq 2B(s,x)$ follows.

From (11.2) and (11.21) we get

$$|Q_{n+1}(s,x) - Q_n(s,x)|$$

$$\leq 4Z_\beta^{-2}(s^\alpha x)\int_x^\infty Q_n(s,y) - Q_{n-1}(s,y)\ B(s,y)Z_\beta^2(s^\alpha y)dy.$$

This inequality and (11.21) readily imply that the bounded sequence

$$P_n(s,x) = s^{-\alpha}(1 + s^\alpha x)^{-\beta/2}Q_n(s,x)$$

is uniformly convergent. Consequently, the limit

$$\lim_{m\to\infty} Q_m(s,x) = Q(s,x)$$

exists, and the function $Q(s,x)$ satisfies equation (11.13) and

$$|Q(s,x)| \leq 2s^\alpha(1 + s^\alpha x)^{\beta/2}. \tag{11.22}$$

We now conclude easily from (11.22) and (11.18) that the function

$$\tilde{y}_s(x) = y_s(x)\exp\left\{\int_0^x Q(s,y)dy\right\}$$

is indeed a solution of problem (11.8). This function is bounded for each fixed s $(0 < s < s_0)$. This implies the assertion of the theorem, for by (11.22) and (11.11), we have

$$\tilde{y}_s(x) \sim 1 - c_\alpha x s^\alpha \qquad (s \to 0).$$

COROLLARY. *Assume that the function $p(x)$ is integrable at zero and*

$$\lim_{x\to\infty} [p(x)x^{-\beta}] = A.$$

Then the conclusion of Theorem 11.2 *holds true.*

PROOF. We set $\Psi(x) = x^{-\beta}p(x) - A$. Since in this case

$$W(s,x) = s^\alpha\int_{xs^\alpha}^\infty y^\beta\Psi(ys^{-\alpha})Z_\beta^2(y)dyZ_\beta^{-2}(s^\alpha x) = s^\alpha I(xs^\alpha,s),$$

it will suffice to show that

$$|I(y,s)| < |y+1|^{\beta/2}\varphi(s)$$

holds for some function $\varphi(s)$ such that $\varphi(s) \to 0$ as $s \to +0$. To do this, we set $\Phi(x) = \sup_{y>x} |\Psi(y)|$. Evidently $\Phi(x) \to 0$ as $x \to \infty$. Consider two cases. If $y > s^{\alpha-\varepsilon}$ $(\varepsilon > 0)$, then it is easily seen that

$$|I(y,s)| < \Phi(s^{-\varepsilon})(y+1)^{\beta/2}. \tag{11.23}$$

But if $y < s^{\alpha-\varepsilon}$, then

$$|I(y,s)| \leqslant \left|\left(\int_y^{s^{\alpha-\varepsilon}} + \int_{s^{\alpha-\varepsilon}}^{\infty}\right) x^{\beta}\Psi(s^{\lambda\alpha}x)Z_{\beta}^{2}(x)Z_{\beta}^{-2}(y)\,dx\right|.$$

The integral from $s^{\alpha-\varepsilon}$ to ∞ is estimated in the same way as (11.23); to estimate the other integral, we observe first that $0 < Z_{\beta}(y) < Z_{\beta}(x)$ for $x < y$. Since $p(x)$ is integrable at zero, we have

$$\begin{aligned}\int_y^{s^{\alpha-\varepsilon}} x^{\beta}|\Psi(xs^{-\alpha})|\,dx &\leqslant c_1 s^{\alpha\beta+\alpha}\left(1 + \int_1^{s^{-\varepsilon}} \Phi(x)x^{\beta}\,dx\right)\\ &\leqslant c_2 s^{\alpha(\beta+1)}(1 + s^{-\varepsilon(\beta+1)})\\ &\leqslant c_3 s^{(\alpha-\varepsilon)(\beta+1)}.\end{aligned}$$

These inequalities imply the desired result whenever $\alpha > \varepsilon > 0$.

REMARK 1. For $\beta = 0$, Theorem 11.2 was proved by Eskin [1] under the additional assumption that $|p(x) - A| < A/4$.

REMARK 2. The value of the function $W(s,x)$ for $x = x_0$ does not depend on the values of $p(x)$ for $x = x_0$. Hence it follows by Theorem 11.2 that if the assumptions of Theorem 11.2 hold for some function $p(x)$, then they remain valid for any function $p_1(x)$ such that the difference $p_1 - p$ has compact support and $p_1(x)$ is integrable at zero. Consequently, under the assumptions of Theorem 11.2, the asymptotic formula (11.15) remains valid for solutions of problem (11.8) in the domain $x > x_0$, where $x_0 > 0$. A similar assertion is of course true for solutions of the problem on the left half-line.

REMARK 3. Suppose that the assumptions of Theorem 11.2 hold on both the left and the right half-line (generally with different constants $A_i \geqslant 0$ such that $A_1 + A_2 > 0$). Then one can easily use Theorem 11.2 to determine the principal term of the asymptotic expansion of Green's function for equation (11.8). This function is bounded on the entire real axis as $s \to +0$. To this end one must use the method of Eskin [1], where the expansion is determined for $\beta = 0$ (see also Eskin [2]).

As mentioned above, Theorem 11.2 yields sufficient conditions in terms of the coefficient $\sigma^2(x)$ for (11.1) to hold.

THEOREM 11.3. *Let $f(x)$ be a function such that the integral* (10.10) *is absolutely convergent and non-zero. Assume that the following conditions hold for $x > 0$, $x < 0$, respectively:*

$$\left.\begin{aligned} &\left| s Z_\beta^{-2}(A_1^\alpha s^\alpha x) \int_x^\infty (p(y) - A_1 y^\beta) Z_\beta^2(A_1^\alpha s^\alpha y)\,dy \right| \\ &\qquad\qquad < s^\alpha (1 + s^\alpha x)^{\beta/2} \varphi(s), \\ &\left| s Z_{\bar\beta}^{-2}(A_2^\alpha s^\alpha |x|) \int_{-\infty}^x (p(y) - A_2 |y|^\beta) Z_\beta^2(A_2^\alpha s^\alpha |y|)\,dy \right| \\ &\qquad\qquad < s^\alpha (1 + s^\alpha |x|)^{\beta/2} \varphi(s), \end{aligned}\right\} \quad (11.24)$$

$$(\varphi(s) \to 0 \quad as \quad s \to 0, \quad A_i \geqslant 0, \quad A_1 + A_2 > 0).$$

Then the process $X(t)$ defined by the operator (10.6) *satisfies* (11.2) *and* (11.3). *The constant c in these formulas is given in terms of A_1 and A_2 by*

$$c = \frac{\alpha^{2\alpha}(A_1^\alpha + A_2^\alpha)}{\Gamma(1 + \alpha)} .$$

PROOF. From (10.9), (11.24) and Theorem 11.2, we get that

$$\mathbf{E} e^{-s\xi_n} \sim 1 - c_\alpha (A_1^\alpha + A_2^\alpha) s^\alpha$$

for $s \to +0$. Hence, by Lemma 10.2, we see that

$$\mathbf{P}\{\xi_n > T\} \sim \frac{c_\alpha (A_1^\alpha + A_2^\alpha)}{\Gamma(1 - \alpha)} T^{-\alpha} = \frac{\alpha^{2\alpha}(A_1^\alpha + A_2^\alpha)}{\Gamma(1 + \alpha)} T^{-\alpha}$$

as $T \to \infty$. To complete the proof it suffices to apply Theorem 10.1.

COROLLARY 1. Theorem 11.3 can be used to investigate the asymptotic behavior of solutions of the Cauchy problem for a parabolic equation. Indeed, it follows from Lemmas 3.6.1 and 3.6.3 that the solution $u(t,x)$ of the problem

$$\frac{\partial u}{\partial t} = \frac{\sigma^2(x)}{2}\frac{\partial^2 u}{\partial x^2} + f(x), \qquad u(0,x) = \varphi(x) \tag{11.25}$$

can be written as

$$u(t,x) = \mathbf{E}_x\int_0^t f(X(s))ds + \mathbf{E}_x\varphi(X(t)).$$

This representation and Theorem 11.3 imply the following: Let $\varphi(x)$ be a bounded function and suppose that the assumptions of Theorem 11.3 hold. Then the solution of problem (11.25) satisfies the relation

$$u(x,t) \sim \frac{\bar{f}t^{\alpha}}{(A_1^{\alpha} + A_2^{\alpha})\alpha^{2\alpha}\varphi(1 - \alpha)}$$

as $t \to \infty$.

COROLLARY 2. Starting from (11.4) one can prove still simpler sufficient conditions for the conclusions of Theorem 11.3 and Corollary 1 to hold. Thus, if $\alpha = \frac{1}{2}$ $(\beta = 0)$, we have $Z_0(x) = e^{-x}$ by (11.10). Conditions (11.24) may then be given the form

$$\sup_{x\geqslant 0}\left|\sqrt{s}\int_x^{\infty}(p(y) - A_1)e^{-2A_1^{\alpha}\sqrt{s}(y-x)}dy\right| \to 0,$$

$$\sup_{x\leqslant 0}\left|\sqrt{s}\int_{-\infty}^{x}(p(y) - A_2)e^{2A_2^{\alpha}\sqrt{s}(y-x)}dy\right| \to 0$$

as $s \to +0$. Let $p(y)$ be a bounded function such that

$$\frac{1}{T}\sup_{x>0}\left|\int_x^{x+T}(p(y) - A_1)dy\right| = \frac{1}{T}\,\varphi(T) \to 0,$$

$$\frac{1}{T}\sup_{x<0}\left|\int_{x-T}^{x}(p(y) - A_2)dy\right| \to 0$$

as $T \to \infty$. Then clearly conditions (11.24) are also satisfied. In fact we conclude from (11.26) that, for example,

$$\left|\sqrt{s}\int_x^{\infty} p(y) - A\ )e^{-2A_1^{\alpha}\sqrt{s}(y-x)}dy\right| = \tag{Contd}$$

$$\text{(Contd)} \quad = \left|\sqrt{s}\int_0^{\infty} 2A_1^{\alpha}\sqrt{s}e^{-2A_1^{\alpha}\sqrt{s}z}dz\int_x^{z+x}(p(u) - A_1)du\right|$$

$$\leqslant \sqrt{s}\int_0^{\infty} 2A_1^{\alpha}\sqrt{s}e^{-2A_1^{\alpha}\sqrt{s}z}\varphi(z)dz$$

$$= 2A_1^{\alpha}\left[\int_0^{s^{-1/2}} se^{-2A_1^{\alpha}\sqrt{s}z}\varphi(z)dz + \int_{s^{-1/2}}^{\infty} se^{-2A_1^{\alpha}\sqrt{s}z}\varphi(z)dz\right].$$

The integral from 0 to $s^{-\frac{1}{2}}$ is of order $O(s^{\frac{1}{2}})$. By (11.26), the other integral satisfies the estimate

$$\int_{s^{-1/2}} se^{-2A_1^{\alpha}\sqrt{s}z}\varphi(z)dz \leqslant \sup_{z>s^{-1/2}} \frac{\varphi(z)}{z}\int_0^{\infty} ue^{-2A_1^{\alpha}u}du \to 0.$$

Thus conditions (11.26) are sufficient for the conclusions of Theorem 11.2 and its corollary to hold. These conditions are of course satisfied if $p(x)$ (or $\sigma^2(x)$) is periodic or converges to a periodic function as $x \to \infty$ or $x \to -\infty$.

CHAPTER V

STABILITY OF SYSTEMS OF STOCHASTIC EQUATIONS

1. Statement of the problem

In Chapter I we studied problems of stability under random perturbations of the parameters. We noted there that no significant results can be expected unless the random perturbations possess sufficiently favorable mixing properties. Fortunately, in practical applications one may often assume that the "noise" has a "short memory interval." The natural limiting case of such noise is of course white noise. Thus it is very important to study the stability of solutions of Itô equations since this is equivalent to the study of stability of systems perturbed by white noise.

Any result concerning conditions for the stability of stochastic systems is apparently itself "stable", in the sense that it carries over to systems perturbed by noise which is "almost white" in a well-defined sense. Some relevant results were obtained by the author in [8]. However, as yet, no complete investigation has been made of conditions under which the stability (instability) of a system of Itô equations implies the stability (instability) of a "similar" system.

Below we present a theory of stability for Itô equations. The main stimulus for setting up this theory was the extremely suggestive paper of Kac and Krasovskiĭ (1). These authors investigate the stability of the solution $X(t) = 0$ of the equation $dX/dt = f(X,t,Y(t))$, where $Y(t)$ is a homogeneous Markov chain with finitely many states. They solve the problem in terms of Ljapunov functions, but instead of the derivative d^0V/dt along the sample path they consider, roughly speaking, the expectation LV of this derivative. The paper also contains important results, amenable to generalization in various directions, concerning the stability of linear systems and stability in the first approximation.

In our paper [2] we remarked that by suitable interpretation these results yield information about the stability of our invariant set (in this case, the hyperplane $x = 0$) of the many-dimensional Markov process $(X(t),Y(t))$. This interpretation will be given in Section 7.6 below.

A brief survey of Chapters V-VII follows.

Consider the system

$$dX(t) = b(t,X)dt + \sum_{r=1}^{k} \sigma_r(t,X)d\xi_r(t). \tag{1.1}$$

We shall assume that $X(t)$, $b(t,x)$ and $\sigma_r(t,x)$ are vectors in E_ℓ, and $\xi_r(t)$ are independent Wiener processes. We assume moreover that the coefficients of (1.1) satisfy the following Lipschitz condition in every domain which is bounded in x, i.e.,

$$\sum_{r=1}^{k} |\sigma_r(t,x) - \sigma_r(t,y)| + |b(t,x) - b(t,y)| < B|x - y|. \tag{1.2}$$

In some cases the Lipschitz constant B will be assumed to be independent of the domain, i.e. inequality (1.2) will be assumed to hold throughout $E = \{t > 0\} \times E_\ell$. We shall limit ourselves to conditions for stability of the trivial solution $X(t) = 0$. Accordingly, we assume that

$$b(t,0) \equiv 0, \qquad \sigma_r(t,0) \equiv 0. \tag{1.3}$$

In the present chapter, the solution of stability problems for systems of stochastic differential equations will be given primarily in the same terms as above. That is, we shall use the same methods, as employed for the qualitative behavior of solutions of such systems in Chapter III.

Theorems 3.1, 4.1 and 4.3 proved below are natural generalizations to stochastic systems of Ljapunov's second method. One feature of these theorems should be mentioned. They all require that the Ljapunov function is sufficiently smooth in t and x in a neighborhood of $x = 0$, except possibly at the point $x = 0$ itself. Unlike for a deterministic system, for a stochastic system there often does not even exist a Ljapunov function which is smooth at the origin. This will be clear from the example at the end of Section 3.

In this connection, we introduce the following definition.

Let U be a domain with closure $\tilde{U}$ in the space $E = I \times E_\ell$, and set $U^\varepsilon(0) = \{(t,x): |x| < \varepsilon\}$. We shall say that a function $V(t,x)$ is in class $\mathbf{C}_2^0(U)$ ($V(t,x) \in \mathbf{C}_2^0(U)$) if it is twice continuously differentiable with respect to x and continuously differentiable with respect to t throughout U, except possibly for the set $x = 0$, and continuous in the closed set $\tilde{U} \setminus U^\varepsilon(0)$ for any $\varepsilon > 0$.

As in Chapter I, we shall consider the stability of the moment of different orders (p-stability) and stability in probability. However, whereas in Chapter I we were able to derive conditions only for weak stability in probability (see Section 1.5), here we shall consider stability in probability in a stronger sense. To be precise, we shall present conditions under which not only does $|X(t)|$ tend to zero in probability uniformly in t, but also $\sup_{t>0} |X(t)|$ tends to zero in probability as $|X(t)| \to 0$. With this definition, the stability or instability of the equilibrium position is determined by the behavior of the coefficients of the equation only in a neighborhood of the equilibrium position. It is therefore natural to expect that for a broad range of cases the full system (1.1) will be stable provided the first-approximation system

$$dX(t) = \frac{\partial b(t,0)}{\partial x} X dt + \sum_{r=1}^{k} \frac{\partial \sigma_r(t,0)}{\partial x} X d\xi_r(t) \tag{1.4}$$

is stable.

The question as to when stability of the system (1.4) implies that of (1.1) will be answered in Chapter VII. We shall prove that if $\partial b/\partial x$ and $\partial \sigma_r/\partial x$ are independent of t, then it suffices that (1.4) is asymptotically stable in probability. If the coefficients depend on time, one must also assume that the coefficients are temporally uniform in a certain sense.

Of major importance in this connection is the problem of stability of linear stochastic systems. In the general case, this problem is extremely difficult. We shall solve it in Chapter VI, though in a rather ineffective way. We shall prove there that the stability or instability of a linear stochastic system with time-independent coefficients is determined by the sign of the conditional expectation of a certain function, given the stationary distribution of a certain Markov process on the ℓ-dimensional sphere. For $\ell = 2$ the density of this distribution can be computed by quadratures, so that the answer is expressible in closed form.

The question of instability conditions is even more complicated. The analogs of the instability theorems of Ljapunov and Četaev do not hold in for stochastic systems. Roughly speaking, the reason for this is that the sample paths of stochastic systems may leave the "instability set" because of purely random forces. This is made particularly clear by the second example in Section 7.3. In this example, the unstable deterministic system $dx_1/dt = x_1$, $dx_2/dt = -x_2$ is "impaired" by the addition of a small drift and a small diffusion

$$\left.\begin{aligned} dX_1(t) &= (X_1 + b(X_1,X_2))dt + \sigma(X_1,X_2)d\xi_1(t), \\ dX_2(t) &= -X_2 dt + \sigma(X_1,X_2)d\xi_2(t). \end{aligned}\right\} \tag{1.5}$$

In this case, for any small number $\varepsilon > 0$, the functions b and σ may be chosen so that

$$|b(x_1,x_2)| + |\sigma(x_1,x_2)| < \varepsilon|x|,$$

and the system (1.5) is asymptotically stable in the large. It is well known that this phenomenon is impossible if $\sigma \equiv 0$.

In Chapter V (Theorem 5.4.2) we shall present a sufficient condition for instability which at first glance is relevant only for very special cases. However, it will follow from the discussion in Chapters VI and VII that if the diffusion of the system is nondegenerate (in a fairly weak sense), then it is often possible to construct a function satisfying the assumptions of Theorem 5.4.2.

On this basis, we shall consider in Chapter VII the problem of instability in the first approximation. In the same chapter we shall also consider the problem of stability under damped random perturbations and some applications to statistics. At the end of Chapter VII we shall discuss the possibility of generalizing the results to a wider class of processes.

2. Some auxiliary results

This chapter will make systematic use of properties of martingales and supermartingales. The following definitions of these important classes of stochastic process are due to Doob.

Let $(\Omega,\mathfrak{A},\mathbf{P})$ be a probability space, $\mathcal{M}_t \subset \mathfrak{A}$ a family of σ-algebras of events in Ω, defined for each $t \geqslant 0$, such that $\mathcal{M}_s \subset \mathcal{M}_t$ for $s < t$. Let $y(t,\omega)$, $t \geqslant 0$, be a stochastic process with finite expectation $\mathbf{E}y(t,\omega)$, such that $y(t,\omega) = y(t)$ is an $\mathcal{M}_t$-measurable random variable for each t. The family $(y(t,\omega),\mathcal{M}_t)$ is called a *supermartingale* if for any $s < t$

$$\mathbf{E}(y(t)/\mathcal{M}_s) \leqslant y(s) \qquad (P\text{-a.s.}). \tag{2.1}$$

If we replace the inequality sign in (2.1) by equality, we get the definition of a *martingale*.

The following examples of martingales and supermartingales are important for the sequel.

EXAMPLE 1. The Wiener process $\xi(t)$ is a martingale with respect to the system of σ-algebras $\mathcal{N}_t$, since

$$\mathbf{E}(\xi(t)/\mathcal{N}_s) = \mathbf{E}([\xi(s) + (\xi(t) - \xi(s))]/\mathcal{N}_s) = \xi(s) \qquad (\text{a.s.}).$$

An analogous argument shows that the more general process

$$y(t) = \int_0^t \sigma(s)d\xi(s)$$

is also a martingale.

EXAMPLE 2. Let $V(t,x)$ be a function twice continuously differentiable with respect to x and continuously differentiable with respect to t in $I \times U$, where $U \subset E_\ell$ is a bounded closed domain. Suppose that in this domain

$$LV(t,x) = \frac{\partial V}{\partial t} + \frac{1}{2}\sum_{i,j=1}^{\ell} a_{ij}(t,x)\frac{\partial^2 V}{\partial x_i \partial x_j} + \sum_{i=1}^{\ell} b_i(t,x)\frac{\partial V}{\partial x_i}$$

$$\equiv \frac{\partial V}{\partial t} + \frac{1}{2}\sum_{r=1}^{k}\left(\sigma_r(t,x), \frac{\partial}{\partial x}\right)^2 V + \left(b(t,x), \frac{\partial}{\partial x}\right)V \leqslant 0$$

holds. Set $\tau(t) = \min(\tau,t)$, where τ is the first exit time from U at the sample path of the process $X(t)$ defined by equation (1.1). Then the process $y(t) = V(\tau(t),X(\tau(t)))$ is a supermartingale with respect to the system $\mathcal{N}_t$. In fact, under the above assumptions it follows from Lemma 3.3.1 that

$$\mathbf{E}[V(\tau(t),X(\tau(t)))/\mathcal{N}_s] \leqslant V(s,X(s)) \qquad \text{(a.s.)}.$$

Hence condition (2.1) is satisfied for almost all paths such that $\tau > s$, and consequently $X(\tau(s)) = X(s) \in U$. For almost all paths such that $\tau \leqslant s$ we have $\mathbf{E}(y(t)/\mathcal{N}_s) = y(s)$, since then $\tau(s) = \tau(t) = \tau$. If $LV \leqslant 0$ for all $x \in E_\ell$, $t \geqslant 0$, and $E_{s,x}V(t,X(t))$ exists, an analogous argument shows that the process $V(t,X(t))$ is also a supermartingale.

These properties generally fail to hold if the condition $LV \leqslant 0$ is not satisfied on some set (even at one point). In certain cases, however, the supermartingale property of the process remains valid even when this happens. Let us call the random variable $\tau^\Gamma = \inf\{t : X(t) \in \Gamma\}$ the first time at which the set Γ is reached. A closed set Γ is said to be *inaccessible* to a process $X(t)$ if $\mathbf{P}\{\tau^\Gamma < \infty\} = 0$. Since the sample paths of the process are continuous, a set Γ is inaccessible if and only if

$$\mathbf{P}\{\tau_{U_\delta(\Gamma)} \longrightarrow \infty \quad \text{as} \quad \delta \to 0\} = 1.$$

Here $U_\delta(\Gamma)$ is the δ-neighborhood of the set Γ.

LEMMA 2.1. *Let $V(t,x)$ be a function twice continuously differentiable with respect to x, continuously differentiable with respect to t on the set $I \times \{U \setminus \Gamma\}$ and bounded in $I \times U$, where U is a bounded domain in E_ℓ and $\Gamma \subset U$ is a set inaccessible to the process $X(t)$ defined by equation* (1.1). *Assume that $LV \leqslant 0$ on the set $I \times (U \setminus \Gamma)$. Then the process $V(\tau_U(t),X(\tau_U(t)))$ is a supermartingale.*

PROOF. Let $\tau_{U,\delta}$ denote the first exit time from the set $U \setminus U_\delta(\Gamma), \tau_{U,\delta}(t) = \min(\tau_{U,\delta},t)$. Since Γ is inaccessible, it follows that for all t, we have

$$\tau_{U,\delta}(t) \to \tau_U(t) \qquad \text{(a.s.)} \tag{2.2}$$

as $\delta \to 0$. On the other hand, it is clear from Example 2 that

$$\mathbf{E}(V(\tau_{U,\delta}(t), X(\tau_{U,\delta}(t)))/\mathcal{N}_s) \leqslant V(\tau_{U,\delta}(s), X(\tau_{U,\delta}(s))) \qquad \text{(a.s.)}.$$

Letting $\delta \to 0$ in this inequality and using (2.2) and the fact that V is bounded, we get the required assertion.

LEMMA 2.2.[1]. *Let the coefficients b and σ_r of equation* (1.1) *satisfy condition* (1.3). *Assume further that condition* (1.2) *holds throughout $E = I \times E_\ell$. Then for any real β, $t \geqslant s$, $x \neq 0$,*

$$\mathbf{E}|X^{s,x}(t)|^\beta \leqslant |x|^\beta \exp\{k(t-s)\}, \tag{2.3}$$

where k is a constant depending only on β and the constant B of (1.2).

PROOF. The function $V(x) = |x|^\beta$ is twice continuously differentiable in the domain $|x| > \delta$ for any $\delta > 0$. Applying Itô's formula (3.3.8) in this domain, we get for $Y^{s,x}(t) = |X^{s,x}(t)|^\beta$ the formula

$$\begin{aligned} Y^{s,x}(\tau_\delta(t)) = Y^{s,x}(s) &+ \beta\int_s^{\tau_\delta(t)} |X^{s,x}(u)|^{\beta-2}\Big[(b(u,X^{s,x}(u)),X^{s,x}(u))du \\ &+ \frac{1}{2}\sum_{i=1}^{\ell} a_{ii}(u,X^{s,x}(u))du \\ &+ \sum_{r=1}^{k}(\sigma_r(u,X^{s,x}(u)),X^{s,x}(u))d\xi_r(u)\Big] \\ &+ \frac{1}{2}\beta(\beta-2)\int_s^{\tau_\delta(t)} |X^{s,x}(u)|^{\beta-4}(A(u,X^{s,x}(u)) \\ &\times X^{s,x}(u),X^{s,x}(u))du, \end{aligned} \tag{2.4}$$

where τ_δ denotes the first exit time from the set $|x| > \delta$ and $\tau_\delta(t) = \min(\tau_\delta, t)$. It is obvious that the random variable $Y^{s,x}(\tau_\delta(t))$ has an expectation. (If $\beta \leqslant 0$, this follows from the

[1] See Nevel'son and Has'minskiĭ [2].

fact that it is bounded, and if $\beta > 0$ we may conclude this from Example 1 in Section 3.4). Calculating expectations in (2.4) and using (1.2) and (1.3), we easily obtain that

$$\mathbf{E}Y^{s,x}(\tau_\delta(t)) \leqslant |x|^\beta + k\mathbf{E}\int_s^{\tau_\delta(t)} Y^{s,x}(u)du \tag{2.5}$$

for some $k = k(\beta,B,1)$. Since $\tau_\delta(u) + u$ for $u < \tau_\delta(t)$, it follows from (2.5) that

$$EY^{s,x}(\tau_\delta(t)) \leqslant |x|^\beta + k\mathbf{E}\int_s^{\tau_\delta(t)} Y^{s,x}(\tau_\delta(u))du$$

$$\leqslant |x|^\beta + k\int_s^t \mathbf{E}Y^{s,x}(\tau_\delta(u))du.$$

Applying the Gronwall-Bellman lemma to this inequality, we get the estimate

$$E|X^{s,x}(\tau_\delta(t))|^\beta \leqslant |x|^\beta \exp\{k(t - s)\}. \tag{2.6}$$

Setting $\beta = -1$ in (2.6) and using Čebyšev's inequality, we get

$$\mathbf{P}_{s,x}\{\tau_\delta(t) < t\} < \frac{\delta}{|x|} e^{k(t-s)}.$$

This implies

$$P_{s,x}\{\tau_\delta < t\} \to 0 \quad \text{as} \quad \delta \to 0 \tag{2.7}$$

for every set. Letting $\delta \to 0$ in (2.6) and using (2.7), we get (2.3).

REMARK 1. It follows from (2.7) that under the assumptions of the lemma the point $x = 0$ is inaccessible to the process $X^{s,x}(t)$. For this assertion to hold it suffices that condition (1.2) is satisfied in every cylinder $I \times K$, where $K \subset E$ is compact and that $X(t)$ is regular. This is intuitively obvious (and easily proved rigorously), since whether the path of a regular process can hit $x = 0$ depends only on the behavior of the coefficients of the equation in the neighborhood of that point. For further reference, we state this result as a lemma.

LEMMA 2.3. *Suppose that the coefficients of equation* (1.1) *satisfy* (1.3), *condition* (1.2) *holds in every domain bounded with respect to* x, *and the process* $X^{s,x_0}(t)$ *is regular. Then the point* $x = 0$ *is inaccessible to any sample path of the process if* $x_0 \neq 0$.

Lemmas 2.1 and 2.3 imply.

LEMMA 2.4. *Let* $V(t,x)$ *be a function in class* $\mathbf{C}_2^0((t > 0) \times U)$, *bounded in the domain* $(t > 0) \times U$, *where* U *is a neighborhood of the origin, and suppose that* $LV(t,x) \leqslant 0$ *in this domain. Then the process* $V(\tau_U(t), X(\tau_U(t)))$ *is a supermartingale, so that*

$$\mathbf{E}V(\tau_U(t), X^{s,x}(\tau_U(t))) \leqslant V(s,x)$$

for $x \in U$.

REMARK 2. By virtue of (2.7) we have

$$P_{s,x}\{\tau_\delta(t) \to t \text{ as } \delta \to 0\} = 1,$$

and hence, letting $\delta \to 0$ in (2.4), we see that Itô's formula (3. 3.8) is applicable to the function $|x|^\beta$ on the whole of E_ℓ, despite the fact that if $\beta < 2$, this function does not satisfy the assumption of Theorem 3.3.1 at zero. This conclusion holds true for any function $V(t,x) \in \mathbf{C}_2^0(E)$ such that $0 \leqslant V(t,x) < k|x|^\beta$.

Our subsequent applications of martingale theory to stability problems are based on the following theorem which we give here without proof.

THEOREM 2.1. *(Doob* [3]*). If* $(y(t,\omega), \mathcal{M}_t, t \geqslant 0)$ *is a positive supermartingale, then the limit* $y_\infty = \lim_{t\to\infty} y(t,\omega)$ *almost surely exists and is finite. Moreover* $Ey_\infty = \lim_{t\to\infty} Ey(t,\omega)$.

In Chapter VI we shall need also the following

THEOREM 2.2. *(Doob* [3]*). If* $(y(t,\omega), \mathcal{M}_t, t \geqslant 0)$ *is a martingale, then for any* $k > 0$

$$P\left\{\sup_{t_0 \leqslant t \leqslant T} |y(t,\omega)| > k\right\} \leqslant \frac{E|y(T,\omega)|}{k}.$$

3. Stability in probability

A solution $X(t,\omega)$ of equation (1.1) is said to be *stable in probability* for $t \geqslant 0$ if for any $s \geqslant 0$ and $\varepsilon > 0$

$$\lim_{x\to 0} P\{\sup_{t>s} |X^{s,x}(t)| > \varepsilon\} = 0.$$

This definition is considerably stronger[2] than (1.5.3). It says that the sample path of the process issuing from a point x

[2] The relation between this definition and definition (1.5.3) is discussed in Section 6.11.

at time s will always remain within any prescribed neighborhood of the origin with probability tending to one as $x \to 0$. The importance of the definition will be clear from Theorem 3.1 below.

Before we state the theorem, we recall that a function $V(t,x)$ is said to be *positive definite* (in Ljapunov's sense) in a neighborhood of the set $x = 0$ if $V(t,0) = 0$ and in this neighborhood $V(t,x) > W(x)$, where $W(x) > 0$ for $x \neq 0$.

Theorem 3.1 is analogous to the well-known theorem of Ljapunov for deterministic systems. For nondegenerate processes, it was first proved by the author in [2]. Subsequently it has been generalized in various directions[3].

THEOREM 3.1. *Let* $\{t > 0\} \times U = U_1$ *be a domain containing the line* $x = 0$, *and assume there exists a function* $V(t,x) \in \mathbf{C}_2^0(U_1)$ *which is positive definite in Ljapunov's sense and satisfies*

$$LV = \frac{\partial V}{\partial t} + \sum_{i=1}^{\ell} b_i(t,x) \frac{\partial V}{\partial x_i} + \frac{1}{2} \sum_{i,j=1}^{\ell} a_{ij}(t,x) \frac{\partial^2 V}{\partial x_i \partial x_j} \leqslant 0$$

for $x \neq 0$. *Then the trivial solution of equation* (1.1) *is stable in probability.*

PROOF. Let r be a number such that the r-neighborhood U_r of the point $x = 0$ is contained in U together with its boundary. We set $V_r = \inf_{x \in U \setminus U_r} V(t,x)$ ($V_r > 0$ by assumption). By Lemma 2.4, we have

$$\mathbf{E}V(\tau_{U_r}(t), X^{s,x}(\tau_{U_r}(t))) \leqslant V(s,x)$$

for $|x| < r$. Using this and Čebyšev's inequality, we get

$$\mathbf{P}\{\sup_{s \leqslant u \leqslant t} |X^{s,x}(u)| > r\} \leqslant \frac{\mathbf{E}V(\tau_{U_r}(t), X^{s,x}(\tau_{U_r}(t)))}{V_r}$$

$$\leqslant \frac{V(s,x)}{V_r} .$$

Letting $t \to \infty$, we finally have

$$\mathbf{P}\{\sup_{u \geqslant s} |X^{s,x}(u)| > r\} \leqslant \frac{V(s,x)}{V_r} .$$

Since $V(s,0) = 0$ and the function $V(s,x)$ is continuous, this implies the desired assertion.

---

[3] Gihman [2,3] has given another proof, making no use of nondegeneracy, but using only Ljapunov functions which are smooth at zero. A similar result was obtained by Kushner in [2].

REMARK. We shall say that the solution $X(t) \equiv 0$ of equation (1.1) is *uniformly stable in probability* for $t > 0$ if for any $\varepsilon > 0$ the function $\mathbf{P}\{\sup_{t>s} |X^{s,x}(t)| \geqslant \varepsilon\}$ tends to zero as $x \to 0$, uniformly in $s \geqslant 0$. An examination of the proof of Theorem 3.1 immediately reveals that a sufficient condition for uniform stability in probability is that the function $V(t,x)$ satisfies the assumptions of Theorem 3.1 and that it has an infinitesimal upper limit, i.e.,

$$\lim_{x \to 0} \sup_{t>0} V(t,x) = 0.$$

A question of major theoretical and practical interest is whether there exists for every system which is stable in probability, a Ljapunov function satisfying the assumptions of Theorem 3.1. For simplicity's sake, we shall confine ourselves to the temporally homogeneous case, assuming moreover that the "noise" is nondegenerate everywhere, except possibly at $x = 0$.

THEOREM 3.2[4]. *Assume that the coefficients b and σ_r of equation* (1.1) *are independent of time, and that its solution $X(t) \equiv 0$ is stable in probability. Suppose that in a neighborhood of $x = 0$ condition* (1.2) *holds and also the nondegeneracy condition*

$$\sum_{i,j=1}^{\ell} a_{ij}\lambda_i\lambda_j > m(x) \sum_{i=1}^{\ell} \lambda_i^2, \tag{3.1}$$

where $m(x)$ is a continuous function such that $m(x) > 0$ for $x \neq 0$.

Then in a neighborhood of $x = 0$ there exists a positive definite function $V(x)$, twice continuously differentiable except perhaps at $x = 0$, such that $LV = 0$.

PROOF. Let $U_r = \{|x| < r\}$ be a sufficiently small neighborhood of $x = 0$. Let $u_\delta(x)$ denote a solution in the domain $U_r \smallsetminus U_\delta$ of the problem

$$Lu = 0; \qquad u\big|_{|x|=\delta} = 1; \qquad u\big|_{|x|=\delta} = 0.$$

It follows from Lemma 3.6.2 that

$$u_\delta(x) = \mathbf{P}\{|X^x(\tau_{r,\delta})| = r\},$$

where $\tau_{r,\delta}$ is the first time at which the sample path reaches the set $\{|x| = r\} \cup \{|x| = \delta\}$.

It is clear that the sequence $u_\delta(x)$ of L-harmonic functions is monotone increasing as $\delta \to 0$. Its limit $V(x)$ is also L-harmonic. Let τ_0 denote the first time at which the path of the process reaches the point 0. Then it follows from the obvious relations

[4] See Has'minskiĭ [2], Kushner [5].

between events that

$$\{\sup_{t>0} |X^x(t)| \geqslant r\} \subset \bigcup_{\delta>0} \{|X^x(\tau_{r,\delta})| = r\} \cup \{\tau_0 < \infty\},$$

$$\bigcup_{\delta>0} \{|X^x(\tau_{r,\delta})| = r\} \subset \{\sup_{t>0} |X^x(t)| \geqslant r\},$$

and from Lemma 2.3 we obtain

$$\mathbf{P}\{\sup_{t>0} |X^x(t)| \geqslant r\} = \lim_{\delta\to 0} \mathbf{P}\{|X^x(\tau_{r,\delta})| = r\} = V(x).$$

Since the solution $X \equiv 0$ is stable in probability, it follows from this equality that $V(x) \to 0$ as $x \to 0$. Finally, the strong maximum principle implies that the function $u_\delta(x)$ and hence also $V(x)$, is positive for $|x| > \delta_1 > \delta$. Thus the function $V(x)$ is positive definite in Ljapunov's sense, and $LV = 0$. This we wished to prove.

REMARK 1. Malkin [1] showed that the analog of Theorem 3.2 for deterministic systems does not hold. It follows that the nondegeneracy condition (3.1) cannot be dropped (though it can be weakened).

REMARK 2. The Ljapunov function constructed in Theorem 3.2 is continuous only at zero. It is readily shown that in general a Ljapunov function which is smooth at zero may not exist. This will be clear from the following example.

Let $X(t)$ be a one-dimensional process, described by the equation

$$dX = bXdt + \sigma X d\xi(t), \tag{3.2}$$

where b and σ are constants. The differential generator of this process is

$$L = \frac{\partial}{\partial t} + \frac{1}{2}\sigma^2 x^2 \frac{\partial^2}{\partial x^2} + bx\frac{\partial}{\partial x}.$$

If $b < \sigma^2/2$, the solution $X(t) \equiv 0$ of the system (3.2) is stable, since the function $V(x) = |x|^{1-2b/\sigma^2}$ satisfies the assumptions of Theorem 3.1. If $b \geqslant 0$, this function is not differentiable at zero. Using the maximum principle for elliptic equations, one readily shows that any function $V_1(x)$ such that $V_1(0) = 0$, $V_1(\varepsilon) \geqslant \delta$, satisfies $V_1(x) \geqslant \delta(|x|/|\varepsilon|)^{1-2b/\sigma^2}$ in the domain $0 < x < \varepsilon$. Hence it is clear that when $b > 0$, then there can be no Ljapunov function smooth at the origin and independent of t. A similar argument shows that there does not even exist a Ljapunov function smooth at zero which depends on t but has an infinitesimal upper limit.

4. Asymptotic stability in probability and instability[5]

The solution $X(t) \equiv 0$ of equation (1.1) is said to be *asymptotically stable in probability* if it is stable in probability and moreover

$$\lim_{x\to 0} \mathbf{P}\{\lim_{t\to\infty} X^{s,x}(t) = 0\} = 1. \tag{4.1}$$

In this section we shall frequently assume that the following condition is satisfied[6]: **D**. *Any solution of equation* (1.1), *beginning in the domain* $\varepsilon < |x| < r$, *almost surely reaches the boundary of this domain in a finite time, for any sufficiently small* r *and* $\varepsilon > 0$.

It follows from Theorem 3.7.1 that condition **D** is satisfied if there exists in the domain $0 < |x| < r$ a function $W(t,x) \in C_2^0(\{t > 0\} \times U_r)$, such that for any ε, $0 < \varepsilon < r$,

$$W(t,x) \geqslant 0, \qquad LW(t,x) < -c_\varepsilon < 0, \quad \text{if} \quad |x| > \varepsilon. \tag{4.2}$$

In the following theorem, $U \subset E_\ell$ is some neighborhood of the origin.

THEOREM 4.1. *Suppose that there exists a positive definite function* $V(t,x) \in C_2^0(\{t > 0\} \times U)$, *which has an infinitesimal upper limit and satisfies* $LV \leqslant 0$. *Let condition* **D** *hold. Then the solution* $X(t) \equiv 0$ *of equation* (1.1) *is asymptotically stable in probability.*

PROOF. By Lemma 2.4, the stochastic process $V(\tau_U(t), X^{s,x}(\tau_U(t)))$ is a supermartingale. By Theorem 2.1, this implies that almost surely:

$$\lim_{t\to\infty} V(\tau_U(t), X^{s,x}(\tau_U(t))) = \xi. \tag{4.3}$$

Let B_x denote the set of sample paths of $X^{s,x}(t)$ such that $\tau_U = \infty$. Since the function V satisfies the assumptions of Theorem 3.1, the solution $X(t) \equiv 0$ is stable in probability, and consequently

$$P(B_x) \to 1 \quad \text{as} \quad x \to 0. \tag{4.4}$$

---

[5] The conditions for asymptotic stability and instability in Theorems 4.1 and 4.2 generalize corresponding results of Has'minskiĭ [2]. Conditions for stability in the large have been considered by Nevel'son [1], to whom, in particular, Theorem 4.4 is due. Stability in the large of stochastic systems in a different setting has been investigated by Kac [2].

[6] An analogous condition for deterministic systems was considered by Krasovskiĭ [1, p. 23] in connection with the inversion of Ljapunov's theorems on asymptotic stability and instability.

It follows from condition **D** that for all paths contained in the set B_x, except for a set of paths of probability zero, we have $\inf_{t>0} |X^{s,x}(t)| = 0$, and in view of Lemma 2.3 we have also the stronger relation

$$\varlimsup_{t\to\infty} |X^{s,x}(t)| = 0.$$

Since the function V has an infinitesimal upper limit, it follows that also $\varlimsup_{t\to\infty} V(t,X^{s,x}(t)) = 0$. But by (4.3) the limit

$$\lim_{t\to\infty} V(\tau_U(t),X^{s,x}(\tau_U(t))) = \lim_{t\to\infty} V(t,X^{s,x}(t))$$

exists for almost all paths in B_x. By the above discussion this limit is equal to zero. Since the function $V(t,x)$ is positive definite for paths in B_x, this implies that

$$\lim_{t\to\infty} |X(t)| = 0.$$

The assertion of the theorem follows now from this relation and (4.4).

COROLLARY 1. As it was mentioned above, condition **D** may be replaced by the requirement that there exists a function $W(t,x)$ satisfying the inequalities (4.2). The function $V(t,x)$ itself satisfies these inequalities if LV is negative definite. We have thus proved the following generalization of Ljapunov's theorem on asymptotic stability of deterministic systems: *The solution $X(t) \equiv 0$ of equation (1.1) is asymptotically stable in probability if there exists in the domain $\{t > 0\} \times U$ a positive definite function $V(t,x) \in C_2^0(\{t > 0\} \times U$, which has an infinitesimal upper limit, such that the function LV is negative definite in this domain.*

COROLLARY 2. Condition **D** always holds if the matrix $A(t,x)$ satisfies the nondegeneracy condition (3.1). Indeed, then the function $W = k - |x|^n$ satisfies conditions (4.2) for a suitable choice of k and n. This means that, if condition (3.1) holds, then the existence of a function $V(t,x)$ satisfying the assumptions of Theorem 3.1 and having an infinitesimal upper limit is also sufficient for asymptotic stability in probability of the solution $X(t) + 0$ of equation (1.1). This fact and Theorem 3.2 yield the following proposition: Assume that the coefficients b and σ_r are independent of t and that the nondegeneracy condition (3.1) is satisfied. Then, if the solution of equation (1.1) is stable in probability, it is also asymptotically stable in probability. This proposition can be generalized to temporally nonhomogeneous systems. The example of deterministic systems shows that condition (3.1) cannot be dropped (though it can be weakened).

As before, we let U_r denote the subset $\{|x| < r\}$ of E_ℓ.

THEOREM 4.2. *Assume that there exists a function* $V(t,x) \in \mathbf{C}_2^0(\{t > 0\} \times U_r)$ *such that*

$$LV \leqslant 0 \quad \text{as} \quad x \in U_r, \qquad x \neq 0, \tag{4.5}$$

$$\liminf_{x\to 0} \inf_{t>0} V(t,x) = \infty. \tag{4.6}$$

Let condition **D** *hold. Then the solution* $X(t) \equiv 0$ *of equation* (1.1) *is not stable in probability. Moreover, in this case the event*

$$\{\sup_{t>0} |X^{s,x}(t)| < r\}$$

has probability zero for all $s > 0$, $x \in U_r$.

PROOF. Let $\tau_{r,\varepsilon}$ denote the first time of reaching the set $\{|x| = r\} \cup \{|x| = \varepsilon\}$, $\tau_{r,\varepsilon}(t) = \min(\tau_{r,\varepsilon}, t)$. By (4.5) and Lemma 3.3.1

$$\mathbf{E}V(\tau_{r,\varepsilon}(t), X^{s,x}(\tau_{r,\varepsilon}(t))) \leqslant V(s,x)$$

holds in the domain $U_r \setminus U_\varepsilon$ for any $\varepsilon < r$. Letting $t \to \infty$ and using condition **D**, we conclude that $\mathbf{E}V(\tau_{r,\varepsilon}, X^{s,x}(\tau_{r,\varepsilon})) \leqslant V(s,x)$. Čebyšev's inequality implies now the estimate

$$\inf_{|x|<\varepsilon, t>0} V(t,x)\mathbf{P}\{\sup_{0<t<\tau^\varepsilon} |X^{s,x}(t)| < r\} < V(s,x),$$

where τ^ε is the first time the set $|x| = \varepsilon$ is reached. Since by Lemma 2.3, $\tau^\varepsilon \to \infty$ almost surely as $\varepsilon \to 0$, we infer the required assertion from the last inequality and (4.6), letting $\varepsilon \to 0$.

REMARK. Arguments similar to those used to deduce the corollary from Theorem 4.1 yield the following sufficient conditions for instability.

(1) The solution $X(t) \equiv 0$ of equation (1.1) is unstable if conditions (4.5), (4.6) and (3.1) hold in the domain $\{t > 0\} \times U_r$.

(2) The solution $X(t) \equiv 0$ of equation (1.1) is unstable if condition (4.6) holds and moreover $\sup_{\varepsilon<|x|<r} LV < 0$ for any $\varepsilon > 0$.

DEFINITION. The solution $X(t) \equiv 0$ of equation (1.1) is said to be *(asymptotically) stable in the large* if it is stable in probability and also for all s,x

$$\mathbf{P}\{\lim_{t\to\infty} X^{s,x}(t) = 0\} = 1.$$

THEOREM 4.3. *A sufficient condition for the solution* $X(t) \equiv 0$ *of equation* (1.1) *to be stable in the large is that it is uniformly stable in probability, and moreover the process* $X(t)$ *is recurrent relative to the domain* $|x| < \varepsilon$ *for any* $\varepsilon > 0$.

PROOF. Since the solution $X(t) \equiv 0$ is uniformly stable in probability, it follows that for any $\varepsilon > 0$ there exists a $\delta > 0$ such that

$$\sup_{s>0,|y|<\delta} \mathbf{P}\{\sup_{t>s} |X^{s,y}(t)| > \varepsilon\} < \varepsilon.$$

Let τ_δ denote the first time at which the path of the process reaches the set $|x| \leqslant \delta$. By assumption, $\tau_\delta < \infty$ almost surely. Using the strong Markov property of the process and choosing $\delta > 0$ such that $|x| > \delta$, we get

$$\mathbf{P}\{\overline{\lim_{t\to\infty}} |X^{s,x}(t)| > \varepsilon\}$$

$$= \int_{u=s}^{\infty}\int_{|y|=\delta} \mathbf{P}\{\tau_\delta \in du, X^{s,x}(\tau_\delta) \in dy\}\mathbf{P}\{\overline{\lim_{t\to\infty}} |X^{u,y}(t)| > \varepsilon\}$$

$$\leqslant \int_{u=s}^{\infty}\int_{|y|=\delta} \mathbf{P}\{\tau_\delta \in du, X^{s,x}(\tau_\delta) \in dy\}\mathbf{P}\{\sup_{t>u} |X^{u,y}(t)| > \varepsilon\}$$

$$\leqslant \varepsilon.$$

This implies the required assertion.

From Theorem 4.3 one readily derives various sufficient conditions for stability in the large in terms of Ljapunov functions. The following theorem generalizes to stochastic equations a well-known theorem of Barbasin and Krasovskiĭ [1].

THEOREM 4.4. *A sufficient condition for the solution* $X(t) \equiv 0$ *of equation* (1.1) *to be stable in the large is that there exists a positive definite function* $V(t,x) \in \mathbf{C}_2^0(E)$ *with an infinitesimal upper limit such that the function* LV *is negative definite and*

$$\inf_{t>0} V(t,x) \to \infty \quad \text{as } |x| \to \infty.$$

PROOF. We observe that under these assumptions the solution $X(t) \equiv 0$ is uniformly stable in probability, by virtue of the remark following Theorem 3.1. Moreover, by Lemma 3.8.1 and Theorem 3.7.1, this solution is recurrent relative to the domain $|x| < \varepsilon$ for any $\varepsilon > 0$.

THEOREM 4.5. *The following conditions are sufficient for the solution* $X(t) \equiv 0$ *of equation* (1.1) *to be stable in the large:*

(1) *the process* $X(t)$ *is regular;*

(2) *there exists a nonnegative function* $V_1(t,x) \in \mathbf{C}_2^0(E)$ *such that the function* LV_1 *is negative definite;*

(3) *there exists a positive definite function* $V_2(t,x) \in \mathbf{C}_2^0(E)$, *having an infinitesimal upper limit, such that* $LV_2 \leqslant 0$.

PROOF. The proof follows from Theorem 4.3 and from the above-mentioned theorems of Chapters III and V. Note that by Theorem 3.3.1 we can replace condition (1) by the following condition (1'): There exists a non-negative function $V_3(t,x) \in \mathbf{C}_2^0(E)$ such that $LV_3 < kV_3$ for some positive constant k and $\lim_{R\to\infty} \inf_{|x|>R} V_3 = \infty$. Similarly, (2) can be replaced by (2'): The nondegeneracy condition (3.1) holds in $U_R \setminus U_\varepsilon$ for any R and $\varepsilon < R$. (2) may be also replaced by the even weaker condition that $a_{ii}(t,x) > a_{R,\varepsilon} > 0$ for some i.

5. Examples

EXAMPLE 1. Consider the one-dimensional process described by the following Itô equation in E_ℓ:

$$dX(t) = b(t,X)dt + \sigma(t,X)d\xi(t). \tag{5.1}$$

Here the differential generator is

$$L = \frac{\partial}{\partial t} + b(t,x)\frac{\partial}{\partial x} + \frac{1}{2}\sigma^2(t,x)\frac{\partial^2}{\partial x^2}. \tag{5.2}$$

Suppose that the expansions

$$b(t,x) = b(t)x + o(|x|); \qquad \sigma(t,x) = \sigma(t)x + o(|x|) \tag{5.3}$$

hold in a neighborhood of $x = 0$ where, in accordance with the conditions of Section 1, the functions $b(t)$ and $\gamma(t)$ are bounded and the relations (5.3) hold uniformly in $t > 0$.

Assume that

$$\int_0^t \left[b(s) - \frac{\sigma^2(s)}{2} + \varepsilon\right] ds < k \tag{5.4}$$

holds for some $\varepsilon > 0$, $k > 0$ and all $t > 0$. Then for sufficiently small $\gamma > 0$ the auxiliary function

$$V_1(t,x) = |x|^\nu \exp\left\{ -\nu\int_0^t (b(s) - \frac{\sigma^2(s)}{2} + \varepsilon)ds \right\} = |x|^\nu V(t)$$

satisfies all the assumptions of Theorem 3.1. Indeed, that $V_1(t,x)$ is positive definite follows from (5.4). Moreover, by (5.2) and (5.3),

$$LV_1(t,x) = \nu|x|^\nu V(t)[-\varepsilon + \tfrac{1}{2}\nu\sigma^2(t)] + o(|x|^\nu).$$

Thus, if $\nu < \varepsilon/\sup_{t>0} \sigma^2(t)$, then the function $LV_1(t,x)$ is negative definite in a sufficiently small neighborhood of $x = 0$. Consequently, *the solution* $X \equiv 0$ *is stable in probability if condition* (5.4) *holds.*

Let us now assume that

$$\int_0^t \left[b(s) - \frac{\sigma^2(s)}{2} - \varepsilon\right]ds > -k. \tag{5.5}$$

holds for some $\varepsilon > 0$, $k > 0$ and all $t > 0$. Then the auxiliary function

$$V_2(t,x) = -\ln|x| + \int_0^t \left[b(s) - \frac{\sigma^2(s)}{2} - \varepsilon\right]ds$$

obviously satisfies condition (4.6). Moreover

$$LV_2(t,x) \leqslant -\varepsilon + o(1) \qquad (x \to 0).$$

Hence, by Remark 2 to Theorem 4.2, it follows that the trivial solution of equation (5.1) is unstable if condition (5.5) holds.

In the cases considered above, the stability (or instability) of the linearized system

$$dX(t) = b(t)Xdt + \sigma(t)Xd\xi(t)$$

implies the stability (or instability) of the full system (5.1). In the general case, however, this is not so. In Chapter VII we shall consider in greater detail the question of conditions under which the theorem on stability in the first approximation is valid.

We mention one other peculiar consequence of Example 1. Condition (5.4) is satisfied, in particular, by a system (5.1) in which the function $b(s)$ is positive, provided the difference $b(s) - \sigma^2(s)/2$ is smaller than a negative constant. Thus, the system $dx/dt = b(t,x)$, which is unstable (even in the linear approximation!) can be "stabilized" by introducing an additive stochastic term $\sigma(t,x)d\xi(t)$, if the "intensity" of the noise $\sigma^2(t,x)$ is sufficiently high. For example, the linear system with constant coefficients $dX = bXdt + \sigma X d\xi(t)$ is stable for $b < \sigma^2/2$. Writing this equation as

$$\dot{X} = (b + \sigma\xi)X, \tag{5.6}$$

one is tempted to interpret the results as follows: An unstable first-order deterministic system is stabilized if white noise of sufficiently high intensity is superimposed on its coefficient. This assertion is in conflict with physical intuition. Neither is it correct, if we define a solution of equation (5.6) to be the limit of a sequence of solutions $x_n(t)$ of the equations $\dot{x}_n = (b + \sigma\xi_n(t))x_n$, where $\xi_n(t)$ is a sequence of Gaussian processes whose autocorrelation functions converge to a δ-function. It can be shown (this follows, in particular, from certain results of Stratonovič [3] and Has'minskiĭ [8]) that in this case (and also under much more general conditions) the above limit procedure leads not to an Itô equation but to the analogous stochastic equation

$$dX(t) = bX(t)dt + \sigma X(t)d^*\xi(t),$$

where the stochastic differential $d^*\xi(t)$ is to be understood in the sense of Stratonovič [2].

The solution of the Itô equation (1.1) can be constructed as the limit in mean square as $h \to 0$ of solutions of the stochastic difference equations

$$X(t_n + h) - X(t_n)$$
$$= b(t_n, X(t_n))h + \sum_{r=1}^{k} \sigma_r(t_n, X(t_n))[\xi_r(t_n + h) - \xi_r(t_n)].$$

On the other hand, the solution of the analogous Stratonovič equation

$$dX(t) = b(t, X(t))dt + \sum_{r=1}^{k} \sigma_r(t, X(t))d^*\xi_r(t) \tag{5.7}$$

is defined as the limit in mean square as $h \to 0$ of the solutions of the finite-difference equations

$$X(t_n + h) - X(t_n)$$
$$= b(t_n, X(t_n))h + \sum_{r=1}^{k} \sigma_r\left(t_n, \frac{X(t_n) + X(t_n + h)}{2}\right)$$

$$\times(\xi_r(t_n + h) - \xi_r(t_n)). \qquad (5.7')$$

Stratonovič [2] showed that if the function $\sigma_r(t,x)$ is continuously differentiable with respect to x, then the above implicit difference scheme is equivalent, as $h \to 0$, to the explicit scheme

$$X(t_n + h) - X(t_n)$$

$$= \left[b(t_n, X(t_n)) + \frac{1}{2}\sum_{r=1}^{k}\frac{\partial\sigma_r(t_n, X(t_n))}{\partial x}\,\sigma_r(t_n, X(t_n))\right]h$$

$$+ \sum_{r=1}^{k}\sigma_r(t_n, X(t_n))[\xi_r(t_n + h) - \xi_r(t_n)]. \qquad (5.7'')$$

Consequently, the stochastic equation (5.7) is equivalent to the Itô equation

$$dX(t) = \left[b(t, X(t)) + \frac{1}{2}\sum_{r=1}^{k}\frac{\partial\sigma_r}{\partial x}\,\sigma_r(t, X(t))\right]dt$$

$$+ \sum_{r=1}^{k}\sigma_r(t, X(t))d\xi_r(t). \qquad (5.8)$$

Thus, the differential generator of the process defined by equation (5.7) is

$$L = \frac{\partial}{\partial t} + \left(b(t,x) + \frac{1}{2}\sum_{r=1}^{k}\frac{\partial\sigma_r}{\partial x}\,\sigma_r(t,x),\ \frac{\partial}{\partial x}\right)$$

$$+ \frac{1}{2}\sum_{r=1}^{k}\left(\sigma_r(t,x),\ \frac{\partial}{\partial x}\right)^2$$

$$= \frac{\partial}{\partial t} + \left(b(t,x),\ \frac{\partial}{\partial x}\right)$$

$$+ \frac{1}{2}\sum_{r=1}^{k}\left(\sigma_r(t,x),\ \frac{\partial}{\partial x}\left(\sigma_r(t,x),\ \frac{\partial}{\partial x}\right)\right). \qquad (5.9)$$

The reason for the equivalence of the difference schemes (5.7') and (5.7") may be explained briefly as follows. Since the vectors $\sigma_r(t,x)$ are continuously differentiable with respect to x, it follows that

$$\sigma_r\left(t_n, \frac{X(t_n) + X(t_n + h)}{2}\right)$$

$$= \sigma_r(t_n,X(t_n)) + \frac{1}{2}\left(\frac{\partial\sigma_r(t_n,X(t_n))}{\partial x} + o(1)\right)(X(t_n + h) - X(t_n))$$

as $h \to 0$. Substituting this relation into (5.7') and using the equality

$$\left(J - \sum_{r=1}^{k} \frac{\partial\sigma_r}{\partial x} \Delta\xi_r(t)\right)^{-1} = J + \sum_{r=1}^{k} \frac{\partial\sigma_r}{\partial x} \Delta\xi_r(t) + o(\Delta\xi_r(t)),$$

$$\text{as} \quad \Delta\xi_r(t) \to 0$$

(where J is the unit matrix and $\Delta\xi_r(t) = \xi_r(t + h) - \xi_r(t)$), we conclude that

$$X(t_n + h) - X(t_n)$$

$$= b(t_n,X(t_n))h + \sum_{r=1}^{k} \sigma_r(t_n,X(t_n))\Delta\xi_r(t_n)$$

$$+ \sum_{r=1}^{k} \sum_{j=1}^{k} \frac{\partial\sigma_r}{\partial x} (t_n,X(t_n))\sigma_j(t_n,X(t_n))\Delta\xi_r(t_n)\Delta\xi_j(t_n)$$

$$+ o(h). \tag{5.10}$$

It can be shown further that the terms corresponding to $j \neq r$ in the last sum are in a certain sense small quantities of higher order than h as $h \to 0$, owing to the mutual independence of $\xi_r(t)$ and $\xi_j(t)$ for $r \neq j$. Moreover, as $h \to 0$, the expression $(\Delta\xi_r(t_n))^2$ in (5.10) can be replaced by its expectation h. It should now be clear from (5.10) and these relations why (5.7') and (5.7") are equivalent as $h \to 0$.

It follows from the above considerations that in dealing with physical problems in which white noise is an idealization of a real process with small time correlation, it is often natural to regard the equation as a stochastic equation of type (5.7).

In particular, we see from (5.9) that in the one-dimensional case considered above the operator associated with equation (5.6), regarded as a Stratonovič equation, is

$$L = \frac{\partial}{\partial t} + bx \frac{\partial}{\partial x} + \frac{\sigma^2}{2} x \frac{\partial}{\partial x}\left(x \frac{\partial}{\partial x}\right)$$

$$= \frac{\partial}{\partial t} + \left(b + \frac{\sigma^2}{2}\right)x \frac{\partial}{\partial x} + \frac{\sigma^2}{2} x^2 \frac{\partial^2}{\partial x^2} .$$

Combining this with Example 5.1, we get the result that the process $X(t)$ defined by the equation

$$dX = bXdt + \sigma X d^{*}\xi$$

is stable for $b < 0$ and unstable for $b > 0$. This accords with physical intuition.

Thus, the unstable one-dimensional deterministic system $\dot{x} = bx$ ($b > 0$ is a constant) cannot be stabilized by a "physically feasible"[7] perturbation of its parameter. This was noticed by Leibowitz [1], who also conjectured that an analogous result holds in the many-dimensional case. However, the example presented below in Section 6.9 shows that this is not so.

EXAMPLE 2. Let the deterministic system

$$\frac{dx}{dt} = b(t,x) \tag{5.11}$$

be exponentially stable in the sense of Remark 1.6.1, and suppose that the function $b(t,x)$ has bounded first and second derivatives with respect to the space variables. Then, slightly modifying the proof of Theorem 11.1 in Krasovskiĭ [1], we easily see that there exists a function $W(t,x)$ for the system (5.11) such that

$$k_1|x|^2 < W(t,x) < k_2|x|^2,$$

$$\frac{d^0W}{dt} = \frac{\partial W}{\partial t} + \sum_{i=1}^{\ell} b_i(t,x)\frac{\partial W}{\partial x_i} < -k_3|x|^2;$$

$$\left|\frac{\partial W}{\partial x}\right| < k_4|x|; \qquad \left|\frac{\partial^2 W}{\partial x_i \partial x_j}\right| < k_5.$$

Using this Ljapunov function to investigate the stability of the "perturbed" system

$$dX(t) = b(t,X)dt + \sum_{r=1}^{k} \sigma_r(t,x)d\xi_r(t) + F(t,X)dt, \tag{5.12}$$

we get

$$LW = \frac{d^0W}{dt} + \frac{1}{2}\sum_{i,j=1}^{\ell} a_{ij}(t,x)\frac{\partial^2 W}{\partial x_i \partial x_j} + \sum_{i=1}^{\ell} F_i(t,x)\frac{\partial W}{\partial x_i}$$

[7] We shall continue to use this term in this sense in the sequel, though it is somewhat vague. For example, if a stochastic process $X(t)$ is a "continuous approximation" to a discrete Markov chain which is the solution of a finite-difference equation, it is natural to use an Itô equation (see Skorohod [1]).

$$\leq - k_3|x|^2 + k_5\|A(t,x)\| + k_4|x||F(t,x)|.$$

Hence it follows by Theorem 4.1 that the solution $X(t) \equiv 0$ of equation (5.12) is asymptotically stable in probability if the system (5.11) is exponentially stable in some neighborhood of the origin, and in this neighborhood

$$\sum_{r=1}^{k} |\sigma_r(t,x)| + |F(t,x)| < \varepsilon|x| \tag{5.13}$$

for sufficiently small ε. If the system (5.11) is exponentially stable in the large and condition (5.13) holds everywhere, it follows from Theorem 4.3 that the system (5.12) is stable in the large. Therefore in this example sufficiently strong stability of the deterministic system (5.11) implies stability of the system (5.12), provided condition (5.13) holds for $\varepsilon < \varepsilon_0$. It is not difficult to find effective estimates for ε_0. In particular, the above conclusion holds true for "physically feasible" random perturbations of the parameters of the system by white noise.

EXAMPLE 3. In our discussion of Example 1 we saw that the one-dimensional system $x' = bx$ $(b < 0)$ remains stable when its parameter is perturbed by white noise of arbitrary intensity. This holds true both for perturbations "of Itô's kind" and for physically feasible perturbations. We shall now show that if the dimension of the space is greater than 2, or in the case of physically feasible perturbations, greater than 1, then sufficiently strong isotropic noise will destroy the stability. To this end it will suffice to show that if σ is a sufficiently large constant, then for the systems

$$dX_i(t) = b_i(t,X)dt + \sigma \sum_{j=1}^{\ell} X_j d\xi_{(i-1)\ell+j}(t) \qquad (i = 1,\ldots,\ell),$$

$$d\tilde{X}_i(t) = b_i(t,\tilde{X})dt + \sigma \sum_{j=1}^{\ell} \tilde{X}_j d^*\xi_{(i-1)\ell+j}(t) \qquad (i = 1,\ldots,\ell)$$

(where $\xi_1(t),\ldots,\xi_{2\ell}(t)$ are independent Wiener processes), there exists a function $V(x)$ satisfying the assumptions of Theorem 4.2, provided the functions $b_i(t,x)$ satisfy conditions (1.2), (1.3). It is readily seen that the differential generators of the processes $X(t)$ and $\tilde{X}(t)$ are

$$L = \frac{\partial}{\partial t} + \sum_{i=1}^{\ell} b_i(t,x)\frac{\partial}{\partial x_i} + \frac{\sigma^2}{2}|x|^2 \sum_{i=1}^{\ell} \frac{\partial^2}{\partial x_i^2},$$

$$\tilde{L} = L + \frac{\sigma^2}{2} \sum_{i=1}^{\ell} x_i \frac{\partial}{\partial x_i} .$$

Considering the auxiliary function

$$V(x) = -\ln|x|^2 = -\ln(x_1^2 + \ldots + x_\ell^2)$$

and assuming that the constant σ is sufficiently large, we see by (1.2) and (1.3) that

$$LV = -\frac{2(x,b(t,x))}{|x|^2} - \sigma^2(\ell - 2) < 0 \quad \text{for} \quad \ell > 2,$$

$$\tilde{L}V = -\frac{2(x,b(t,x))}{|x|^2} - \sigma^2(\ell - 1) < 0 \quad \text{for} \quad \ell > 1.$$

Applying Theorem 4.2 and the subsequent Remark 2, we get the above assertions.

We shall now prove that the "Itô perturbations" considered in this example do not destroy the stability of the asymptotically stable system $dx_i = b_i x_i dt$ ($b_i < 0$; $i = 1,2$) for any value of σ. In this case,

$$L = \frac{\partial}{\partial t} + b_1 x_1 \frac{\partial}{\partial x_1} + b_2 x_2 \frac{\partial}{\partial x_2} + \frac{\sigma^2}{2} |x|^2 \sum_{i=1}^{2} \frac{\partial^2}{\partial x_i^2} .$$

Considering the auxiliary function

$$V(x) = |x|^\alpha = (x_1^2 + x_2^2)^{\alpha/2}$$

for sufficiently small positive α, we readily obtain the inequality

$$LV(x) = \alpha|x|^{\alpha-2}\left\{b_1 x_1^2 + b_2 x_2^2 + \frac{\alpha\sigma^2}{2} |x|^2\right\} < 0.$$

Hence, it follows by Theorems 4.1 and 4.3 that the system is asymptotically stable in the large.

EXAMPLE 4. Consider the system

$$dX_1 = X_2 dt + \sigma(X_1,X_2) d\xi_1(t);$$

$$dX_2 = -X_1 dt + \sigma(X_1,X_2) d\xi_2(t).$$

It is clear that in the absence of random perturbations ($\sigma \equiv 0$) the equilibrium position of this system is stable, but not asymptotically stable. The differential generator is

$$L = x_2 \frac{\partial}{\partial x_1} - x_1 \frac{\partial}{\partial x_2} + \frac{1}{2} \sigma^2(x)\left(\frac{\partial^2}{\partial x_1^2} + \frac{\partial^2}{\partial x_2^2}\right) .$$

It is obvious that the function $W(x) = -\ln(x_1^2 + x_2^2)$ satisfies conditions (4.5), (4.6). Consequently, the system is unstable if $\sigma(x) \neq 0$ for $x \neq 0$. This example shows that a non-asymptotically stable deterministic system may become unstable when driven by white noise whose intensity tends arbitrarily fast to zero as $x \to 0$.

6. Differentiability of solutions of stochastic equations with respect to the initial conditions

We have already seen (Section 3) that Ljapunov functions satisfying the assumptions of the stability theorems can be constructed as expectations of certain functionals of the relevant processes. However, only in the nondegenerate case can one guarantee, using the theory of partial differential equations, the necessary smoothness of these expectations. Gihman [1] and Blagoveščenskiĭ and Freĭdlin [1] have demonstrated an alternative approach: One first proves that the solution $X^{s,x}(t)$ of the stochastic equation is smooth with respect to s,x, and then the smoothness of the corresponding expectations follows as a corollary. This approach is applicable to processes with diffusion of arbitrary degree of degeneracy; for this reason it imposes stringent restrictions on the smoothness of the coefficients b and σ_r.

In this section we shall present Gihman's theorem on the differentiability of the solutions of stochastic equations with respect to the initial conditions; we shall then establish certain auxiliary relations which will be used in Section 7 to prove the existence of Ljapunov functions for certain stable systems.

THEOREM 6.1. *Let the coefficients of the equation*

$$dX^{s,x}(t) = b(t,X^{s,x})dt + \sigma(t,X^{s,x})d\xi(t) \tag{6.1}$$

in E_ℓ be continuous in t,x and with continuous bounded derivatives of order up to and including 2 *with respect to $x_1 .\,.\,.\,.\,. x_\ell$. Then the solution $X^{s,x}(t)$ of equation* (8.1) *is twice continuously differentiable in mean square with respect to x* [8]. *The*

[8] If $\Phi(x_1,\ldots,x_\ell,t)$ is a random function depending on the parameters $x_1,\ldots,x_\ell,t$, its partial derivative in mean square with respect to x_i is defined as the random variable $\frac{\partial \Phi}{\partial x_i}(x_1,\ldots,x_\ell,t)$ such that

$$\mathbf{E}\left\{\frac{1}{\Delta x_i}\left[\Phi(x_1,\ldots,x_i + \Delta x_i,\ldots,x_\ell,t) - \Phi(x_1,\ldots,x_\ell,t)\right]\right. \text{(Contd)}$$

derivatives

$$\frac{\partial}{\partial x_i} X^{s,x}(t), \qquad \frac{\partial^2}{\partial x_i \partial x_j} X^{s,x}(t)$$

are then continuous in x in mean square. They are defined by the system obtained by formally differentiating (6.1) *with respect to x.*

We shall not give here all the details of the proof of Theorem 6.1; the interested reader can find them in Gihman and Skorohod [1]. We describe the idea of the proof, incidentally explaining some of the arguments in Gihman and Skorohod [1] which will be needed later.

To avoid cumbersome notation, we limit ourselves to the case in which the dimension ℓ of the space E_ℓ is 1. It is easy to see that the stochastic process

$$Y_{x,\Delta x}(t) = \frac{1}{\Delta x}[X^{s,x+\Delta x}(t) - X^{s,x}(t)]$$

is a solution of the equation

$$Y_{x,\Delta x}(t) = 1 + \int_s^t A(x,\Delta x,u) Y_{x,\Delta x}(u)du$$

$$+ \int_s^t B(x,\Delta x,u) Y_{x,\Delta x}(u)d\xi(u), \tag{6.2}$$

where

$$A(x,\Delta x,t) = \frac{b(t,X^{s,x+\Delta x}(t)) - b(t,X^{s,x}(t))}{X^{s,x+\Delta x}(t) - X^{s,x}(t)},$$

$$B(x,\Delta x,t) = \frac{\sigma(t,X^{s,x+\Delta x}(t)) - \sigma(t,X^{s,x}(t))}{X^{s,x+\Delta x}(t) - X^{s,x}(t)}.$$

By the assumptions of the theorem, the functions A and B are almost surely bounded by some constant k.

For any $n \geqslant 1$, we apply Itô's formula (3.3.8) to the process $Z(x,\Delta x,t) = [Y_{x,\Delta x}(t)]^{2n}$ and thus get from (6.2) the relation

---

[8] (Contd)

$$- \frac{\partial \Phi}{\partial x_i}(x_1,\ldots,x_\ell,t)\Big\}^2 \to 0$$

as $\Delta x_i \to 0$.

$$Z(x,\Delta x,t) = 1 + n\int_s^t Z(x,\Delta x,u)[2A(x,\Delta x,u) + (2n-1)B^2(x,\Delta x,u)]du$$

$$+ 2n\int_s^t Z(x,\Delta x,u)B(x,\Delta x,u)d\xi(u).$$

As in the proof of Lemma 2.2, we now calculate expectations on both sides of this equality and apply the Gronwall-Bellman lemma, to get the inequality

$$\mathbf{E}[Y_{x,\Delta x}(t)]^{2n} \leqslant e^{k(t-s)}, \tag{6.3}$$

where the constant k depends only on the lowest upper bounds of σ_x' and b_x' and the number n. In particular it follows from (6.3) that $X^{s,x+\Delta x}(t) \to X^{s,x}(t)$ in probability as $\Delta x \to 0$. Hence the coefficients A and B of equation (6.2) converge in probability as $\Delta x \to 0$ to the functions $b_x'(u,X^{s,x}(u))$ and $\sigma_x'(u,X^{s,x}(u))$, respectively. Since the functions A,B,b_x' and σ_x' are also bounded, it follows that all moments of the differences $A - b_x'$ and $B - \sigma_x'$ converge to zero. Hence, as before, we readily conclude that $Y_{x,\Delta x}(t)$ converges in mean square as $\Delta x \to 0$ to a solution of the equation

$$\zeta_x(t) = 1 + \int_s^t b_x'(u,X^{s,x}(u))\zeta_x(u)du$$

$$+ \int_s^t \sigma_x'(u,X^{s,x}(u))\zeta_x(u)d\xi(u). \tag{6.4}$$

By definition, the process $\zeta_x(t)$ is equal to $\frac{\partial}{\partial x} X^{s,x}(t)$. It is also easy to see on the basis of (6.2), (6.3) and (6.4) that for any integer $n \geqslant 1$

$$\left.\begin{aligned} &M[\zeta_x(t)]^{2n} \leqslant e^{k(t-s)}, \\ &M[Y_{x\Delta x}(t) - \zeta_x(t)]^{2n} \to 0 \quad \text{as} \quad \Delta x \to 0. \end{aligned}\right\} \tag{6.5}$$

Similar arguments prove the existence and continuity of the second derivatives.

LEMMA 6.1. *Let the coefficients of equation* (6.1) *be continuous in* t,x *and satisfy the conditions*

$$\sigma(t,0) \equiv 0, \qquad b(t,0) \equiv 0. \tag{6.6}$$

Suppose also that they have continuous bounded first and second partial derivatives with respect to $x_1,\dots,x_\ell$. Then for any real β the function $u(s,x) = \mathbf{E}|X^{s,x}(t)|^\beta$ is twice continuously differentiable with respect to $x_1,\dots,x_\ell$, except perhaps at $x = 0$. We also have then

$$\left.\begin{aligned} \left|\frac{\partial u(s,x)}{\partial x}\right| &\leqslant k_1|x|^{\beta-1}e^{k_2(t-s)}, \\ \left|\frac{\partial^2 u(s,x)}{\partial x_i \partial x_j}\right| &\leqslant k_1|x|^{\beta-2}e^{k_2(t-s)} \end{aligned}\right\} \tag{6.7}$$

for some $k_1 > 0$, $k_2 > 0$.

PROOF. As before, we consider only the case $\ell = 1$, leaving it to the reader to extend the argument to more dimensions. By formal differentiation, we get

$$u'_x(s,x) = \beta\mathbf{E}\left[|X^{s,x}(t)|^{\beta-2}X^{s,x}(t)\,\frac{\partial X^{s,x}(t)}{\partial x}\right]. \tag{6.8}$$

The existence of the expectation on the right of (6.8) follows from Lemma 2.2, Theorem 6.1 and the estimate

$$|X^{s,x}(t)|^{\beta-1}\left|\frac{\partial X^{s,x}(t)}{\partial x}\right| < |X^{s,x}(t)|^{2\beta-2} + \left|\frac{\partial X^{s,x}(t)}{\partial x}\right|^2.$$

We first consider the case $\beta = 2$. Then

$$\left|\frac{u(s,x+\Delta x) - u(s,x)}{\Delta x} - 2\mathbf{E}\left[X^{s,x}(t)\,\frac{\partial X^{s,x}(t)}{\partial x}\right]\right|$$

$$\leqslant \left(\mathbf{E}\left[\frac{(X^{s,x+\Delta x}(t))^2 - (X^{s,x}(t))^2}{\Delta x} - 2X^{s,x}(t)\,\frac{\partial X^{s,x}(t)}{\partial x}\right]^2\right)^{1/2}$$

$$= \left\{\mathbf{E}\left\{2X^{s,x}(t)\left(Y_{x,\Delta x}(t) - \frac{\partial X^{s,x}(t)}{\partial x}\right) + [Y_{x,\Delta x}(t)]^2\Delta x\right\}^2\right\}^{1/2}$$

$$\leqslant \left\{32\left[\mathbf{E}(X^{s,x}(t))^4\mathbf{E}\left(Y_{x,\Delta x}(t) - \frac{\partial X^{s,x}(t)}{\partial x}\right)^4\right]^{1/2} + 2\mathbf{E}[Y_{x,\Delta x}(t)]^4(\Delta x)^2\right\}^{1/2}.$$

Hence, by the estimates (6.3) and (6.5), it follows that the derivative $\frac{\partial}{\partial x}\mathbf{E}|X^{s,x}(t)|^2$ exists, and also that

$$\sup_{s<t\leqslant s+T} \mathbf{E}\left(\frac{\Delta y}{\Delta x}\right)^2 < K,$$

$$\mathbf{E}\left[\frac{\Delta y}{\Delta x} - 2X^{s,x}(t)\,\frac{\partial X^{s,x}(t)}{\partial x}\right]^2 \to 0 \quad \text{as} \quad \Delta x \to 0, \tag{6.9}$$

where T is any positive number and we have set

$$y = [X^{s,x}(t)]^2; \qquad y + \Delta y = [X^{s,x+\Delta x}(t)]^2. \tag{6.10}$$

Using (6.3), (6.5) and Lemma 2.2, we readily obtain

$$\mathbf{E}(\Delta y)^4 \to 0 \quad \text{as} \quad \Delta x \to 0. \tag{6.11}$$

Now let $\beta \neq 2$. Then, again using the notation (6.10), we get

$$\left|\frac{u(s,x+\Delta x) - u(s,x)}{\Delta x} - \beta\mathbf{E}\left[|X^{s,x}(t)|^{\beta-2}X^{s,x}(t)\,\frac{\partial X^{s,x}(t)}{\partial x}\right]\right|$$

$$= \left|\mathbf{E}\left\{\left[\frac{(y+\Delta y)^{\beta/2} - y^{\beta/2}}{\Delta y} - \frac{\beta}{2}\,y^{\beta/2-1}\right]\frac{\Delta y}{\Delta x}\right.\right.$$

$$\left.\left. + \frac{\beta}{2}\,y^{\beta/2-1}\left[\frac{\Delta y}{\Delta x} - 2X^{s,x}(t)\,\frac{\partial X^{s,x}(t)}{\partial x}\right]\right\}\right|$$

$$\leqslant \left\{\mathbf{E}\left[\frac{(y+\Delta y)^{\beta/2} - y^{\beta/2}}{\Delta y} - \frac{\beta}{2}\,y^{\beta/2-1}\right]^2 \mathbf{E}\left(\frac{\Delta y}{\Delta x}\right)^2\right\}^{1/2}$$

$$+ \frac{\beta}{2}\left\{\mathbf{E}y^{\beta-2}\mathbf{E}\left(\frac{\Delta y}{\Delta x} - 2X^{s,x}(t)\,\frac{\partial X^{s,x}(t)}{\partial x}\right)^2\right\}^{1/2}. \tag{6.12}$$

By Lemma 2,2 and (6.9), the second term on the right of (6.12) tends to zero as $\Delta x \to 0$. It also follows from (6.9) that the function $E(\Delta y/\Delta x)^2$ is bounded. We claim that the function

$$J = \frac{(y+\Delta y)^{\beta/2} - y^{\beta/2}}{\Delta y} - \frac{\beta}{2}\,y^{\beta/2-1}$$

converges to zero in mean square.

Since $y > 0$, $y + \Delta y > 0$, it follows that

$$|J| < k|\Delta y|\,[(y+\Delta y)^{\beta/2-2} + y^{\beta/2-2}],$$

for some $k > 0$ [9], and hence the relation

[9] This estimate follows from the obvious inequality ($u > 0$)
(Contd)

$$EJ^2 \leqslant k_1^2(\mathbf{E}|\Delta y|^4)^{1/2}\{[\mathbf{E}(y + \Delta y)^{2\beta-8}]^{1/2} + (\mathbf{E}y^{2\beta-8})^{1/2}\}$$

holds. The last estimate, (6.11) and Lemma 2.2, imply that $\lim_{\Delta x \to 0} \mathbf{E}J^2 = 0$. Hence, by (6.12), we conclude that the first derivative of $u(x)$ with respect to x exists and is given by formula (6.8). Using (6.5) and (2.3), we see from (6.8) that also

$$|u'_x(s,x)| < |\beta|\mathbf{E}\left\{|X^{s,x}(t)|^{\beta-1}\left|\frac{\partial X^{s,x}(t)}{\partial x}\right|\right\}$$

$$\leqslant |\beta|(\mathbf{E}|X^{s,x}(t)|^{2\beta-2})^{1/2}\left(\mathbf{E}\left|\frac{\partial X^{s,x}(t)}{\partial x}\right|^2\right)^{1/2}$$

$$\leqslant k_1|x|^{\beta-1}e^{k_2(t-s)}.$$

Similarly one proves the existence and continuity of the second derivative u''_{xx} and the estimate (6.7).

LEMMA 6.2. *Under the assumptions of Lemma* 6.1, *the function* $u(s,x) = \mathbf{E}|X^{s,x}|^\beta$ *is differentiable with respect to* s. *Moreover, for* $x \neq 0$

$$Lu = \frac{\partial u}{\partial s} + \frac{1}{2}\sum_{i,j=1}^{\ell} a_{ij}(s,x)\frac{\partial^2 u}{\partial x_i \partial x_j} + \sum_{i=1}^{\ell} b_i(s,x)\frac{\partial u}{\partial x_i} = 0.$$

PROOF. The proof differs only in details from that of Theorem 5.1, Chapter VIII of Gihman and Skorohod [1]. We shall therefore confine the present discussion to those parts of the proof which are new. As before, we consider only the case $\ell = 1$. Expressing the difference $u(s + \Delta s, X^{s,x}(s + \Delta s)) - u(s + \Delta s, x)$ by means of Itô's formula and using the identity $\mathbf{E}u(s + \Delta s, X^{s,x}(s + \Delta s)) = u(s,x)$, we get

$$u(s,x) - u(s + \Delta s,x)$$

$$= \mathbf{E}\int_s^{s+\Delta s}\left[\frac{\partial}{\partial x}u(s + \Delta s, X^{s,x}(t))b(t,X^{s,x}(t))\right.$$

---

9 (Contd)

$$\frac{|u^{\beta/2} - 1 - (\beta/2)u^{\beta/2-1}(u - 1)|}{(u - 1)^2} < k(u^{\beta/2-2} + 1),$$

if we set $u = (y + \Delta y)/y$.

$$+ \tfrac{1}{2}\sigma^2(t,X^{s,x}(t))\frac{\partial^2}{\partial x^3} u(s + \Delta s,X^{s,x}(t))\Bigg] dt. \tag{6.13}$$

Next, using the explicit expressions for u, $\partial u/\partial x$ and $\partial^2 u/\partial x^2$ and proceeding as in the case of (6.7), we derive the estimates

$$|u_x^{(i)}(s + \Delta s,x) - u_x^{(i)}(s,x)| \leqslant k\Delta s|x|^{\beta-i} \qquad (i = 0,1,2).$$

Applying these estimates and (6.7), we easily deduce the assertion of the lemma from (6.13) by letting $\Delta s \to 0$.

COROLLARY. The function

$$V(s,x) = \int_s^{s+T} \mathbf{E}|X^{s,x}(t)|^\beta dt$$

is in class $\mathbf{C}_2^0(E)$, and

$$LV(s,x) = \mathbf{E}|X^{s,x}(s + T)|^\beta - |x|^\beta.$$

PROOF. We set $u_t(s,x) = \mathbf{E}|X^{s,x}(t)|^\beta$. Differentiating V with respect to s and applying Lemma 6.2, we get

$$LV(s,x) = \mathbf{E}|X^{s,x}(s + T)|^\beta - \mathbf{E}|X^{s,x}(s)|^\beta + \int_s^{s+T} Lu_t(s,x)dt$$

$$= \mathbf{E}|X^{s,x}(s + T)|^\beta - |x|^\beta,$$

as required.

7. Exponential p-stability and q-instability[10]

The solution $X(t) \equiv 0$ of the system

$$dX(t) = b(t,X)dt + \sum_{r=1}^{k} \sigma_r(t,X)d\xi_r(t) \tag{7.1}$$

in E_ℓ is said to be

[10] Theorems 7.1 and 7.2 are due to Nevel'son and Has'minskiĭ [2] (the second theorem is proved there under slightly more restrictive conditions).

(1) *p-stable* ($p > 0$) for $t \geq 0$, if

$$\sup_{|x| \leq \delta, t \geq s} \mathbf{E}|X^{s,x}(t)|^p \to 0 \quad \text{as} \quad \delta \to 0 \qquad (s \geq 0);$$

(2) *asymptotically p-stable*, if it is p-stable and moreover $\mathbf{E}|X^{s,x}(t)|^p \to 0$ as $t \to \infty$;

(3) *exponentially p-stable*, if for some positive constants A and α

$$\mathbf{E}|X^{s,x}(t)|^p \leq A|x|^p \exp\{-\alpha(t - s)\}. \tag{7.2}$$

The case most frequently considered in the literature to date is that of p-stability for $p = 1$ (stability in the mean) and for $p = 2$ (stability in mean square).

The following two theorems give necessary and sufficient conditions for exponential p-stability of stochastic systems in terms of Ljapunov functions. They may be regarded as generalizations of well-known theorems for deterministic systems (see Krasovskiĭ [1, Section 11]).

THEOREM 7.1. *The trivial solution of the system* (7.1) *is exponentially p-stable for* $t \geq 0$ *if there exists a function* $V(t,x)$ *of class* $\mathbf{C}_2^0(E)$ *such that*

$$k_1|x|^p \leq V(t,x) \leq k_2|x|^p, \tag{7.3}$$

$$LV(t,x) \leq -k_3|x|^p \tag{7.4}$$

for certain positive constants k_1, k_2, k_3.

PROOF. Conditions (7.3) and (7.4) are sufficient for the process $X(T)$ to be regular, since the function $V(t,x)$ satisfies the assumptions of Theorem 3.4.1. It follows from the same theorem that $EV(t,X^{s,x}(t))$ exists for all $t > s$. Expressing the difference $V(t,X^{s,x}(t)) - V(s,x)$ by means of Itô's formula (3.3.13), calculating expectations and using conditions (7.3) and (7.4), we get

$$\mathbf{E}V(t,X^{s,x}(t)) - V(s,x) = \int_s^t \mathbf{E}LV(u,X^{s,x}(u))\,du.$$

Differentiating this equality with respect to t and using (7.3), (7.4), we see that

$$\frac{d}{dt}\mathbf{E}V(t,X^{s,x}(t)) \leq -\frac{k_3}{k_2}\mathbf{E}V(t,X^{s,x}(t))$$

This implies the estimate

$$\mathbf{E}V(t, X^{s,x}(t)) \leqslant V(s,x)\exp\left\{-\frac{k_3}{k_2}(t-s)\right\}.$$

Together with (7.3), this estimate yields (7.2). The proof is complete.

THEOREM 7.2. *If the solution* $X(t) \equiv 0$ *of the system* (7.1) *is exponentially p-stable and the coefficients* b *and* σ_r *have continuous bounded derivatives with respect to* x *up to second order, then there exists a function* $V(t,x) \in \mathbf{C}_2^0(E)$ *satisfying inequalities* (7.3), (7.4) *and also*

$$\left|\frac{\partial V}{\partial x_i}\right| < k_4|x|^{p-1}, \qquad \left|\frac{\partial^2 V}{\partial x_i \partial x_j}\right| < k_4|x|^{p-2} \tag{7.5}$$

for some $k_4 > 0$.

PROOF. We claim that the function

$$V(t,x) = \int_t^{t+T} \mathbf{E}|X^{t,x}(u)|^p du \tag{7.6}$$

satisfies all the conditions of the theorem for suitable choice of the constant $T > 0$. Indeed, by (7.2),

$$V(t,x) \leqslant \int_t^{t+T} A|x|^p \exp\{-\alpha(u-t)\}du = k_1|x|^p.$$

Since the coefficients b and σ_r have bounded partial derivatives with respect to x_i, while $\sigma_r(t,0) = 0$, $b(t,0) = 0$, we have

$$|a_{ij}(t,x)| < k_5|x|^2, \qquad |b_i(t,x)| < k_5|x|.$$

Hence it follows that

$$|L(|x|^p)| < k_6|x|^p. \tag{7.7}$$

Applying Itô's formula to the function $|x|^p$ and using (7.7), we get

$$\mathbf{E}|X^{t,x}(t+T)|^p - |x|^p = \int_t^{t+T} \mathbf{E}L(|X^{t,x}(u)|^p)du$$

$$\geqslant -k_6\int_t^{t+T} \mathbf{E}|X^{t,x}(u)|^p du = -k_6V(t,x).$$

Choosing T so that

$$\mathbf{E}|X^{t,x}(t+T)|^p < \frac{1}{2}|x|^p, \tag{7.8}$$

we thus get the inequality $V(t,x) > |x|^p/(2k_6)$. This proves (7.3). To prove the required smoothness of $V(t,x)$ and to verify (7.4), we apply the Corollary to Lemma 6.2 and (7.8). Finally, using (6.7), we derive the estimate

$$\left|\frac{\partial V(t,x)}{\partial x_i}\right| = \int_t^{t+T} \frac{\partial}{\partial x_i} \mathbf{E}|X^{t,x}(u)|^p du$$

$$\leqslant \int_t^{t+T} k_1|x|^{p-1}\exp\{k_2(u-t)\}du = k_4|x|^{p-1}.$$

The second part of (7.5) is proved in similar fashion. This completes the proof.

The next lemma is useful in investigations of the stabilization of stochastic systems (see Chapter VIII).

LEMMA 7.1. *Assume that the coefficients* $b(t,x)$ *and* $\sigma_r(t,x)$ *satisfy the conditions of Theorem* 6.2, *and moreover that*

$$\int_s^\infty \mathbf{E}|X^{s,x}(t)|^p dt < \infty. \tag{7.9}$$

Then

$$\lim_{t\to\infty} \mathbf{E}|X^{s,x}(t)|^p = 0. \tag{7.10}$$

PROOF. By Remark 2 in Section 2, we can apply Itô's formula (3.3.8) to the function $|x|^p$. Doing this and using the estimate (7.7), which follows from the assumptions of our lemma, we see that

$$|\mathbf{E}|X^{s,x}(t+h)|^p - \mathbf{E}|X^{s,x}(t)|^p| < k\int_t^{t+h} \mathbf{E}|X^{s,x}(u)|^p du$$

holds for some constant $k > 0$. Thus,

$$\left|\frac{\partial}{\partial t}\mathbf{E}|X^{s,x}(t)|^p\right| \leqslant k\mathbf{E}|X^{s,x}(t)|^p. \tag{7.11}$$

Inequalities (7.9) and (7.11) obviously imply (7.10).

We now consider the concept of q-instability.

The trivial solution of the system (7.1) is said to be *exponentially q-unstable* ($q > 0$) if

$$\mathbf{E}|X^{s,x}(t)|^{-q} < A|x|^{-q}\exp\{-\alpha(t-s)\}$$

for some positive constants A and α. Similarly, we modify the other definitions at the beginning of this section; we replace p by $-q$ and a neighborhood of 0 by a neighborhood of the point at infinity, to get the definitions of *q-instability* and *asymptotic q-instability*.

It is clear that asymptotic q-instability for some $q > 0$ implies instability in probability, since, by Čebyšev's inequality,

$$\mathbf{P}\{|X^{s,x}(t)| < R\} < R^q\mathbf{E}|X^{s,x}(t)|^{-q}$$

for any $R > 0$. In order to avoid the difficulties created by possible irregularity of the process $X(t)$, we shall assume till the end of this section that the coefficients b and σ_r of equation (7.1) have bounded derivatives with respect to the space variables.

Later, in connection with the problem of instability in the first approximation, we shall be especially interested in the investigation of exponential q-instability. The proofs of the following two theorems are almost word-for-word repetitions of those of Theorems 7.1 and 7.2.

THEOREM 7.3. *The solution* $X(t) \equiv 0$ *of the system* (7.1) *is exponentially q-unstable for* $t \geq 0$ *if there exists a function* $V(t,x)$ *of class* $\mathbf{C}_2^0(E)$ *such that*

$$k_1|x|^{-q} \leq V(t,x) \leq k_2|x|^{-q};$$

$$LV(t,x) \leq -k_3|x|^{-q}. \tag{7.12}$$

THEOREM 7.4. *If the coefficients* b *and* σ_r *have continuous bounded derivatives with respect to* x *up to second order, and the solution* $X(t) \equiv 0$ *of the system* (7.1) *is exponentially q-unstable, then there exists a function* $V(t,x)$ *satisfying inequalities* (7.12) *and the inequalities*

$$\left|\frac{\partial V}{\partial x_i}\right| \leq k_4|x|^{-q-1}; \qquad \left|\frac{\partial^2 V}{\partial x_i \partial x_j}\right| < k_4|x|^{-q-2}.$$

REMARK 1. It follows from Theorems 7.2 and 4.4 that the solution $X(t) \equiv 0$ of the system (7.1) is asymptotically stable in the large if it is exponentially p-stable for some $p > 0$, and the functions $b(t,x)$ and $\sigma_r(t,x)$ have continuous bounded derivatives with respect to x of order up to 2 inclusive.

REMARK 2. Let the function $V(t,x) \in \mathbf{C}_2^0(E)$ be positive definite and such that $V(t,x) < k|x|^p$. Suppose moreover that $LV(t,x) \geqslant 0$. Then the system (7.1) is not asymptotically p-stable. Indeed, we deduce from Lemma 2.4 and Remark 2 at the end of Section 2, by applying Itô's formula and then taking expectations, that

$$k\mathbf{E}|X^{s,x}(t)|^p > \mathbf{E}V(t,X^{s,x}(t)) \geqslant V(s,x).$$

8. Almost sure exponential stability

Kozin [3] raises the question of finding conditions under which almost all solutions of the system (7.1) are exponentially stable. He proves that a sufficient condition is that the trivial solution be exponentially stable in mean square. Using a different method, we shall prove both this and a more general result.

THEOREM 8.1. *Under the assumptions of Theorem 7.1, there exists a constant* $\gamma > 0$ *such that, if* $x \in E_\ell$, $s \geqslant 0$, *the inequality* $|X^{s,x}(t)| < K_{s,x}e^{-\gamma t}$ *holds almost surely for* $t \geqslant s$, *where the random variable* $K_{s,x}$ *is almost surely finite.*

PROOF. Setting

$$W(t,x) = V(t,x)\exp\left\{\frac{k_3 t}{k_2}\right\}$$

we see by (7.3) and (7.4) that for $x \neq 0$

$$LW = \frac{k_3}{k_2}\exp\left\{\frac{k_3 t}{k_2}\right\}V + \exp\left\{\frac{k_3 t}{k_2}\right\}LV \leqslant 0.$$

Hence the process $W(t,X^{s,x}(t))$ is a supermartingale. Since it is positive, it follows from Theorem 2.1 that for all s,x the process $W(t,X^{s,x}(t))$ converges almost surely to a finite limit as $t \to \infty$. Consequently,

$$\sup_t W(t,X^{s,x}(t)) = A_{s,x} < \infty$$

almost surely. Therefore,

$$V(t,X^{s,x}(t)) \leqslant A_{s,x}e^{-\gamma t}.$$

This together with condition (7.3) implies the required assertion. In exactly the same way one proves the following

THEOREM 8.2. Under the assumptions of Theorem 7.3, there

exists a constant $\gamma > 0$ such that for any $x \neq 0$, $s \geq 0$

$$|X^{s,x}(t)| > K_{s,x}e^{\gamma t}$$

holds almost surely for $t \geq s$, where the random variable $K_{s,x}$ is almost surely positive.

CHAPTER VI

SYSTEMS OF LINEAR STOCHASTIC EQUATIONS

1. One-dimensional systems

In this chapter we shall study a linear homogeneous system of equations whose coefficients are perturbed by Gaussian white noise $\dot{\eta}_i^j(t)$. A system of this type can be written as

$$\frac{dX_i}{dt} = \sum_{j=1}^{\ell} (b_i^j(t) + \dot{\eta}_i^j(t))X_j(t). \tag{1.1}$$

The white noise processes $\dot{\eta}_i^j(t)$ figuring here are generalized Gaussian stochastic processes with zero mean and covariance matrix

$$\mathbf{E}[\dot{\eta}_i^j(s)\dot{\eta}_n^m(t)] = k_{ij}^{mn}(t)\delta(t - s),$$

where $\delta(t)$ is the Dirac δ-function, $i,j,m,n = 1,\ldots,\ell$. It is well known that the dependent white noise processes $\dot{\eta}_i^j(t)$ may be replaced by linear combinations of at most ℓ^2 independent processes.

We append a nonrigorous justification of the last statement (it can be made rigorous within the framework of the theory of generalized stochastic processes of Gel'fand [2] and Itô [3]). Let $\dot{\eta}_1(t),\ldots,\dot{\eta}_n(t)$ be Gaussian white noise processes with covariance matrix $k_{ij}(t)$ $(t - s)$. Let $\lambda_1(t),\ldots,\lambda_n(t)$ denote the eigenvalues and $f_1(t),\ldots,f_n(t)$ the associated normalized eigenvectors of the matrix $\|k_{ij}(t)\|$. Since the matrix $\|k_{ij}(t)\|$ is symmetric, the vectors $f_i(t)$ are orthogonal. Now let $\xi_1(t),\ldots,\xi_n(t)$ be independent Gaussian white noise processes with unit spectral density, so that

$$\mathbf{E}[\xi_i(s)\xi_j(t)] = \delta_{ij}\delta(t - s).$$

Define new (generalized) stochastic processes by

$$\tilde{\eta}_i(t) = \sum_{k=1}^{n} \sqrt{\lambda_k(t)} f_k^{(i)}(t)\xi_k(t),$$

where $f_k^{(i)}(t)$ is the i-th component of the vector $f_k(t)$. Then, using the relations

$$\sum_{i=1}^{n} f_i^{(k)}(t) f_i^{(\ell)}(t) = \sum_{i=1}^{n} f_k^{(i)}(t) f_\ell^{(i)}(t) = \delta_{k\ell},$$

$$\lambda_k(t) f_k^{(i)}(t) = \sum_{j=1}^{n} k_{ij}(t) f_k^{(i)}(t),$$

which follow from the definition of $\lambda_k(t)$ and $f_k(t)$, we get

$$\mathbf{E}[\tilde{\eta}_i(s)\tilde{\eta}_j(t)] = \sum_{k=1}^{n} \sqrt{\lambda_k(s)\lambda_k(t)} f_k^{(i)}(s) f_k^{(j)}(t)\mathbf{E}[\xi_k(s)\xi_k(t)]$$

$$= k_{ij}(t)\delta(t - s).$$

Thus, the correlation matrices, and hence also the probability distributions of the processes $\dot{\eta}_1(t),\ldots,\dot{\eta}_n(t)$ and $\tilde{\eta}_1(t),\ldots,$ $\tilde{\eta}_n(t)$ coincide. We may therefore set $\|\dot{\eta}_i^j(t)\| = \sum_{r=1}^{k} \sigma_r(t)\xi_r(t)$, $k \leqslant \ell^2$, in equation (1.1).

We shall treat (1.1) as a system of Itô equations

$$dX(t) = B(t)X(t)dt + \sum_{r=1}^{k} \sigma_r(t)X(t)d\xi_r(t). \tag{1.2}$$

Here $B(t)$ and $\sigma_r(t) = \|\sigma_{ir}^j(t)\|$ are $\ell \times \ell$ matrices, $\ell^2 \geqslant k$, and

$$\mathbf{E}\left[\sum_{r=1}^{k} \sigma_{ir}^j(s)\xi_r(s) \sum_{r=1}^{k} \sigma_{nr}^m(t)\xi_r(t)\right] = k_{ij}^{mn}(t)\delta(t - s).$$

In accordance with (5.1.2), we shall assume that $\|B(t)\|$, $\|\sigma_r(t)\|$ are bounded functions of time on any finite interval.

A linear stochastic equation with stochastic differentials in the sense of Stratonovič (see Chapter V) can also be reduced to the form (1.2). To be precise, it follows from (5.5.8) that a linear system with Stratonovič differentials $d\xi(t)$ is equivalent to a system involving Itô differentials. Both systems have the same coefficients $\sigma_r(t)$ and the new drift coefficients are related to the old ones by

$$\tilde{B}(t) = B(t) + \frac{1}{2}\sum_{r=1}^{k} \sigma_r^2(t).$$

We mention some properties of solutions of the system (1.2), which follow easily from the properties of stochastic integrals and the uniqueness of the solution.

1. If $X^{(1)}(t),\ldots,X^{(\ell)}(t)$ is a solution of the system (1.2), then the function

$$Y(t) = \sum_{i=1}^{\ell} k_i X^{(i)}(t), \tag{1.3}$$

is also a solution of the system (1.2) for any constants $k_1,\ldots,$ k_ℓ.

2. If $X^{(i)}(t)$ are solutions such that the determinant of the matrix $(X_j^{(i)}(t_0))$ does not vanish, then the solution of the system (1.2) with initial condition $X(t_0) = x_0$ can be expressed as a sum (1.3) for suitable constants k_i (the system of solutions $X^{(1)}(t),\ldots,X^{(\ell)}(t)$ is fundamental).

One consequence of these properties is that a linear stochastic system which is asymptotically stable in probability is asymptotically stable in the large. We leave the verification to the reader.

It is well known that the solution of a deterministic linear system for $\ell = 1$ can be determined by quadratures. An analogous statement holds for the one-dimensional stochastic system

$$dX(t) = b(t)X(t)dt + \sigma(t)X(t)d\xi(t). \tag{1.4}$$

Indeed, a direct check shows that the function[1]

$$X(t) = x_0\exp\left\{\int_0^t \left[b(s) - \frac{\sigma^2(s)}{2}\right]ds + \int_0^t \sigma(s)d\xi(s)\right\}, \tag{1.5}$$

satisfies equation (1.4) and the initial condition $X(0) = x_0$. (To verify this one calucates $dX(t)$ using Itô's formula (3.3.8), treating $X(t)$ as a function of t and the process $y(t) = \int_0^t \sigma(s)d\xi(s)$.)

---

[1] The representation of solutions of equation (1.4) in the form (1.5) is well known; see, e.g., Stratonovič [3], Gihman [2,3].

The representation (1.5) enables us to obtain conditions (for the stability of solutions of equation (1.4) which are an improvement on those derived above (see (5.5.4) and (5.5.5)). First, it follows from (1.5) that the solution $X(t) \equiv 0$ of the system (1.4) is asymptotically stable if the process

$$\eta(t) = \int_0^t [b(s) - \sigma^2(s)]ds + \int_0^t \sigma(s)d\xi(s) \tag{1.6}$$

satisfies the condition $\mathbf{P}\{\eta(t) \to -\infty \text{ as } t \to \infty\} = 1$.

Similarly, the solution $X(t) \equiv 0$ is stable if $\overline{\lim\limits_{t\to\infty}}\, \eta(t) < \infty$ almost surely, and unstable if $\overline{\lim\limits_{t\to\infty}}\, \eta(t) = \infty$ with positive probability.

The process $\eta(t)$ is clearly Gaussian and has independent increments. The study of its growth as $t \to \infty$ can be reduced to that of the growth of the Wiener process, by means of the following simple lemma.

LEMMA 1.1. *Let* $\int_{t_0}^t \sigma(s)d\xi(s)$ *be an Itô stochastic integral with respect to a Wiener process. Then there exists another Wiener process* $\tilde{\xi}(t)$ *such that*

$$\int_{t_0}^t \sigma(s)d\xi(s) = \xi\left(\int_{t_0}^t \sigma^2(s)ds\right) \qquad (a.s.) \tag{1.7}$$

for all $t \geq 0$.

PROOF. Let $t(\tau)$ denote the smallest number such that $\tau = \int_{t_0}^{t(\tau)} \sigma^2(s)ds$. Let us investigate some properties of the process $\xi(\tau) = \int_{t_0}^{\hat{t}(\tau)} \sigma(s)d\xi(s)$. By the properties of stochastic integrals it is easy to see that this process has independent increments and is Gaussian and moreover $\mathbf{E}\xi(\tau) = 0$, $\mathbf{E}[\xi(\tau)]^2 = \tau$. This means that $\tilde{\xi}(\tau)$ is a Wiener process. The assertion of the lemma now follows easily, since for any t we have almost surely:

$$\int_{t_0}^t \sigma(s)d\xi(s) = \int_{t_0}^{\hat{t}} \sigma(s)d\xi(s);$$

$$\xi\left(\int_{t_0}^t \sigma^2(s)ds\right) = \xi\left(\int_{t_0}^{\hat{t}} \sigma^2(s)ds\right),$$

where $\hat{t}$ is the smallest number such that

$$\int_{t_0}^{t} \sigma^2(s)ds = \int_{t_0}^{\hat{t}} \sigma^2(s)ds.$$

In the sequel we shall need the following theorem of Hinčin [1].

Law of the iterated logarithm. If $\xi(t)$ is a Wiener process, then almost surely

$$\overline{\lim_{t\to\infty}} \frac{\xi(t)}{\sqrt{2t\ln\ln t}} = 1.$$

We set

$$\tau(t) = \int_0^t \sigma^2(s)ds,$$

$$J(t) = \frac{\int_0^t [b(s) - \frac{1}{2}\sigma^2(s)]ds}{\left[2\int_0^t \sigma^2(s)ds \ln\ln\left(\int_0^t \sigma^2(s)ds\right)\right]^{1/2}}.$$

THEOREM 1.1. *If* $\tau(\infty) < \infty$, *then the inequality*

$$\overline{\lim_{t\to\infty}} \int_0^t b(s)ds < \infty$$

is a necessary and sufficient condition for the trivial solution of the system (1.4) *to be stable; and the condition*

$$\lim_{t\to\infty} \int_0^t b(s)ds = -\infty$$

is necessary and sufficient for asymptotic stability.

On the other hand, if $\tau(\infty) = \infty$, *then* $\overline{\lim\limits_{t\to\infty}} J(t) < -1$ *is a sufficient condition for asymptotic stability and* $\underline{\lim\limits_{t\to\infty}} J(t) > -1$ *a sufficient condition for instability (of the trivial solution).*

PROOF. By Lemma 1.1, we can write the process $\eta(t)$ as

$$\eta(t) = \int_{t_0}^{t} [b(s) - \tfrac{1}{2}\sigma^2(s)]ds + \xi\left(\int_{t_0}^{t} \sigma^2(s)ds\right).$$

Hence, by virtue of the fact that almost surely

$$\sup_{0 \leqslant \tau < \tau_0} \tilde{\xi}(\tau) < \infty,$$

we at once obtain the first part of the theorem.

Now let $\tau(\infty) = \infty$. Then

$$\underline{\lim}_{t\to\infty} J(t) + \overline{\lim}_{t\to\infty} \frac{\tilde{\xi}(\tau(t))}{(2\tau(t)\ln\ln\tau(t))^{1/2}}$$

$$\leqslant \overline{\lim}_{t\to\infty} \frac{\eta(t)}{[2\tau(t)\ln\ln\tau(t)]^{1/2}}$$

$$\leqslant \overline{\lim}_{t\to\infty} J(t) + \overline{\lim}_{t\to\infty} \frac{\tilde{\xi}(\tau(t))}{[2\tau(t)\ln\ln\tau(t)]^{1/2}} .$$

Hence, using the law of the iterated logarithm, we obtain the second part of the theorem.

REMARK 1. It is readily seen from the relation (5.5.8) between Ito and Stratonovič stochastic equations, that the assertions of Theorem 1.1 for the case $\tau(\infty) < \infty$ remain valid for the Stratonovič variant

$$dX(t) = b(t)X(t)dt + \sigma(t)X(t)d^{*}\xi(t).$$

The assertions for $\tau(\infty) = \infty$ also remain valid if the function $J(t)$ is replaced by

$$J_1(t) = \frac{\int_{t_0}^{t} b(s)ds}{[2\tau(t)\ln\ln\tau(t)]^{1/2}} .$$

Hence, in particular, it follows that the unstable solution $X \equiv 0$ of the equation $\dot{x} = bx$ cannot be stabilized by physically feasible (see Section 5.5) perturbations of the parameter b. For a constant b this was demonstrated in Section 5.5.

REMARK 2. It follows from Theorem 1.1 and Remark 1 that random noise does not affect the stability properties of the system $\dot{x} = bx$ if $\int_0^\infty \sigma^2(s)ds < \infty$.

The representation (1.5) may also be used to derive conditions for p-stability and q-instability in the one-dimensional case. In fact, for any real α one easily sees from (1.5) that

$$\mathrm{E}|X(t)|^\alpha = |x_0|^\alpha \exp\left\{\alpha\int_{t_0}^{t}\left[b(s) - \frac{\sigma^2(s)}{2}\right]ds + \frac{\alpha^2}{2}\int_{t_0}^{t}\sigma^2(s)ds\right\} . \tag{1.8}$$

For example, let b and σ be constants. Then, as we already have mentioned in Section 5.5, the necessary and sufficient condition for asymptotic stability is $b < \sigma^2/2$, and formula (1.8) can be rewritten as

$$\mathbf{E}|X(t)|^{\alpha} = |x_0|^{\alpha}\exp\left\{\alpha\left[\left(b - \frac{\sigma^2}{2}\right) + \frac{\alpha}{2}\sigma^2\right](t - t_0)\right.$$

Hence we see that an asymptotically stable one-dimensional linear system with constant coefficients is p-stable for sufficiently small p. Later, in Section 4 we shall see that this important property carries over to many-dimensional systems. The analogous statement for instability does not hold. For example, if $b = \sigma^2/2 \neq 0$, the system is unstable, but it is not q-unstable for any $q > 0$.

2. Equations for moments[2]

It is well known that the solution of a linear homogeneous deterministic system with constant coefficients can be determined from the roots of an auxiliary algebraic equation. Unfortunately, there is apparently no analogous reduction procedure for stochastic systems. However, as has been pointed out by many authors (see Leibowitz [1], Gihman [2,3] and others), the problem of determining the moments of orders 1,2,3,... can be reduced to solving an auxiliary deterministic system of linear differential equations.

Let us examine the situation more closely. Expressing the system (1.1) in integral form and calculating the conditional expectation, given $X(t_0) = x_0$, we easily derive the following equation for the vector $m_1(t) = \mathbf{E}X^{t_0,x_0}(t)$:

$$\frac{dm_1}{dt} = B(t)m_1 \qquad (B(t) = \|b_i^j(t)\|) \tag{2.1}$$

with initial condition

$$m_1(t_0) = x_0.$$

Systems of equations for the second, third, etc. moments can be obtained by using Itô's formula (3.3.8).

Applying this formula to the function $x_i x_j$, we get

$$d(X_i(t)X_j(t)) = X_i(t)dX_j(t) + X_j(t)dX_i(t) + \qquad \text{(Contd)}$$

[2] The results of this section are derived from Leibowitz [1], Gihman [2,3].

(Contd) $$+ \sum_{r=1}^{k} (\sigma_r(t)X(t))_i(\sigma_r(t)X(t))_j dt$$

$$= [X_i(t)(B(t)X(t))_j + X_j(t)(B(t)X(t))_i]dt$$

$$+ \sum_{r=1}^{k} \{[X_i(t)(\sigma_r(t)X(t))_j + X_j(t)(\sigma_r(t)X(t))_i]d\xi_r(t)$$

$$+ (\sigma_r(t)X(t))_i(\sigma_r(t)X(t))_j dt\}.$$

Expressing this relation in integral form and taking expectations, we get the system of differential equations:

$$\frac{dm_{ij}}{dt} = \sum_{n=1}^{\ell} \Big\{b_i^n(t)m_{jn}(t) + b_j^n(t)m_{in}(t)$$

$$+ \sum_{s,n=1}^{\ell} \sum_{r=1}^{k} \sigma_{ir}^s(t)\sigma_{jr}^n(t)m_{sn}(t)\Big\}$$

$$(i,j = 1,2,\ldots,\ell), \tag{2.2}$$

with the unknowns

$$m_{ij}(t) = \mathbf{E}[X^{t_0,x_0}(t)X^{t_0,x_0}(t)].$$

This system contains $n(n + 1)/2$ independent equations, since $m_{ij}(t) = m_{ji}(t)$.

The same method yields equations for

$$m_{i_1 i_2 i_3}(t) = \mathbf{E}[X_{i_1}^{t_0,x_0}(t)X_{i_2}^{t_0,x_0}(t)X_{i_3}^{t_0,x_0}(t)]$$

and so on.

REMARK 1. Comparing equations (2.1) and (1.2), we see that if the system (1.2) is stable in the mean, then the deterministic system obtained by suppressing the "fluctuation" terms is stable.

REMARK 2. Since

$$\mathbf{E}|X^{t_0,x_0}(t)|^2 = m_{11}(t) + \ldots + m_{ii}(t),$$

it follows that the system (1.2) is stable in mean square (asymptotically or exponentially) if and only if the deterministic system (2.2) is stable in the corresponding sense. If the coefficients b_i^j and σ_{ir}^j are constants, the system (1.1) is asymptotically (exponentially) stable in mean square if and only if the roots of the equation

$$|\lambda J - \tilde{B}| = 0 \tag{2.3}$$

have negative real parts. (Here $\tilde{B}$ is the matrix of the system (2.2), J is the $\ell^2 \times \ell^2$ unit matrix). Necessary and sufficient conditions for the real parts of the roots of the algebraic equation (2.3) to be negative can be found in the form of inequalities for the elements of the matrix B (see Gantmaher [1], p. 486). However, these conditions (the Routh-Hurwitz conditions) are quite cumbersome, since they involve computation of ℓ^2 determinants of orders up to ℓ^2. In Section 10 we shall see that these conditions can be simplified in a special case of practical importance.

REMARK 3. The roots of equation (2.3) are continuous functions of the coefficients b_i^j and σ_{ir}^j; thus, if the system (1.2) with constant coefficients is asymptotically stable in mean square, the same holds for a system with coefficients deviating slightly from those of (1.2). A more general result can be proved by the method of Ljapunov functions (see Chapter VII).

3. Exponential p-stability and q-instability[3]

In this section we shall give some improvements and applications of the theorems of Chapter V, proving necessary and sufficient conditions for p-stability and q-instability relative to the linear system

$$dX(t) = B(t)X(t)dt + \sum_{r=1}^{k} \sigma_r(t)X(t)d\xi_r(t). \tag{3.1}$$

We shall assume throughout that the functions $\|B(t)\|$, $\|\sigma_r(t)\|$ are bounded.

THEOREM 3.1. *The solution $X(t) \equiv 0$ of the system* (3.1) *is exponentially p-stable if and only if there exists a function $V(t,x)$, homogeneous of degree p in x, such that for some constants $k_i > 0$*

$$k_1|x|^p \leqslant V(t,x) \leqslant k_2|x|^p; \qquad LV(t,x) \leqslant -k_3|x|^p, \tag{3.2}$$

$$\left|\frac{\partial V}{\partial x_i}\right| \leqslant k_4|x|^{p-1}; \qquad \left|\frac{\partial^2 V}{\partial x_i \partial x_j}\right| \leqslant k_4|x|^{p-2} \quad (i,j = 1,\ldots,\ell).$$

[3] The main theorems of this section are due to Nevel'son and Has'minskiĭ [2].

PROOF. Sufficiency follows from Theorem 5.7.1. To prove necessity we proceed as in the proof of Theorem 5.7.2, using the function

$$V(t,x) = \int_t^{t+T} \mathbf{E}|X^{t,x}(u)|^p du. \tag{3.3}$$

Since the coefficients of the system (3.1) have bounded derivatives of arbitrary order with respect to x, it follows from Theorem 5.7.2 that this function satisfies (3.2). We claim that $V(t,x)$ is homogeneous of degree p. In fact, (1.3) implies that the solution $X^{t,x}(u)$ admits the representation

$$X^{t,x}(u) = \sum_{i=1}^{\ell} x_i X^{(i)}(u), \tag{3.4}$$

where $(x_1,\ldots,x_\ell) = x$ and $(X^{(1)}(u),\ldots,X^{(\ell)}(u))$ is a fundamental system of solutions of equations (3.1) satisfying the initial conditions $X^i_j(t) = \delta^i_j$ (δ^i_j is the Kronecker symbol). Substituting (3.4) into (3.3), we see that $V(t,x)$ is homogeneous of degree p.

In applications of Theorem 3.1, one would like to be sure that the p-stability of a stochastic system can be "detected" with the aid of homogeneous functions from a relatively small class. For the general case we have unfortunately no results of this kind. However, for even p the following theorem is valid.

THEOREM 3.2. *A necessary condition for exponential p-stability of even order* ($p = 2,4,\ldots$) *of the system* (3.1) *is that for every positive definite form* $W(t,x)$ *of degree p whose coefficients are continuous bounded functions of time there exist a positive definite form* $V(t,x)$ *of the same degree such that*

$$LV = - W.$$

The same condition, with the phrase "for every ... " replaced by "for some ... ", is also sufficient.

The proof is analogous to that of Theorem 3.1; the only difference is that instead of the function (3.3) one considers the function

$$V_1(t,x) = \mathbf{E}\int_t^{\infty} W(u,X^{t,x}(u))du.$$

The infinite upper limit of integration here causes no difficulties, since the required smoothness of $V_1(t,x)$ as a function of x follows from the fact that this function is a form in x.

If the coefficient matrices B and σ_r in (3.1) are constant, so that we have a linear stationary system

$$dX(t) = BX(t)dt + \sum_{r=1}^{k} \sigma_r X(t)d\xi_r(t), \tag{3.5}$$

then the forms $V(t,x)$ and $W(t,x)$ in the statement of Theorem 3.2 may be replaced by forms $V(x)$ and $W(x)$ with constant coefficients. In this case, Theorem 3.2 yields the following algorithm for a construction of algebraic criteria for p-stability for even p. Given some positive definite form $W(x)$ of degree p, we look for a form $V(x)$ of the same degree such that $LV = -W(x)$. Comparing the coefficients at the monomials $x_1^{k_1}\ldots x_\ell^{k_\ell}$ on both sides of this equation ($k_1 + \ldots + k_\ell = p$), we get a system of linear equations for the coefficients of $V(x)$. It follows from Theorem 3.2 that the system is p-stable if and only if the function $V(x)$ turns out to be positive definite.

This procedure is well known for deterministic systems (see Četaev [1]). Its applicability to stability in mean square of stochastic systems was first indicated by Kac and Krasovskiĭ [1].

THEOREM 3.3. *The solution* $X(t) \equiv 0$ *of the system* (3.1) *is exponentially* q*-unstable if and only if there exists a function* $V(t,x)$*, homogeneous in* x *of degree* $-q$*, such that for some constants* $k_i > 0$ *we have*

$$k_1|x|^{-q} \leqslant V(t,x) \leqslant k_2|x|^{-q}; \qquad LV(t,x) \leqslant -k_3|x|^{-q};$$

$$\left|\frac{\partial V}{\partial x_i}\right| \leqslant k_4|x|^{-q-1}; \qquad \left|\frac{\partial^2 V}{\partial x_i \partial x_j}\right| \leqslant k_4|x|^{-q-2}$$

$$(i, j, = 1, 2, \ldots, \ell).$$

The reader should have no difficulty in proving this theorem, using the Theorems 5.7.3 and 5.7.4.

A deterministic stationary linear system $dx/dt + Bx$ is q-unstable for any q if and only if all the roots λ_i of the characteristic equation $|\lambda J - B| = 0$ have positive real parts. This is easily checked directly, but we shall view this as one of the consequences of Theorem 3.3.

It is well known that the existence of a positive definite quadratic form $W(x)$ such that the form $d^0W/dt = LW$ is also positive definite is a necessary and sufficient condition for the numbers $\mathrm{Re}\lambda_i$, $i = 1,\ldots,\ell$, to be positive. Set $V(x) = [W(x)]^{-q/2}$. It is readily seen that this function satisfies all the assumptions of Theorem 3.3, and this implies the above assertion.

The following two examples illustrate applications of the theorems proved in this section.

EXAMPLE 1. The differential generator of the one-dimensional system

$$dX(t) = bX(t)dt + \sigma X(t)d\xi(t)$$

with constants b and c is equal to

$$L = bx \frac{\partial}{\partial x} + \frac{1}{2}\sigma^2 x^2 \frac{\partial^2}{\partial x^2}.$$

Setting $V(x) = |x|^p$, we get

$$LV = p|x|^{p-1}\left[b + \frac{\sigma^2}{2}(p - 1)\right].$$

It follows from Theorem 3.2 that in this case the inequality $b + \frac{1}{2}\sigma^2(p - 1) < 0$ is a necessary and sufficient condition for exponential p-stability. Of course, this conclusion also follows from the explicit expression (1.5)

EXAMPLE 2. Recall that the expression

$$L = \frac{\partial}{\partial t} + \left(Bx, \frac{\partial}{\partial x}\right) + \frac{1}{2}\left(A(x)\xi\frac{\partial}{\partial x}, \frac{\partial}{\partial x}\right)$$

$$\equiv \frac{\partial}{\partial t} + \left(Bx, \frac{\partial}{\partial x}\right) + \frac{1}{2}\sum_{r=1}^{k}\left(\sigma_r x, \frac{\partial}{\partial x}\right)^2$$

defines the differential operator of the system (3.5). We now consider some sufficient conditions for p-stability of this system, restricting ourselves to Ljapunov functions which are powers of a positive definite quadratic from $(Wx,x) = \sum W_{ij}x_i x_j$. We introduce the notation: $V(x) = (Wx,x)^{p/2}$, B^* is the matrix adjoint to B, $\lambda^D_{min} = \lambda^D_1 < \lambda^D_2 < \ldots < \lambda^D_\ell = \lambda^D_{max}$ are the eigenvalues of a symmetric $\ell \times \ell$ matrix D. In addition, we set

$$m = \inf_{|x|=1} \lambda^{A(x)}_{min}; \qquad M = \sup_{|x|=1} \lambda^{A(x)}_{max}.$$

We first observe that for any positive semi-definite symmetric matrices D_1 and D_2 we have

$$\lambda^{D_1}_{min} \mathrm{tr} D_2 < \mathrm{tr}(D_1 D_2) < \lambda^{D_1}_{max} \mathrm{tr} D_2. \tag{3.6}$$

This inequality is easily proved by reducing D_1 to diagonal form.

Next,

$$LV = p(Wx,x)^{p/2-1} \qquad \text{(Contd)}$$

$$\times\left[\frac{1}{2}\,\mathrm{tr}(A(x)W) + ((WB + B^*W)x,x) + \left(\frac{p}{2} - 1\right)\mathrm{tr}(A(x)F(x))\right] =$$

$$\text{(Contd)} = p(Wx,x)^{p/2-1}\Phi(x) \tag{3.7}$$

holds, where $F(x) = ((f_{ij}(x)))$ is the matrix with the elements

$$f_{ij}(x) = \frac{(Wx_i)(Wx_j)}{(Wx,x)} .$$

It follows from the preceding theorem that a sufficient condition for the linear system (3.5) to be exponentially p-stable is that the expression $\Phi(x)$ (in square brackets in (3.7)) is negative for $x \neq 0$. But if $\Phi(x) \geqslant 0$, then (see Remark 2 at the end of Section 5.7) the system is not exponentially p-stable.

It follows from (3.6) and (3.7) that for $p \geqslant 2$

$$\Phi(x) \leqslant \frac{1}{2} \lambda_{\max}^{A(x)} \mathrm{tr}W + ((WB + B^*M)x,x) + \left(\frac{p}{2} - 1\right)\lambda_{\max}^{A(x)} \frac{(W^2x,x)}{(Wx,x)}$$

$$\leqslant \left[M\left(\frac{1}{2} \mathrm{tr}W + \left(\frac{p}{2} - 1\right)\lambda_{\max}^{W}\right) + \lambda_{\max}^{WB+B^*W}\right](x,x), \tag{3.8}$$

$$\Phi(x) \geqslant \left[m\left(\frac{1}{2} \mathrm{tr}W + \left(\frac{p}{2} - 1\right)\lambda_{\min}^{W}\right) + \lambda_{\min}^{WB+B^*W}\right](x,x). \tag{3.9}$$

Hence we see that if for some $p \geqslant 2$ there exists a positive definite matrix W such that

$$M\left[\frac{1}{2} \mathrm{tr}W + \left(\frac{p}{2} - 1\right)\lambda_{\max}^{W}\right] + \lambda_{\max}^{WB+B^*W} < 0, \tag{3.10}$$

then the system is exponentially p-stable for this p. On the other hand, if

$$m\left[\frac{1}{2} \mathrm{tr}W + \left(\frac{p}{2} - 1\right)\lambda_{\min}^{W}\right] + \lambda_{\min}^{WB+B^*W} \geqslant 0, \tag{3.11}$$

then the system is not exponentially p-stable.

Similarly, if for some $p \leqslant 2$ there exists a positive definite matrix W such that

$$\frac{1}{2} M\mathrm{tr}W + m\left(\frac{p}{2} - 1\right)\lambda_{\min}^{W} + \lambda_{\max}^{WB+B^*W} < 0, \tag{3.12}$$

then the system (3.1) is exponentially p-stable, while if

$$\frac{1}{2} m\mathrm{tr}W + M\left(\frac{p}{2} - 1\right)\lambda_{\max}^{W} + \lambda_{\min}^{WB+B^*W} \geqslant 0, \tag{3.13}$$

the system is not exponentially p-stable.

Although the conditions furnished by (3.10) through (3.13) are quite weak, we shall show now that in a certain special case they yield necessary and sufficient conditions for exponential p-stability. Let $A(x) = \delta|x|^2 J$ (where J is the unit matrix). Then $M = m = \delta$.

Now suppose that $B + B^* = -\lambda J$, and set $W = J$. It then follows from (3.8) - (3.13) that this system is asymptotically p-stable if and only if $\frac{1}{2}\delta(\ell + p - 2) < \lambda$. An analogous argument shows that the system is q-unstable if and only if $\frac{1}{2}\delta(\ell - q - 2) > \lambda$ for some positive q.

4. Exponential p-stability and q-instability (continued)

It is well known that an asymptotically stable linear deterministic system with constant coefficients is exponentially stable. This statement holds true for a system with variable coefficients which is uniformly (in time) asymptotically stable. In this section we shall prove analogs of these properties for linear stochastic systems with constant coefficients. Systems with variable coefficients will be considered in the next section.

LEMMA 4.1. *If a linear system with constant coefficients*

$$dX(t) = BX(t)dt + \sum_{r=1}^{k} \sigma_r X(t) d\xi_r(t) \tag{4.1}$$

is stable in probability, then it is p-stable for sufficiently small p.

PROOF. Since the system (4.1) is stable in probability, there exists $\alpha > 0$ such that

$$\sup_{|x| \leqslant 2^{-\alpha}} \mathbf{P}\{\sup_{t \geqslant 0} |X^x(t)| > 1\} \leqslant \frac{1}{2}.$$

Further, since the system is linear, we have

$$X^{\gamma x}(t) = \gamma X^x(t). \tag{4.2}$$

Therefore, for any k,

$$\sup_{|x| \leqslant 2^{k\alpha}} \mathbf{P}\{\sup_{t \geqslant 0} |X^x(t)| > 2^{\alpha(k+1)}\} \leqslant \frac{1}{2}. \tag{4.3}$$

Let τ denote the first time at which the path of the process reaches the set $|x| = 2^\alpha$. Using the strong Markov property of the process $X(t)$ and (4.3), we get

$$\sup_{|x| \leqslant 1} \mathbf{P}\{\sup_{t \geqslant 0} |X^x(t)| > 2^{2\alpha}\} = \qquad \text{(Contd)}$$

(Contd) $$= \sup_{|x|\leqslant 1} \int_{u=0}^{\infty}\int_{|y|=2^{\alpha}} \mathbf{P}\{\tau \in du, X^{x}(u) \in dy\}$$

$$\times \mathbf{P}\{\sup_{t\geqslant 0} |X^{x}(t)| > 2^{2\alpha}\}$$

$$\leqslant \frac{1}{2} \sup_{|x|\leqslant 1} \mathbf{P}\{\tau < \infty\} = \frac{1}{2} \sup_{|x|\leqslant 1} \mathbf{P}\{\sup_{t>0} |X^{x}(t)| > 2^{\alpha}\}$$

$$< \frac{1}{2^2}$$

. . . .

$$\sup_{|x|\leqslant 1} \mathbf{P}\{\sup_{t\geqslant 0} |X^{x}(t)| > 2^{k\alpha}\} < \frac{1}{2^k}. \tag{4.4}$$

Now let $x \in E_{\ell}$ be such that $|x| = 1$. Then, using (4.4) with $p < 1/\alpha$, we have

$$\mathbf{E}[\sup_{t>0} |X^{x}(t)|^{p}] \leqslant \sum_{k=1}^{\infty} 2^{k\alpha p}\mathbf{P}\{2^{(k-1)\alpha} < \sup_{t>0} |X^{x}(t)| < 2^{k\alpha}\}$$

$$\leqslant \sum_{k=1}^{\infty} 2^{k\alpha p}2^{-(k-1)} = 2 \sum_{k=1}^{\infty} 2^{-k(1-\alpha p)} = K(p)$$

$$< \infty. \tag{4.5}$$

It follows from (4.5) and (4.2) that

$$\sup_{|x|<\delta} \mathbf{E}[\sup_{t>0} |X^{x}(t)|^{p}] \leqslant \delta^{p}K(p),$$

and this inequality implies the assertion.

LEMMA 4.2. *If the system* (4.1) *is asymptotically stable in probability, then it is asymptotically p-stable for sufficiently small p.*

PROOF. Let x be such that $|x| = 1$. Then (4.5) implies that

$$\mathbf{E}\{\sup_{t\geqslant 0} |X^{x}(t)|^{p}\} < K \tag{4.6}$$

holds for some $p > 0$.

Moreover, as was mentioned in Section 1, a linear system is

asymptotically stable in probability if and only if it is asymptotically stable in the large. Therefore, almost surely

$$X^x(t) \to 0 \quad \text{as} \quad t \to \infty. \tag{4.7}$$

It now follows from (4.6) and (4.7) by Lebesgue's bounded convergence theorem (Halmos [1], p. 110) that

$$\mathbf{E}|X^x(t)|^p \to 0 \quad \text{as} \quad t \to \infty.$$

This and Lemma 4.1 imply the assertion.

LEMMA 4.3. *If the system* (4.1) *is asymptotically p-stable for some p, then it is also exponentially p-stable.*

PROOF[4]. We first show that, under the assumptions of the lemma, for every $Q < 1$ there exists a $T > 0$ such that

$$\sup_{|x|=1} \mathbf{E}|X^x(T)|^p \leqslant Q. \tag{4.8}$$

To do this we use the representation (3.4). It follows from the assumptions that for each $i = 1,\dots,\ell$ and sufficiently large T

$$\mathbf{E}|X^{(i)}(T)| < \frac{Q}{\ell^{p+1}}.$$

Next, using the inequality

$$|A_1 + \dots + A_\ell|^p \leqslant \ell^p(|A_1|^p + \dots + |A_\ell|^p)$$

and (3.4), we see that if $|x| = 1$, then

$$\mathbf{E}|X^x(T)|^p \leqslant \ell^p \sum_{i=1}^{\ell} \mathbf{E}|X^{(i)}(T)|^p \leqslant Q.$$

We now choose T so that (4.8) holds with $Q = e^{-1}$ and, in view of (4.2), we rewrite (4.8) as

$$\mathbf{E}|X^x(T)|^p \leqslant e^{-1}|x|^p. \tag{4.9}$$

Then

$$\mathbf{E}|X^x(2T)|^p = \int_{E_\ell} P(x,T,dy)\mathbf{E}|X^y(T)|^p \leqslant \qquad \text{(Contd)}$$

[4] The proof of Lemma 4.3 essentially uses the same idea as that of Theorem 6.1 in Kac and Krasovskiĭ [1].

(Contd) $$\leqslant e^{-1}\int_{E_\ell} P(x,T,dy)|y|^p = e^{-1}\,|X^x(T)|^p$$

$$\leqslant e^{-2}|x|^p \tag{4.10}$$

. . . .

$$\mathbf{E}|X^x(kT)|^p \leqslant e^{-k}|x|^p.$$

Let $t = nT + t_1$ $(0 \leqslant t_1 < T)$, and let

$$K = \sup_{t>0,|x|=1} \mathbf{E}|X^x(t)|^p.$$

Here $K < \infty$ by virtue of Lemma 4.1. This and (4.10) imply

$$\mathbf{E}|X^x(t)|^p = \int_{E_\ell} P(x,nT,dy)\mathbf{E}|X^y(t_1)|^p$$

$$\leqslant K\mathbf{E}|X^x(nT)|^p \leqslant K|x|^p e^{-n}$$

$$\leqslant Ke|x|^p e^{-t/T} = K_1|x|^p e^{-t/T}.$$

This completes the proof.

Lemma 4.2 and 4.3 immediately imply

THEOREM 4.1. *If the linear system* (4.1) *with constant coefficients is asymptotically stable in probability, then it is exponentially p-stable for all sufficiently small positive p.*

Similar arguments hold for q-instability. Thus one easily proves the following

THEOREM 4.2. *If the solutions of the linear system* (4.1) *with constant coefficients satisfy the relation*

$$P\{|X^x(t)| \to \infty \text{ as } t \to \infty\} = 1$$

for $x \neq 0$, then the system is exponentially q-unstable for all sufficiently small positive q.

Theorems 4.1 and 4.2 fail to hold for linear systems with variable coefficients. For example, the deterministic system $dx/dt = -x(t+1)$ is asymptotically stable but not exponentially stable. However, if certain additional assumptions are made, then one can prove analogs of Lemmas 4.1 and 4.3, hence also of Theorem 4.1. But first we have to introduce new definitions of stability and instability and then study the properties of systems satisfying these definitions. This we do in the next section.

5. Uniform stability in the large

The solution $X(t) \equiv 0$ of the system

$$dX(t) = b(t,X)dt + \sigma(t,X)d\xi(t) \tag{5.1}$$

is said to be *stable in the large uniformly in* $t > 0$, if it is uniformly stable in probability and moreover for any $x \in E_\ell$, $\varepsilon > 0$,

$$\sup_{s>0} \mathbf{P}\{ \sup_{u>s+T} |X^{s,x}(u)| > \varepsilon\}_{T\to\infty} \to 0. \tag{5.2}$$

Let us give a few comments on this definition.

1. It follows from (5.2) that the system (5.1) is stable in the large in the sense of the definition of Section 5.4. In fact, the equivalence of the events

$$\left\{\bigcap_{n=1}^{\infty} \sup_{u>s+n} |X^{s,x}(u)| > \varepsilon\right\} = \overline{\lim_{t\to\infty}} |X^{s,x}(t)| > \varepsilon\}$$

and (5.2) imply that

$$\mathbf{P}\ \overline{\lim_{t\to\infty}} |X^{s,x}(t)| > \varepsilon\} = \lim_{n\to\infty} \mathbf{P}\{ \sup_{u>s+n} |X^{s,x}(u)| > \varepsilon\} = 0.$$

Since ε is arbitrary, this shows that the system is stable in the large.

2. A sufficient condition for the solution $X(t) \equiv 0$ of the system (5.1) to be stable in the large is that it is uniformly stable in probability and, moreover, for any $x \in E_\ell$, $\varepsilon > 0$,

$$\sup_{s>0} \mathbf{P}\{|X^{s,x}(s + T)| > \varepsilon\}_{T\to\infty} \to 0. \tag{5.3}$$

This evidently follows from the inequalities

$$\mathbf{P}\{ \sup_{u>s+T} |X^{s,x}(u)| > \varepsilon\}$$
$$= \left(\iint_{|y|<\delta} + \int_{|y|\geq\delta}\right) P(s,x,s + T,dy)\mathbf{P}\{ \sup_{u>s+T} |X^{s+T,y}(u)| > \varepsilon\}$$
$$\leq \sup_{s>0,|y|<\delta} \mathbf{P}\{\sup_{u>s} |X^{s,y}(u)| > \varepsilon\} + \mathbf{P}\{|X^{s,x}(s + T)| \geq \delta\},$$

where the first and second terms on the right can be made arbitrarily small by a suitable choice of δ and T.

Yet another sufficient condition for uniform asymptotic stability in the large is given by

LEMMA 5.1. *A sufficient condition for the solution* $X(t) \equiv 0$ *of equation* (5.1) *to be stable in the large uniformly in* $t > 0$ *is that it be uniformly stable in probability, and that the family of processes* $X^{s,x}(t)$ *associated with different values of the parameters* s,x *is for any* $\varepsilon > 0$ *uniformly recurrent relative to the domain* $|x| < \varepsilon$, *in the sense that*

$$\sup_{s>0} \mathbf{P}\{\tau_\varepsilon^{s,x} - s > T\} \to 0 \quad \text{as} \quad T \to \infty, \tag{5.4}$$

where $\tau_\varepsilon^{s,x}$ *is the first time the path of the process* $X^{s,x}(t)$ *reaches the set* $|x| = \varepsilon$.

PROOF. It is clear that for any $\varepsilon > 0$, δ and $T > 0$, we have

$$\{\sup_{u>s+T} |X^{s,x}(u)| > \varepsilon\}$$

$$\subset \{\tau_\delta^{s,x} > s + T\} \cup \{\tau_\delta^{s,x} \leqslant s + T;\ \sup_{u>\tau_\delta^{s,x}} |X^{s,x}(u)| > \varepsilon\}.$$

Hence, applying the strong Markov property, we get

$$\mathbf{P}\{\sup_{u>s+T} |X^{s,x}(u)| > \varepsilon\}$$

$$\leqslant \mathbf{P}\{\tau_\delta^{s,x} - s > T\}$$

$$+ \int_{v=s}^{s+T}\int_{|y|=\delta} \{\tau_\delta^{s,x} \in ds\ X^{s,x}(\tau_\delta^{s,x}) \in dy\}\mathbf{P}\{\sup_{u>v} |X^{v,y}(u)| > \varepsilon\}$$

$$\leqslant \{\tau_\delta^{s,x} - s > T\} + \sup_{v>0,|y|=\delta} \{\sup_{u>v} \mathbf{P}|X^{v,y}(u)| > \varepsilon\}.$$

For any $\varepsilon > 0$, the second term on the right of this inequality can be made arbitrarily small by suitable choice of δ, and the first, by a choice of T (see (5.4)). This proves the lemma.

REMARK. For condition (5.4) to hold, it is sufficient that there exists a positive function $V(t,x)$ in the domain $|x| > \varepsilon$ such that

$$\left.\begin{aligned} &\inf_{|x|>R,t>0} V(t,x) = V_R \to \infty \quad \text{as} \quad R \to \infty, \\ &\sup_{\varepsilon<|x|\delta,t>0} V(t,x) = V^{(\delta)} < \infty, \end{aligned}\right\} \tag{5.5}$$

$$LV < -k \qquad (k > 0). \tag{5.6}$$

In fact, conditions (5.5) and (5.6) imply by Theorem 3.4.1 that the process $X(t)$ is regular. Applying Theorem 3.7.1, we get the inequality

$$\mathbf{E}(\tau_\delta^{s,x} - s) < \frac{V(s,x)}{k} \leqslant \frac{V(|x|)}{k}.$$

Hence, using Čebyšev's inequality, we get (5.4).

This remark and the remark following Theorem 5.3.1 imply

LEMMA 5.2. *A sufficient condition for the solution* $X(t) \equiv 0$ *of equation* (5.1) *to be stable in the large uniformly in* $t > 0$ *is that there exists a positive definite function* $V(t,x) \in \mathbf{C}_2^0(E)$, *with infinitesimal upper limit, such that the function* LV *is negative definite and* (5.5) *holds.*

Comparing this lemma with Theorem 5.4.4, we see that a system is uniformly stable in the large if there exists a Ljapunov function V satisfying the assumptions of Theorem 5.4.4 and such that $\sup_{t>0} V(t,x)$ is bounded in any bounded (with respect to x) domain.

We now state the analogs of Theorem 4.1 and Lemma 4.1 for nonstationary linear systems.

THEOREM 5.1. *If the linear system*

$$dX(t) = B(t)X(t)dt + \sum_{r=1}^{k} \sigma_r(t)X(t)d\xi_r(t) \tag{5.7}$$

is stable in the large uniformly in $t > 0$, *then it is exponentially p-stable for sufficiently small positive p.*

LEMMA 5.3. *If the system* (5.7) *is uniformly stable in probability, then it is p-stable for sufficiently small p, and there exists an* $\alpha > 0$ *such that for all* $k = 1,2,3,\ldots$

$$\sup_{s>0,|x|\leqslant 1} \mathbf{P}\{\sup_{t\leqslant s} |X^{s,x}(t)| > 2^{k\alpha}\} \leqslant \frac{1}{2^k}. \tag{5.8}$$

The proof of inequality (5.8) is an almost word for word repetition of that of inequality (4.4), except that properties of uniform stability are used. From (5.8) we get the first assertion of the lemma, in the same way as Lemma 4.1 was deduced from (4.4).

PROOF OF THEOREM 5.1. It will suffice to verify that there exists a $T > 0$ such that for all positive $p < \alpha$

$$\sup_{s>0,|x|=1} \mathbf{E}|X^{s,x}(s + T)|^p < 1. \tag{5.9}$$

Indeed, the assertion will then follow from (5.9) in the same way as Lemma 4.2 follows from (4.8).

Now, proceeding as in the case of (4.5), we see that for arbitrary $\alpha > 0$, $n > 0$, $p > 0$

$$\mathbf{E}|X^{s,x}(s+T)|^p \leqslant \mathbf{E}[\sup_{t\geqslant s+T} |X^{s,x}(t)|^p]$$

$$\leqslant \sum_{k+-\infty}^{+\infty} 2^{k\alpha p}\mathbf{P}\{2^{(k-1)\alpha} \leqslant \sup_{t\geqslant s+T} |X^{s,x}(t)| < 2^{k\alpha}\}$$

$$< 2^{-\alpha p} + 2^{n\alpha p}\mathbf{P}\{\sup_{t>s+T} |X^{s,x}(t)| > 2^{-\alpha}\}$$

$$\sum_{k=n+1}^{\infty} 2^{k\alpha p}\{\sup_{t>s} |X^{s,x}(t)| > 2^{\alpha(k-1)}\}.$$

If we now select α as in Lemma 5.3 and use (5.8), we get the inequality

$$|X^{s,x}(s+T)|^p$$

$$\leqslant 2^{n\alpha p} + 2^{n\alpha p}\mathbf{P}\{\sup_{t>s+T} |X^{s,x}(t)| > 2^{-\alpha}\} + 2\sum_{k=n+1}^{\infty} 2^{k(\alpha p-1)}.$$

Now choose $p < 1/\alpha$, then let n be large enough, so that

$$2\sum_{k=n+1}^{\infty} 2^{k(\alpha p-1)} < \frac{1}{2}(1-2^{-\alpha p}).$$

Finally take T large enough, so that

$$2^{n\alpha p} \sup_{s>0,|x|=1} \mathbf{P}\{\sup_{t>s+T} |X^{s,x}(t)| > 2^{-\alpha}\} < \frac{1}{2}(1-2^{-\alpha p})$$

(such choice of T is possible by (5.2)). This yields (5.9) and hence the assertion of the theorem.

To conclude this section we state the analogous instability theorem (the proof is similar).

THEOREM 5.2. *If the linear system* (5.7) *is uniformly unstable in the sense that for any* $x \neq 0$, $A > 0$

$$\sup_{s>0} \mathbf{P}\{\sup_{u>s+T} |X^{s,x}(u)| < A\} \underset{\to\infty}{} \to 0,$$

then the system is exponentially q-unstable for sufficiently small positive q.

6. Stability of products of independent matrices

Let $A_n = ((a_{ij}^{(n)}))$ $(i,j = 1,2,\ldots,\ell;\ n = 1,2,\ldots)$ be a sequence of identically distributed[5] $\ell \times \ell$ matrices and let $P(dA)$ be their common probability distribution. It is easy to see that the sequence

$$x_0;\quad x_1 = A_1 x_0;\quad \ldots\ x_n = A_n x_{n-1}\ \ldots, \tag{6.1}$$

where $x_0 \in E_\ell$, is a Markov chain in E_ℓ. In applications (see Kalman [1]) one is often interested to find conditions under which $|x_n| = |A_n A_{n-1} \ldots A_1 x_0| \to 0$ for all $x_0 \in E_\ell$ (or, what is the same, $\|A_n A_{n-1} \ldots A_1\| \to 0$ in some sense) as $n \to \infty$. In Chapter I (Section 1.8, Example 2) we showed that in certain cases the stability theory of linear systems whose coefficients are step functions can be reduced to this problem.

The solution of this problem will provide yet another illustration of how to apply the methods of this chapter to a discrete model.

1. We first consider the trivial case $\ell = 1$.

Then

$$|x_n| = |x_0| \exp\left\{\sum_{i=1}^{\infty} \ln|A_i|\right\}. \tag{6.2}$$

If $a = \mathbf{E}\ln|A_i|$ exists, then it follows from the law of large numbers that $|x_n| \to 0$ almost surely when $a < 0$, and $|x_n| \to \infty$ when $a > 0$. For $a = 0$ we may have either stability (e.g., if $|A_n| = 1$ almost surely) or instability. However, if the random variable $\ln|A_i|$ has finite non-zero variance, one readily shows that

$$\overline{\lim_{n\to\infty}}\ |x_n| = \infty, \quad \text{but} \quad \underline{\lim_{n\to\infty}}\ |x_n| = 0$$

almost surely.

It is also obvious that $\mathbf{E}|x_n|^p \to 0$ if and only if $\mathbf{E}|A_i|^p < 1$. The condition for (exponential) q-instability is similar.

2. Since the matrix A_n and the vector X_{n-1} are independent, it follows from (6.1) that

---

[5] That is to say, the joint distributions of the ℓ^2 random variables $a_{ij}^{(n)}$ $(i,j = 1,\ldots,\ell)$ are independent for different n. The distributions of the elements of the same matrix may of course be dependent.

$$\mathbf{E}x_n = \mathbf{E}A_n\mathbf{E}x_{n-1} = (\mathbf{E}A)^n x_0.$$

Clearly, $\mathbf{E}x_n \to 0$ if all the eigenvalues of the constant matrix $\mathbf{E}A$ are such that $|\lambda_i| < 1$.

Similar reasoning gives recurrence relations for moments of higher orders. This was pointed out by Bellman [3], who also observed that in the determination of the k-th moment, $\mathbf{E}A$ is replaced by the expectation of the k-fold direct product of the matrix A itself. For example, it follows that x_n is asymptotically stable in mean square if and only if the roots of the matrix $\mathbf{E}[A \times A]$ lie inside the unit disk. (Recall that

$$A \times A = \begin{pmatrix} a_{11}A & \ldots & a_{1\ell}A \\ \ldots & \ldots & \ldots \\ a_{\ell i}A & \ldots & a_{\ell\ell}A \end{pmatrix}$$

is an $\ell^2 \times \ell^2$ matrix.)

3. It is not hard to prove analogs of the theorems in Section 3 for products of independent identically distributed matrices $A_1,\ldots,A_n,\ldots$. We present two theorems of this type, where p is any non-zero real number.

THEOREM 6.1. *A sufficient condition for*

$$\mathbf{E}\|A_n\ldots A_1\|^p \to 0 \quad as \quad n \to \infty \tag{6.3}$$

to hold is that there exists a positive definite function $f(x)$, homogeneous of degree p, such that the function $\mathbf{E}f(Ax) - f(x)$ is negative definite.

PROOF. The assumptions imply that there exist positive constants k_i $(k_3 > k_1)$ for which

$$\mathbf{E}f(Ax) - f(x) \leqslant -k_1|x|^p; \qquad k_2|x|^p < f(x) < k_3|x|^p. \tag{6.4}$$

Hence, setting $q = 1 - (k_1/k_3)$, we get

$$f(Ax) \leqslant f(x)\left(1 - \frac{k_1}{k_2}\right) = qf(x). \tag{6.5}$$

Let $P(x,n,\Gamma)$ denote the transition probability of the Markov chain (6.1). Then, applying (6.5), we get

$$\mathbf{E}f(A_nA_{n-1}\ldots A_1x_0) = \mathbf{E}f(A_nx_{n-1}) = \int P(x_0,n-1,dx)\mathbf{E}f(Ax)$$

$$\leqslant q\mathbf{E}f(A_{n-1}\ldots A_1x_0) \leqslant \ldots \leqslant q^n f(x_0),$$

whence, by (6.4) we may conclude that

$$\mathbf{E}|A_n A_{n-1} \ldots A_1 x_0|^p \leqslant \frac{k_3}{k_2} q^n |x_0|^p. \tag{6.6}$$

Thus,

$$\mathbf{E}\|A_n A_{n-1} \ldots A_1\|^p \leqslant \frac{k_3}{k_2} q^n.$$

This proves the theorem.

REMARK. Examining the above proof we easily see that if we replace the condition that the function $\mathbf{E}f(Ax) - f(x)$ be negative definite by the inequality $\mathbf{E}f(Ax) - f(x) \geqslant 0$, we can prove the existence of a constant k such that

$$\mathbf{E}\|A_n A_{n-1} \ldots A_1\|^p > k > 0 \qquad (n = 1,2,\ldots).$$

THEOREM 6.2. *If* $A_1, A_2, \ldots$ *is a sequence of independent identically distributed matrices satisfying condition* (6.3), *then, for any positive definite function* $g(x)$ *which is homogeneous of degree* p, *there exists a positive definite function* $f(x)$, *homogeneous of the same degree, such that*

$$\mathbf{E}f(Ax) - f(x) = -\, g(x). \tag{6.7}$$

The proof is analogous to that of Theorem 3.1. One first applies (6.3) to establish (6.6). Then, using (6.6), one readily shows that equation (6.7) is satisfied by the function

$$f(x) = g(x) + \sum_{i=1}^{\infty} \mathbf{E}g(A_i A_{i-1} \ldots A_1 x) = g(x) + \sum_{i=1}^{\infty} \mathbf{E}g(x_i)$$

which is homogeneous of degree p. Setting $f(x) = (Fx,x)$, $g(x) = (Gx,x)$ in Theorems 6.1 and 6.2, we get

COROLLARY 1. *A necessary condition for the process* X_n *defined by* (6.1) *to be asymptotically stable in mean square is that, for every positive definite matrix* G, *the solution* F *of the equation*

$$\mathbf{E}(A^*FA) - F = -\, G$$

be a positive definite matrix. The same condition with the phrase "for every ... " replaced by "for some ... " is also sufficient.

COROLLARY 2. *If the assumptions of Theorem* 6.1 *hold with* $p > 0$, *then almost surely*

$$\|A_n A_{n-1} \ldots A_1\| \to 0 \quad \text{as} \quad n \to \infty. \tag{6.8}$$

On the other hand, if the assumptions hold for $p < 0$, then almost surely

$$\|A_n A_{n-1} \dots A_1\| \to \infty \quad as \quad n \to \infty. \tag{6.9}$$

To prove this, we observe that by the assumptions of the theorem the sequence $f(x_n) = f(A_n \dots A_1 x_0)$ is a positive supermartingale. By Theorem 5.2.1 the limit

$$\lim_{n\to\infty} f(A_n A_{n-1} \dots A_1 x_0) = \xi \qquad \text{(a.s.)},$$

is finite and $\mathbf{E}\xi \leqslant \lim_{n\to\infty} \mathbf{E}f(x_n) = 0$. Consequently, $\xi = 0$ almost surely. This at once yields both assertions (6.8) and (6.9).

EXAMPLE. Let G be the group of orthogonal real $\ell \times \ell$ matrices and let $g \in G$, $h \in G$. It is well known (see Halmos [1]) that there exists a unique Borel measure μ (Haar measure) such that $\mu(G) = 1$ and, for any nonempty open set U and any $g \in U$, we have $\mu(U) > 0$, $\mu(gU) = \mu(Ug) = \mu(U)$. The integral with respect to μ is invariant in the sense that

$$\int_G f(g)\mu(dg) = \int_G f(gh)\mu(dg) = \int_G f(hg)\mu(dg). \tag{6.10}$$

We now assume that $A_i = B_1 g_i B_2$, where B_1 and B_2 are fixed nonsingular matrices and $g_i \in G$ are independent matrices distributed over G in accordance with the Haar measure. We shall determine sufficient conditions for (6.8) and (6.9) to hold in this case.

Set $f(x) = |B_2 x|^p$. Then $f(x)$ is obviously a positive definite and homogeneous function of degree p satisfying

$$\mathbf{E}f(Ax) = \int_G |B_2 B_1 g B_2 x|^p \mu(dg).$$

Let e denote some fixed unit vector in E_ℓ, say $e = (1,0,0,\dots,0)$. Then it is obvious that for any $x \neq 0$ there exists a matrix $g_0 \in G$ such that $g_0 e = B_2 x \ |B_2 x|$. Using this and (6.10), we get

$$\mathbf{E}f(Ax) - f(x) = |B_2 x|^p \left(\int_G |B_2 B_1 g g_0 e|^p \mu(dg) - 1\right)$$

$$= |B_2 x|^p \left(\int_G |B_2 B_1 g e|^p \mu(dg) - 1\right).$$

Applying Theorem 6.1 and its Corollary 2, we get the following result.

If for some $p > 0$

$$\int |B_2B_1ge|^p \mu(dg) - 1 < 0, \qquad (6.11)$$

then

$$||A_nA_{n-1}\ldots A_1|| = ||B_1g_nB_2B_1g_{n-1}B_2\ldots B_1g_1B_2|| \to 0 \quad \text{(a.s.)}.$$

Similarly, a sufficient condition for $||A_n\ldots A_1|| \to \infty$ to hold almost surely as $n \to \infty$ is that (6.11) holds for some $p < 0$.

This result may be given a more convenient form. Using the expansion of a^p in powers of p and the inequalities

$$0 < k_1 < |B_2B_1ge| < k_2,$$

we easily see that condition (6.11) holds for sufficiently small $p > 0$ if

$$I = \int_G \ln|B_2B_1ge|\mu(dg) < 0, \qquad (6.12)$$

and for sufficiently small $p < 0$ if $I > 0$. Thus the condition $I < 0$ guarantees that the chain (6.1) of this example is stable; the condition $I > 0$, implies that it is unstable[6].

It is natural to call $I = 0$ the "critical" case. The matrix product is then either unstable or nonasymptotically stable (the latter possibility occurs, for instance, if $B_i \in G$).

It is evident that in this example we may replace G by an subgroup of G or by the group of unitary matrices.

In particular, let us consider the group of rotations of the plane. The element of this group corresponding to the rotation by an angle φ_i is the matrix

$$g_i = \begin{pmatrix} \cos\varphi_i & -\sin\varphi_i \\ \sin\varphi_i & \cos\varphi_i \end{pmatrix}.$$

Let φ_i be independent random variables, uniformly distributed on the interval $[0,2\pi]$, and let $B_2B_1 = \begin{pmatrix} \lambda_1 & 0 \\ 0 & \lambda_2 \end{pmatrix}$. It follows from the foregoing formulas that the product

$$g_n\begin{pmatrix} \lambda_1 & 0 \\ 0 & \lambda_2 \end{pmatrix}g_{n-1}\begin{pmatrix} \lambda_1 & 0 \\ 0 & \lambda_2 \end{pmatrix}\ldots g_1$$

almost surely converges to 0 if

[6] The above example is due to V.N. Tutubalin, who derived condition (6.12) by a different method.

$$I = \frac{1}{2\pi}\int_0^{2\pi} \ln(\lambda_1^2\cos^2\varphi + \lambda_2^2\sin^2\varphi)d\varphi$$

and to infinity if $I > 0$.

In this example we have verified condition (6.12) by letting $p \to +0$ in (6.11). In the next section we shall consider another approach where a sufficient stability condition of type (6.12) will be derived by a different method and in greater generality. Instead of the Haar measure $\mu(dg)$ we shall have to consider a different measure, related in a natural manner to the distribution of the matrices A_i.

7. Asymptotic stability of linear systems with constant coefficients

1. Let us again consider the Markov chain (6.1). Set $\lambda_n = x_n/|x_n|$. The equality $x_n = A_n x_{n-1}$ is equivalent to

$$\lambda_n = \frac{A_n\lambda_{n-1}}{|A_n\lambda_{n-1}|} \,. \tag{7.1}$$

It follows from (7.1) that the sequence $\lambda_0,\lambda_1,\ldots$ is also a temporally homogeneous Markov chain on the sphere $S_\ell = \{|x| = 1\}$ in E_ℓ. It is readily seen that the transition probability $P_0(\lambda,A)$ of this chain has the Feller property (i.e., the function $\int P_0(\lambda,dy)f(y)$ is continuous if $f(\lambda)$ is continuous). Proceeding as in the proof of Theorem 3.2.1, using the compactness of the phase space and the Feller property of the transition function, we easily prove that the chain $\lambda_0,\lambda_1,\ldots$ has a stationary probability distribution. Suppose that the chain is ergodic and let $\nu(d\lambda)$ be its unique stationary distribution. Set $\rho_n = \ln|x_n|$. Obviously

$$\rho_n = \rho_{n-1} + \ln|A_n\lambda_{n-1}| = \rho_0 + \sum_{i=1}^{n} \ln|A_i\lambda_{i-1}|. \tag{7.2}$$

Since the matrices $A_1,A_2,\ldots$ are independent, the pairs $X_n = \{A_n,\lambda_n\}$ $(n = 1,2,\ldots)$ also form a Markov chain. This chain has the stationary distribution $P(dA)\nu(d\lambda)$ in the phase space $\mathcal{A}\times S_\ell$, where $\mathcal{A}$ is the set of real $\ell\times\ell$ matrices. Under fairly broad assumptions about the function $f(A,\lambda)$, $A\in\mathcal{A}$, $\lambda\in S_\ell$, we have the following form of the strong law of large numbers (see Doob [3], Chapter V, Section 6): If

$$\iint |f(A,\lambda)|\mathbf{P}(dA)\nu(d\lambda) < \infty,$$

then the limit

$$\lim_{n\to\infty} \frac{1}{n} \sum_{k=1}^{n} f(A_k,\lambda_k)$$

exists almost surely for any x_0 and is equal to

$$\iint f(A,\lambda)\mathbf{P}(dA)\nu(d\lambda).$$

Suppose that

$$\iint |\ln|A\lambda||\mathbf{P}(dA)\nu(d\lambda) < \infty. \tag{7.3}$$

Then, applying the strong law of large numbers to (7.2), we see that almost surely

$$\lim_{n\to\infty} \frac{\rho_n}{n} = a, \quad \text{where} \quad a = \iint \ln|A\lambda|\mathbf{P}(dA)\nu(d\lambda). \tag{7.4}$$

Since $\rho_n = \ln|A_n A_{n-1}\ldots A_1 x_0|$, it follows from (7.4) that, almost surely,

$$||A_n A_{n-1}\ldots A_1|| \to 0 \quad \text{as} \quad n \to \infty, \quad \text{if} \quad a < 0,$$

$$||A_n A_{n-1}\ldots A_1|| \to \infty \quad \text{as} \quad n \to \infty, \quad \text{if} \quad a > 0.$$

$a = 0$ is the critical case. As can be seen from the example in Section 6, the chain (6.1) could then possibly be stable but not asymptotically so. The typical case here is nevertheless instability.

Let us study this case in more detail. It is well known that under certain additional conditions the function $f(A_n,\lambda_n)$ satisfies the central limit theorem (see Doob [3], Chapter V.7). Assuming that these conditions are satisfied and using (7.2), we get

$$\mathbf{P}\left\{\frac{\rho_n - \mathbf{E}\rho_n}{\sqrt{n}} < \lambda\right\} \to \frac{1}{\sigma\sqrt{2\pi}}\int_{-\infty}^{\lambda} e^{-y^2/2\sigma^2} dy. \tag{7.5}$$

(We are assuming here that $\sigma^2 = \lim_{n\to\infty} \frac{1}{n} D\rho_n \neq 0$. Since $a = 0$, we may apply Lemma 7.2 in Doob [3], Chapter V, to conclude that for some constant

$$|\mathbf{E}\rho_n| = \left|\sum_{i=1}^{n} \mathbf{E}\ln|A_i\lambda_{i-1}|\right| \leqslant k.$$

Hence it follows by (7.5) that the probability of the event $\rho_n > \lambda\sqrt{n}$ does not tend to zero as $n \to \infty$.

In particular, one sees from (7.5) that the approach presented here may also be used to obtain further results on products of random matrices (see Kesten and Furstenberg [1]).

REMARK. The number a can be effectively calculated from (7.4) only when an invariant measure of the chain $\lambda_0, \lambda_1, \ldots$ is known. An integral equation for this measure is easily set up, but its general solution is fraught with difficulties. In some special cases one can easily compute the invariant measure of the chain $\tilde{\lambda}_n$ obtained by "projecting" the chain (6.1) onto the ellipsoid $|Bx| = 1$ instead of the sphere $|x| = 1$. Next one replaces (7.1) and (7.2) by the formulas

$$\tilde{\lambda}_n = \frac{x_n}{|Bx_n|}; \qquad \tilde{\rho}_n = \ln|Bx_n| = \tilde{\rho}_{n-1} + \ln|BA_n\tilde{\lambda}_{n-1}|.$$

This method is also readily applicable to the example considered at the end of Section 6[7].

2. It is natural to ask questions about the behavior of the norms of products of random matrices when each matrix is close to the unit matrix. Limit theorems for this case are discussed in the interesting book of Grenander [1]. Grenander shows that under certain natural assumptions, the limiting distribution of a product of random matrices close to the unit matrix coincides with the distribution of a homogeneous multiplicative matrix-valued stochastic process. This process is governed by a system of linear stochastic equations with constant coefficients and a matrix initial condition. Thus we are again faced with the need to study stability conditions for linear systems of stochastic differential equations

$$dX(t) = BX(t)dt + \sum_{r=1}^{k} \sigma_r X(t) d\xi_r(t) \tag{7.6}$$

with constant coefficients b_i^j and σ_{ir}^j.

To derive necessary and sufficient conditions for asymptotical stability of the system (7.6) we employ the same method as in the case of matrix products.

As before, we set

$$a_{ij}(x) = \sum_{r=1}^{k} \sum_{n,m=1}^{\ell} \sigma_{ir}^n \sigma_{jr}^m x_n x_m; \qquad A(x) = ((a_{ij}(x)));$$

[7] The construction described in this subsection resembles that used by Kesten and Furstenberg in [1]. Sufficient conditions for ergodicity of the Markov chain λ_n on the sphere $|\lambda| = 1$ with identified antipodal points follow from a theorem of Furstenberg [1]. See also the survey article of Sazonov and Tutubalin [1].

$$L = \frac{\partial}{\partial t} + \left(Bx, \frac{\partial}{\partial x}\right) + \frac{1}{2}\left(A(x)\frac{\partial}{\partial x}, \frac{\partial}{\partial x}\right)$$

$$= \frac{\partial}{\partial t} + \left(Bx, \frac{\partial}{\partial x}\right) + \frac{1}{2}\sum_{r=1}^{k}\left(\sigma_r x, \frac{\partial}{\partial x}\right)^2 .$$

It follows from the general properties of the operator L that the matrix $A(x)$ is positive semidefinite. To simplify matters, we first assume that this matrix is also non-degenerate in the sense that there is a constant m such that for any vector $\alpha = (\alpha_1,\ldots,\alpha_\ell)$

$$(A(x)\alpha,\alpha) = \sum_{r=1}^{k} (\sigma_r x,\alpha)^2 \geqslant m|x|^2|\alpha|^2. \tag{7.7}$$

As in Subsection 1, we introduce new variables:

$$\lambda = \frac{x}{|x|}; \qquad \rho = \ln|x|.$$

The process $\Lambda(t) = X(t)/|X(t)|$ on the sphere $S_\ell = \{|x| = 1\}$ is Markovian and temporally homogeneous[8]. To verify this, we need only use Itô's formula (3.3.8). We find it from expressions for $d\Lambda_i(t)$, from which it is clear that the coefficients of dt and $d\sigma_r(t)$ depend only on $\Lambda_1(t),\ldots,\Lambda_\ell(t)$. Condition (7.7) is sufficient for the process $\Lambda(t)$ to be ergodic, since it guarantees that the transition probability has an everywhere positive density (see Chapter IV). Let $\nu(d\lambda)$ denote the unique invariant measure of the process on the sphere.

Next, let $\rho(t) = \ln|X(t)|$. Using Itô's formula, we get

$$d\rho(t) = L\rho(t) + \sum_{r=1}^{k} (\sigma_r\Lambda(t),\Lambda(t))d\xi_r(t)$$

$$= \left[(B\Lambda(t),\Lambda(t)) + \frac{1}{2}\operatorname{tr}A(\Lambda(t)) - (A(\Lambda(t))\Lambda(t),\Lambda(t))\right]$$

$$+ \sum_{r=1}^{k} (\sigma_r\Lambda(t),\Lambda(t))d\xi_r(t). \tag{7.8}$$

As expected in view of the analogy with formula (7.2), the increment of the function $\rho(t)$ is a functional of the process $\Lambda(t)$ and the Wiener processes $\xi_r(t)$. We set

---

[8] It is easy to see that the process obtained from $\Lambda(t)$ by identifying antipodal points $\Lambda(t)$ and $-\Lambda(t)$ is also Markovian.

$$Q(\lambda) = (B\lambda,\lambda) + \frac{1}{2}\,\mathrm{tr}A(\lambda) - (A(\lambda)\lambda,\lambda),$$

$$a = \int_{S_\ell} Q(\lambda)\nu(d\lambda).$$

THEOREM 7.1. *Suppose that condition* (7.7) *is satisfied and* $a < 0$. *Then the solution* $X(t) \equiv 0$ *of the system* (7.6) *is almost surely asymptotically stable. On the other hand, if* $a > 0$, *then for* $x \not\equiv 0$

$$\mathbf{P}\{|X^x(t)| \xrightarrow[t\to\infty]{} \infty\} = 1. \tag{7.9}$$

To prove this we need

LEMMA 7.1. *Let* $\sigma(t,\omega)$ *be a function such that* $\mathbf{E}\sigma^2(t,\omega) < k^2$ *(*$k > 0$ *is a constant). Then the stochastic integral* $\int_0^t [\sigma(s,\omega)\times d\xi(s,\omega)]$ *almost surely satisfies the relation*

$$\frac{1}{T}\int_0^T \sigma(t,\omega)d\xi(t,\omega) \to 0 \quad \text{as} \quad T \to \infty.$$

PROOF. Let $A_{n,m}$ denote the event

$$\left\{\sup_{T>2^n} \frac{1}{T}\left|\int_0^T \sigma(t)d\xi(t)\right| > \frac{1}{m}\right\}, \qquad A^{(m)} = \bigcap_{n=1}^{\infty} A_{n,m}.$$

By virtue of the relation between events

$$B = \left\{\overline{\lim_{T\to\infty}}\, \frac{1}{T}\left|\int_0^T \sigma(t,\omega)d\xi(t)\right| > 0\right\} = \bigcup_{m=1}^{\infty}\bigcup_{n=1}^{\infty} A_{n,m};$$

$$A_{n,m} \supset A_{n+1,m}; \qquad A^{(m+1)} \supset A^{(m)}$$

we get

$$\mathbf{P}(B) = \lim_{m\to\infty}\lim_{n\to\infty} \mathbf{P}(A_{n,m}).$$

The lemma will be proved if we show that $\lim_{n\to\infty} \mathbf{P}(A_{n,m}) = 0$ for any $m > 0$. Since the process $\int_0^t \sigma(s,\omega)d\xi(s)$ is a martingale, it follows from Theorem 5.2.2 that

$$\mathbf{P}\left\{\sup_{2^r \leqslant T \leqslant 2^{r+1}} \left|\int_0^T \sigma(t,\omega)d\xi(t)\right| > \varepsilon\right\} \leqslant \frac{1}{\varepsilon}\mathbf{E}\left|\int_0^{2^{r+1}} \sigma(t,\omega)d\xi(t)\right|$$

$$\leqslant \frac{1}{\varepsilon}\left[\mathbf{E}\left(\int_0^{2^{r+1}} \sigma(t,\omega)d\xi(t)\right)^2\right]^{1/2}$$

$$\leqslant \frac{k2^{(r+1)/2}}{\varepsilon}.$$

Setting $\varepsilon = \varepsilon_0 2^r$, we get the estimate

$$\mathbf{P}\left\{\sup_{2^r \leqslant T \leqslant 2^{r+1}} \frac{1}{T}\left|\int_0^T \sigma(t,\omega)d\xi(t)\right| > \varepsilon_0\right\} < \frac{k\sqrt{2}}{\varepsilon_0} 2^{-t/2}.$$

This estimate implies the inequality

$$\mathbf{P}(A_{n,m}) \leqslant \sum_{r=n}^{\infty} \mathbf{P}\left\{\sup_{2^r \leqslant T \leqslant 2^{r+1}} \frac{1}{T}\left|\int_0^T \sigma(t,\omega)d\xi(t)\right| > \frac{1}{m}\right\}$$

$$\leqslant km\sqrt{2} \sum_{r=n}^{\infty} 2^{-r/2}.$$

Thus $P(A_{n,m}) \to 0$ as $n \to \infty$. Lemma 7.1 is proved.

PROOF OF THEOREM 7.1. We can rewrite (7.8) as

$$\frac{\rho(T) - \rho(0)}{T}$$

$$= \frac{1}{T}\int_0^T Q(\Lambda(t))dt + \frac{1}{T}\sum_{r=1}^{k} \int_0^T (\sigma_r\Lambda(t),\Lambda(t))d\xi_r(t). \tag{7.10}$$

It follows from the strong law of large numbers for the process $\Lambda(t)$, and from Lemma 7.1, that almost surely

$$\lim_{T\to\infty} \frac{\rho(T)}{T} = a. \tag{7.11}$$

This implies both parts of the theorem.

THEOREM 7.2. *Suppose that condition* (7.7) *is satisfied and* $a = 0$. *Then the solution* $X(t) \equiv 0$ *of the system* (7.6) *is neither asymptotically stable nor asymptotically unstable in the sense of* (7.9).

PROOF. Suppose that the solution $X(t) \equiv 0$ is asymptotically stable in probability.

Then, by Theorem 4.1, this solution is exponentially p-stable for all sufficiently small $p > 0$. Hence, by (7.8) and using Jensen's inequality

$$\mathbf{E}(\exp\xi) \geqslant \exp\{\mathbf{E}\xi\}$$

(see Doob [3]), we get

$$|x_0|^p \exp\left\{p\int_0^T \mathbf{E}Q(\Lambda(s))ds\right\} \leqslant \mathbf{E}|X^x(T)|^p < Ae^{-\alpha T}.$$

These inequalities imply that

$$\lim_{T\to\infty} \frac{1}{T}\int_0^T \mathbf{E}Q(\Lambda(s))ds < 0,$$

but this contradicts the assumption $a = 0$. One proves similarly that the solution $X(t) \equiv 0$ cannot satisfy condition (7.9).

8. Systems with constant coefficients (continued)

In the preceeding section we studied conditions implying asymptotical stability of the system (7.6). We assumed the non-degeneracy condition (7.7). We now consider what modifications must be made in the arguments of Section 7 if the diffusion matrix is allowed to degenerate on certain curves, surfaces, or even everywhere.

We first observe that the essential point for all arguments in Section 7 is not so much condition (7.7) as one of its consequences, namely the ergodicity of the Markov process $\Lambda(t)$ on the sphere $|x| = 1$[9]. Now suppose that ergodicity fails to hold and that the path of the Markov process

$$\Lambda^{\lambda_0}(t) = \frac{X^x(t)}{|X^x(t)|} \qquad (\lambda_0 = x/|x|),$$

satisfying the initial condition $\Lambda^{\lambda_0}(t) = \lambda_0$, may belong to different ergodic components A of the process for different λ_0.

---

[9] It is readily seen that the following weaker than (7.7) condition is sufficient: For all vectors λ and μ such that $|\lambda| = |\mu| = 1$, $\lambda \neq \pm\mu$,

$$\sum_{r=1}^{k} (\sigma_r\lambda - (\sigma_r\lambda,\lambda)\lambda,\mu)^2 > 0. \qquad (7.7')$$

Let $\mu_A(d\lambda)$ denote the stationary initial distribution corresponding to the component A.

Applying the strong law of large numbers for the component A and Lemma 7.1, we see as in the proof of Theorem 7.1 that

$$\lim_{t\to\infty} \frac{\ln|X^x(t)|}{t} = a_A = \int Q(\lambda)\mu_A(d\lambda)$$

for μ_A-almost all values of $\lambda = x/|x|$.

Following the proofs of Theorems 7.1 and 7.2, we see that a necessary condition for the system (7.6) to be asymptotically stable is that

$$a_A < 0 \tag{8.1}$$

holds for all ergodic components A. A sufficient condition is that for all $x \in E_\ell$,

$$\lim_{t\to\infty} \frac{\ln|X^x(t)|}{t} < 0$$

holds almost surely. The process $\Lambda(t)$ may have infinitely many ergodic components. For instance, the ergodic components of the process $\Lambda(t)$ for the deterministic system $dx/dt = x$, $dy/dt = y$ are all the points of the circle $x^2 + y^2 = 1$.

Therefore it might seem at first sight that (8.1) may involve an infinite set of conditions. However, as we shall show below, the number a_A can take on at most ℓ distinct values. We shall first prove a simple lemma which is usually employed to investigate the properties of Ljapunov's characteristic numbers (see Malkin [2], p. 328).

LEMMA 8.1. *Let $X_i(t)$ $(i = 1,2)$ be E_ℓ-valued functions such that*

$$\overline{\lim_{t\to\infty}} \frac{1}{t} \ln|X_i(t)| = a_i < \infty.$$

Then

$$\overline{\lim_{t\to\infty}} \frac{1}{t} \ln|x_1X_1(t) + x_2X_2(t)| \leqslant \max(a_1,a_2).$$

PROOF. To be specific, suppose that $a_1 \leqslant a_2$. It follows from the assumptions of the lemma that

$$|X_i(t)| < e^{(a_\ell+\varepsilon)}$$

for any $\varepsilon > 0$ and all $t > T(\varepsilon)$. Therefore

$$|x_1X_1(t) + x_2X_2(t)| \leqslant |x_1|e^{(a_1+\varepsilon)t} + |x_2|e^{(a_2+\varepsilon)t}$$
$$\leqslant ke^{(a_2+\varepsilon)t}.$$

Hence the inequality

$$\overline{\lim_{t\to\infty}} \frac{1}{t} \ln|x_1X_1(t) + x_2X_2(t)| \leqslant a_2.$$

This proves the lemma.

If the process $\Lambda(t)$ is not ergodic, the following lemma may be useful in investigating the stability of the system (7.6).

LEMMA 8.2. *Suppose that there are ℓ linearly independent vectors $\lambda_1,\ldots,\lambda_\ell$ in E_ℓ such that almost surely*

$$\overline{\lim_{T\to\infty}} \frac{1}{T}\int_0^T Q(\Lambda^{\lambda_i}(t))dt < 0. \tag{8.2}$$

Then the system (7.6) *is asymptotically stable.*

PROOF. It follows from (7.10), Lemma 7.1 and (8.2) that

$$\overline{\lim_{t\to\infty}} \frac{1}{t} \ln|X^{\lambda_i}(t)| < 0$$

almost surely. Hence, by Lemma 8.1, we get

$$\overline{\lim_{t\to\infty}} \frac{1}{t} \ln|X^{x}(t)| = \overline{\lim_{t\to\infty}} \frac{1}{t} \ln\left|\sum_{i=1}^{\ell} k_iX^{\lambda_i}(t)\right| < 0,$$

where $x = \sum_{i=1}^{\ell} k_i\lambda_i$. This proves the lemma.

COROLLARY 1. *The process $\Lambda(t)$ has at most ℓ ergodic components A_i, corresponding to the different values of*

$$a_i = \int Q(\lambda)\mu_i(d\lambda).$$

(Here μ_i is the stationary distribution for the component A_i.) Moreover, if $a_1 < a_2 < \ldots < a_\ell$, then $a_\ell < 0$ is a sufficient condition for the system (7.6) *to be asymptotically stable.*

PROOF. Let $A_1,\ldots,A_k$ be the ergodic components of the process $\Lambda(t)$ and assume that the corresponding a_i are monotone increasing: $a_1 < a_2 < \ldots < a_k$. Let λ_i be vectors such that almost surely

$$\lim_{T\to\infty} \frac{1}{T}\int_0^T Q(\Lambda^{\lambda_i}(t))dt = a_i.$$

Then the vectors $\lambda_1,\ldots,\lambda_k$ are linearly independent. In fact, otherwise we have $\lambda_i = c_1\lambda_1 + \ldots + c_{i-1}\lambda_{i-1}$ for some $i \leqslant k$. Therefore

$$X^{\lambda_i}(t) = \sum_{j=1}^{i-1} c_j X^{\lambda_j}(t).$$

Hence, by Lemma 8.1,

$$a_i = \lim_{t\to\infty} \frac{1}{t} \ln|X^{\lambda_i}(t)| \leqslant \max(a_1,\ldots,a_{i-1}) = a_{i-1}.$$

This is a contradiction, and thus $\lambda_1,\ldots,\lambda_k$ are linearly independent. This, together with Lemma 8.2, implies the assertion.

Another obvious consequence of Lemma 8.2 is

COROLLARY 2. *If the process $\Lambda(t)$ has an ergodic component whose stationary distribution $\mu_A(d\lambda)$ is not concentrated on any hyperplane $\sum_{i=1}^{\ell} k_i\lambda_i + k_0 = 0$, then a sufficient condition for asymptotic stability of the system* (7.6) *is that* $a = \int Q(\lambda)\mu_A(d\lambda) < 0$.

Let us consider the case $\ell = 2$ in more detail. Let $\lambda(\varphi)$ denote the vector in the plane with components $\lambda_1 = \cos\varphi$; $\lambda_2 = \sin\varphi$, $\hat{\lambda}(\varphi) = -d\lambda(\varphi)/d\varphi$. As we have already stated, the stochastic process $\varphi(t)$ on the circle generated by the system (7.6) is Markovian. One shows easily that

$$d\varphi(t) = \Phi(\varphi(t))dt + \Psi(\varphi(t))d\xi(t), \tag{8.3}$$

where

$$\Psi^2(\varphi) = (A(\lambda(\varphi))\hat{\lambda}(\varphi),\hat{\lambda}(\varphi));$$

$$\Phi(\varphi) = -\,(B\lambda(\varphi),\hat{\lambda}(\varphi)) + (A(\lambda(\varphi))\lambda(\varphi),\hat{\lambda}(\varphi)),$$

and $\tilde{\xi}(t)$ is a Wiener process with zero mean such that $\mathbf{E}\xi^2(t) = t$.

We first assume that

$$\Psi^2(\varphi) > 0 \qquad (0 \leqslant \varphi < 2\pi). \tag{8.4}$$

(This is implied, for instance, by inequality (7.7)). Then the process $\varphi(t)$ has a stationary distribution which is absolutely continuous and has density $\mu(\varphi)$ relative to the uniform distribution on the circle. This density satisfies the Fokker-Planck-Kolmogorov equation, which in our case is

$$\frac{1}{2}\frac{d^2}{d\varphi^2}(\Psi^2(\varphi)\mu) - \frac{d}{d\varphi}(\Phi(\varphi)\mu) = 0. \tag{8.5}$$

Equation (8.5) has a unique solution satisfying the normalization condition

$$\int_0^{2\pi} \mu(\varphi)d\varphi = 1 \tag{8.6}$$

and the periodicity condition

$$\mu(0) = \mu(2\pi). \tag{8.7}$$

It is easy to see that this solution is given by

$$\mu(\varphi) = k\left[1 + \frac{W(2\pi) - 1}{\int_0^{2\pi} W(s)ds}\int_0^{\varphi} W(u)du\right][W(\varphi)\Psi^2(\varphi)]^{-1}, \tag{8.8}$$

where

$$W(\varphi) = \exp\left\{-2\int_0^{\varphi}\frac{\Phi(v)dv}{\Psi^2(v)}\right\},$$

and the constant k is determined by the normalization (8.6).

Applying Theorems 7.1 and 7.2 to the case $\ell = 2$, we obtain that

$$\int_0^{2\pi} Q(\lambda(\varphi))\mu(\varphi)d\varphi < 0 \tag{8.9}$$

is a necessary and sufficient condition for asymptotic stability, expressed in terms of quadratures.

We shall now allow that the function $\Psi^2(\varphi)$ may vanish. Since

$$\Psi^2(\varphi) = \sum_{r=1}^{k}(\sigma_r\lambda(\varphi),\hat{\lambda}(\varphi))^2,$$

it follows that apart from the trivial case of a deterministic system ($\sigma_r \equiv 0$), there are two possible cases: either

$$\Psi^2(\varphi) = 0 \tag{8.10}$$

only for $\sin\varphi = 0$ and for $\cos\varphi = 0$, or equation (8.19) is equivalent to a fourth-degree equation in $\text{tg}\varphi$ or $\text{ctg}\varphi$. In both cases (in view of the inequality $\Psi^2(\varphi) \geqslant 0$), equation (8.10) is satisfied by at most two values of φ in the interval $0 \leqslant \varphi < \pi$, each of which pairs off with another value differing from it by π. It follows that the process $\varphi(t)$ has at most four ergodic components if $\sigma_{ir}^k \not\equiv 0$.

Let us discuss the possibilities in greater detail.

1. Let φ_1, φ_2, $\varphi_1 + \pi$, $\varphi_2 + \pi$ $(0 \leqslant \varphi_1 < \varphi_2 < \pi)$ be the solutions of equation (8.10). Then one readily sees that the process $\varphi(t)$ is ergodic for $\text{sign}\Phi(\varphi_1) = \text{sign}\Phi(\varphi_2)$ and has two ergodic components if $\text{sign}\Phi(\varphi_1) = -\text{sign}\varphi(\varphi_2)$.

2. Suppose that equation (8.10) has two solutions φ_1 and $\varphi_1 + \pi$ in the interval $[0,2\pi]$. Then the process $\varphi(t)$ is always ergodic, provided $\Phi(\varphi_1) \not\equiv 0$.

3. If $\Phi(\varphi_k) = 0$, the process $\varphi(t)$ has stationary points at $\varphi = \varphi_k$ and $\varphi = \varphi_k + \pi$.

EXAMPLE. Consider the system

$$\left.\begin{aligned} dX_1(t) &= aX_1dt + \sigma_1X_1d\xi_1(t), \\ dX_2(t) &= bX_2dt + \sigma_2X_2d\xi_2(t), \end{aligned}\right\} \tag{8.11}$$

Then

$$\psi^2(\varphi) = (\sigma_1^2 + \sigma_2^2)\cos^2\varphi\sin^2\varphi,$$

$$\Psi(\varphi) = \sin\varphi\cos\varphi(\sigma_1^2\cos^2\varphi - \sigma_2^2\sin^2\varphi - a + b).$$

Thus the points $\varphi_k = k\pi/2$ $(k = 0,1,2,3)$ are stationary points of the process $\varphi(t)$. All the invariant measures of the process $\varphi(t)$ are concentrated at these points. Hence it follows that the system (8.11) is stable if and only if both its components are stable. This conclusion may also be derived directly, since the components of the process (8.11) are independent.

More substantial examples follow in the next section.

9. Two examples

EXAMPLE 1. A much discussed question in the literature (see Samuels [1], Leibowitz [1] and others) is whether a linear deterministic system can be stabilized with artificially disturbing its parameters by white noise. The problem is understood by different authors in different ways, since the stochastic equation obtained by superimposing noise may be made rigorous in various ways. As mentioned in Chapter V, a natural approach is to study the problem for "physically feasible" noise in the sense of Section 5.5. We have already seen that one-dimensional systems cannot be stabilized by physically feasible noise.

We now consider the system

$$\left.\begin{aligned} dX_1(t) &= b_1X_1dt + \sigma(X_1d^*\xi_1(t) + X_2d^*\xi_2(t)), \\ dX_2(t) &= b_2X_2dt + \sigma(X_2d^*\xi_2(t) + X_2d^*\xi_2(t)), \end{aligned}\right\} \quad (9.1)$$

in E_ℓ, where the $d^*\xi_i(t)$ are Stratonovic differentials. The differential generator of this process is obviously (see (5.5.9))

$$L = \frac{\partial}{\partial t} + \left(b_1 + \frac{\sigma^2}{2}\right)x_1\frac{\partial}{\partial x_1} + \left(b_2 + \frac{\sigma^2}{2}\right)x_2\frac{\partial}{\partial x_2} + \frac{\sigma^2}{2}(x_1^2 + x_2^2)\left(\frac{\partial^2}{\partial x_1^2} + \frac{\partial^2}{\partial x_2^2}\right).$$

Hence, using the formulas of Section 8, we easily see that

$$\mu(\varphi) = c\exp\left\{\frac{b_1 - b_2}{\sigma^2}\cos^2\varphi\right\};$$

$$Q(\lambda(\varphi)) = \frac{\sigma^2}{2} + b_1\cos^2\varphi + b_2\sin^2\varphi.$$

Applying Theorems 7.1 and 7.2, we see that the condition

$$a = \int_0^{\pi/2}\left(\frac{\sigma^2}{2} + b_1\cos^2\varphi - b_2\sin^2\varphi\right) \times\exp\left\{\frac{b_1 - b_2}{\sigma^2}\cos^2\varphi\right\}d\varphi < 0 \quad (9.2)$$

is necessary and sufficient for the system (9.1) to be asymptotically stable.

We may easily give another form to condition (9.2) by using the well-known integral representation

$$I_n(z) = \frac{(-1)}{\pi}\int_0^{\pi} e^{-z\cos\vartheta}\cos\vartheta d\vartheta \quad (9.3)$$

for the Bessel function $I_n(z)$ of a pure imaginary argument. In fact, if we set $\varkappa = (b_1 - b_2)/\sigma^2$, formulas (9.2) and (9.3) yield the condition

$$1 + \frac{2b_1}{\sigma^2} < \varkappa\left(1 - \frac{I_1(\varkappa/2)}{I_0(\varkappa/2)}\right) \tag{9.4}$$

which is equivalent to (9.2).

The asymptotic representation for the function $I_n(z)$ as $z \to \infty$ shows that the last inequality is valid if, say, $b_2 < 0$ is chosen with sufficiently large absolute value, and $b_1/\sigma^2 < 3/8$.

We have thus proved that for suitable choice of $b_1 > 0$ and $b_2 < 0$ the unstable deterministic system

$$\frac{dx_1}{dt} = b_1 x_1; \qquad \frac{dx_2}{dt} = b_2 x_2 \tag{9.5}$$

becomes asymptotically stable when its parameters are perturbed by certain physically feasible white noise processes. This result is valid for any deterministic system which is reducible to the canonical form (9.5) by a linear transformation. Indeed, linear transformations do not affect the stability properties of a system.

EXAMPLE 2. Consider a linear system with natural frequency ω, subject to the action of a friction force proportional to velocity with coefficient k. This system is described by the equation

$$\ddot{x} + k\dot{x} + \omega^2 x = 0. \tag{9.6}$$

It is evident that this system is stable for $k > 0$. However, in many problems it seems natural to assume that k is merely the mean value of the frictional coefficient, while its true value is a stochastic process with small correlation interval. It is extremely interesting to study "bifurcation" values of the noise intensity, i.e., values for which the system first becomes unstable.

A limiting case of this problem is to determine a constant σ_0 such that for $\sigma < \sigma_0$ the solution $x = x' \equiv 0$ of the stochastic equation

$$\ddot{x} + (k + \sigma\dot{\xi})\dot{x} + \omega^2 x = 0 \tag{9.7}$$

is asymptotically stable, while for $\sigma > \sigma_0$ it is unstable. The process $\dot{\xi}(t)$ in equation (9.7) is white noise of unit intensity. We shall interpret this equation in the sense described in Section 5.5, as a Stratonovič stochastic equation.

Setting $x_1 = \omega x$, $x_2 = \dot{x}$, we get the system

$$\left.\begin{aligned} dx_1 &= \omega x_2 dt \\ dx_2 &= -(kx_2 + \omega x_1)dt - \sigma x_2 d^* \xi(t) \end{aligned}\right\} \tag{9.8}$$

with differential operator

$$L = \frac{\partial}{\partial t} + \omega x_2 \frac{\partial}{\partial x_1} + \left[\left(\frac{\sigma^2}{2} - k\right)x_2 - \omega x_1\right]\frac{\partial}{\partial x_2} + \frac{1}{2}\sigma^2 x_2^2 \frac{\partial^2}{\partial x_2^2} .$$

We shall first find a sufficient condition for stability in mean square, using the algorithm described in Section 3. Our aim is to find conditions under which a quadratic form

$$W = \frac{A}{2} x_1^2 + Bx_1x_2 + \frac{C}{2} x_2^2,$$

which satisfies equation

$$LW = - x_2^2 - x_2^2$$

is positive definite.

Simple arguments lead to the equalities

$$C = \frac{2}{k - \sigma^2} ; \qquad B = \frac{1}{\omega} ;$$

$$A = \frac{2}{k - \sigma^2} + \frac{1}{\omega^2}\left(k - \frac{\sigma^2}{2}\right) .$$

It is clear that the form W is positive definite if and only if $\sigma^2 < k$. Hence we get a lower bound for the bifurcation value of the noise: $\sigma_0^2 > k$.

To derive an equation for σ_0^2, we use the results of Section 8. According to these, we get that $\sigma = \sigma_0$ satisfies, for a process $\Lambda(t)$ which is ergodic on the circle $|x| = 1$, the equation

$$\int_0^{2\pi} Q(\lambda(\varphi))\mu(\varphi)d\varphi = 0. \tag{9.9}$$

We easily see from (9.8) and the formulas of Section 8 that

$$\Psi^2(\varphi) = \sigma^2\sin^2\varphi\cos^2\varphi,$$

$$\Phi(\varphi) = \left(\frac{\sigma^2}{2} - k\right)\sin\varphi\cos\varphi + \sigma^2\sin^3\varphi\cos\varphi - \omega,$$

$$\begin{aligned} Q(\lambda(\varphi)) &= \left(\frac{\sigma^2}{2} - k\right)\sin^2\varphi - \frac{1}{2}\sigma^2\sin^2\varphi - \sigma^2\sin^4\varphi \\ &= \sin^2\varphi(\sigma^2\cos^2\varphi - k). \end{aligned} \tag{9.10}$$

The function $\Psi^2(\varphi)$ vanishes for $\varphi_k = k\pi/2$ $(k = 0,1,2,3)$. However, Φ does not vanish at these points $(\Phi(\varphi_k) = -\omega)$. Hence the process $\Lambda(t)$ described by equation (8.3) is ergodic. Let us determine the density $\mu(\varphi)$ of its invariant measure. In this case we cannot use formula (8.8), since equation (8.5) has singularities at the points $\varphi = \varphi_k$. Nevertheless, the function $\mu(\varphi)$ satisfies equation (8.5) for $\varphi \neq \varphi_k$. Moreover, it can be shown that μ is bounded and continuous in the neighbourhood of the points $\varphi = \varphi_k$ and that it also satisfies condition (8.6). It is readily verified that the unique solution of equation (8.5) satisfying these additional conditions in the function

$$\mu(\varphi) = \begin{cases} c\int_0^{\varphi} W(u)du[W(\varphi)\Psi^2(\varphi)]^{-1} & \text{for} \quad 0 \leqslant \varphi < \frac{\pi}{2}, \\ c\int_{-\pi/2}^{\varphi} W(u)du[W(\varphi)\Psi^2(\varphi)]^{-1} & \text{for} \quad -\frac{\pi}{2} \leqslant \varphi < 0, \\ \mu(\varphi - \pi) & \text{for} \quad \frac{\pi}{2} \leqslant \varphi < \frac{3\pi}{2}, \end{cases} \tag{9.11}$$

where

$$W(\varphi) = \cos^{-2}\varphi(\mathrm{tg}^2\varphi)^{(2k-\sigma^2)/2\sigma^2}\exp\left\{-\frac{4\omega}{\sigma^2}\,\mathrm{ctg}2\varphi\right\},$$

and the constant c is determined by condition (8.6).

It follows from (9.9), (9.10) and (9.11) that the constant σ_0^2 satisfies the equation

$$F(\sigma^2,k,\omega) \equiv \int_{-\pi/2}^{0} \frac{\sin^2\varphi(\sigma^2\cos^2\varphi - k)}{W(\varphi)\Psi^2(\varphi)} \int_{-\pi/2}^{\varphi} W(v)dvd\varphi \tag{9.12}$$

$$+ \int_0^{\pi/2} \frac{\sin^2\varphi(\sigma^2\cos^2\varphi - k)}{W(\varphi)\Psi^2(\varphi)} \int_0^{\varphi} W(v)dvd\varphi = 0.$$

From this we easily derive our previous estimate $\sigma_0^2 > k$. The parameters k, σ^2 and ω have the same dimension. We can therefore replace them by two dimensionless quantities, say

$$\hat{\sigma}^2 = \frac{\sigma^2}{k}; \qquad \hat{\omega} = \frac{\omega}{k}.$$

When written in these new variables, equation (9.12) becomes

$$F(\hat{\sigma}^2,1,\hat{\omega}) \equiv \hat{F}(\hat{\sigma}^2,\hat{\omega}) = 0. \tag{9.13}$$

This equation is fairly involved. Nevertheless, it enables us to investigate the dependence of the dimensionless critical noise power σ_0^2 on the dimensionless frequency $\hat{\omega}$ for the limiting cases $\hat{\omega} \to 0$ and $\hat{\omega} \to \infty$.

It is easy to see that $\hat{\sigma}_0^2 \to \infty$ as $\hat{\omega} \to 0$; in other words, low-frequency oscillations of the system are stable under very strong perturbations of the friction coefficient.

Let us now investigate in greater detail the other limiting case $\hat{\omega} \to \infty$. The quantity σ_0^2 may now be determined by Laplace approximation of the integrals. To this end, we transform the variables in (9.13) by the formulas

$$-\operatorname{ctg} 2\varphi = z, \qquad -\operatorname{ctg} 2v = u.$$

We introduce the notation

$$R_1(z) = \frac{\sqrt{z^2+1} + zz}{2\sqrt{z^2+1}}, \qquad R_2(z) = \frac{\sqrt{z^2+1} - z}{2\sqrt{z^2+1}},$$

$$R_3(z) = \frac{R_2(z)}{R_1(z)}, \qquad \alpha = \frac{2-\hat{\sigma}^2}{2\hat{\sigma}^2}.$$

Then

$$\hat{F}(\hat{\sigma}^2,\hat{\omega}) = \frac{1}{4\hat{\sigma}^2}\left[\int_{-\infty}^{\infty} \frac{dz(\hat{\sigma}^2 R_1(z) - 1)}{(z^2+1)R_3^{\alpha}(z)} \int_{-\infty}^{z} \frac{du R_3^{\alpha}(u) e^{4\hat{\omega}\hat{\sigma}^{-1}(u-z)}}{(u^2+1)R_1(u)} \right.$$

$$\left. + \int_{-\infty}^{\infty} \frac{dz(\hat{\sigma}^2 R_2(z) - 1)R_3^{\alpha}(z)}{(z^2+1)} \int_{-\infty}^{z} \frac{du e^{4\hat{\omega}\hat{\sigma}^{-2}(u-z)}}{(u^2+1)R_3^{\alpha}(u)R_2(u)}\right].$$

Applying the asymptotic formula

$$\int_{-\infty}^{z} \varphi(u) e^{\lambda(u-z)} dz = \frac{1}{\lambda}\varphi(z) + O\left(\frac{1}{\lambda^2}\right) \qquad (\lambda \to \infty)$$

(Lavrent'ev and Šabat [1], pp. 446-450), we see that

$$\hat{F}(\hat{\sigma}^2,\hat{\omega}) = \frac{1}{16\hat{\omega}}\int_{-\infty}^{\infty} \frac{dz[\hat{\sigma}^2(\sqrt{z^2+1} + z) - 2\sqrt{z^2+1}]}{(z^2+1)^2(\sqrt{z^2+1} + z)}$$

$$+ \frac{1}{16\omega}\int_{-\infty}^{\infty} \frac{dz[\hat{\sigma}^2(\sqrt{z^2+1} - z) - 2\sqrt{z^2+1}]}{(z^2+1)^2(\sqrt{z^2+1} - z)} + O\left(\frac{1}{\hat{\omega}^2}\right) =$$

(Contd)

(Contd) $$= \frac{1}{8\hat{\omega}}\int_{-\infty}^{\infty} \frac{dz}{(z^2+1)^2}(\hat{\sigma}^2 - 2(z^2+1)) + O\left(\frac{1}{\hat{\omega}^2}\right)$$

$$= \frac{\pi}{8\hat{\omega}}\left[\frac{\hat{\sigma}^2}{2} - 2\right] + O\left(\frac{1}{\hat{\omega}^2}\right)$$

as $\hat{\omega} \to \infty$. Hence we get the equality

$$\lim_{\hat{\omega}\to\infty} \hat{\sigma}_0^2(\hat{\omega}) = 4.$$

Therefore the critical noise power for high frequencies is close to $4k$ (Figure 2).

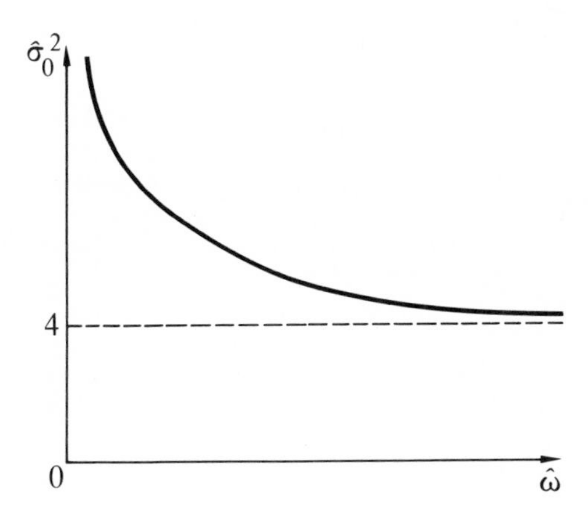

Figure 2

Analogous arguments apply to the investigation of random parametric excitations (i.e., random perturbations of the natural frequency). If the intensity of the noise is low, the investigation is a simple matter; one can apply then the method of averaging. On this subject, see Kolomiec [1,2], Stratonovič and Romanovskiĭ [1], Stratonovič [1], Has'minskiĭ [3] and others.

10. *n*-th order equations[10]

It is well-known that the solution $y = 0$ of the equation

$$y^{(n)} + b_1 y^{(n-1)} + \dots + b_n y = 0 \tag{10.1}$$

is stable if and only if the Routh-Hurwitz conditions

$$\left.\begin{aligned} &\Delta_1 = b_1 > 0; \qquad \Delta_2 = \begin{vmatrix} b_1 & b_3 \\ 1 & b_2 \end{vmatrix} > 0; \\ &\Delta_3 = \begin{vmatrix} b_1 & b_3 & b_5 \\ 1 & b_2 & b_4 \\ 0 & b_1 & b_3 \end{vmatrix} > 0; \ \dots; \ \Delta_n = \begin{vmatrix} b_1 & b_3 & b_5 & 0 \\ 1 & b_2 & b_4 & 0 \\ 0 & b_1 & b_3 & 0 \\ \dots & \dots & \dots & \dots \\ 0 & \dots & \dots & b_n \end{vmatrix} > 0 \end{aligned}\right\} \tag{10.2}$$

are satisfied (see Gantmaher [1]).

We shall derive analogous necessary and sufficient conditions for mean square stability of the system

$$y^{(n)} + (b_1 + \eta_1(t))y^{(n-1)} + \dots + (b_n + \eta_n(t))y = 0.$$

Here $\dot{\eta}_1(t),\dots,\dot{\eta}_n(t)$ are Gaussian white noise processes, generally correlated, so that

$$\mathrm{E}\dot{\eta}_i(s)\dot{\eta}_j(t) = a_{ij}\delta(t - s).$$

Replacing the processes $\dot{\eta}_1(t),\dots,\dot{\eta}_n(t)$ by independent processes as indicated in Section 1, and setting

$$X_1(t) = y(t); \ \dots; \ X_n(t) = y^{(n-1)}(t),$$

we get a system of Itô equations

$$\begin{aligned} dX_1(t) &= X_2(t)dt, \\ dX_2(t) &= X_3(t)dt, \ \dots, \ dX_{n-1}(t) = X_n(t)dt, \\ dX_n(t) &= -\sum_{i=1}^{n} b_i X_{n-i+1}(t)dt - \sum_{i,j=1}^{n} \sigma_{ij} X_{n-i+1} d\xi_j(t), \end{aligned} \tag{10.3}$$

---

[10] See Nevel'son and Has'minskiĭ [1]. Some special cases were considered previously by Caughey and Dienes [1], Rabotnikov [1].

where $((\sigma_{ij}))((\sigma_{ji})) = ((a_{ij}))$.

It is easy to see that the differential generator of the process $X(t)$ is

$$L = \frac{\partial}{\partial t} + \sum_{i=1}^{n-1} x_{i+1} \frac{\partial}{\partial x_i} - \sum_{i=1}^{n} b_i x_{n-i+1} \frac{\partial}{\partial x_n}$$

$$+ \frac{1}{2} \sum_{i,j=1}^{n} a_{ij} x_{n-i+1} x_{n-j+1} \frac{\partial^2}{\partial x_n^2} .$$

Using the methods of Sections 2 and 3 we may now determine necessary and sufficient conditions for the system (10.3) to be stable in mean square. However, the resulting conditions will involve determinants of order n^2. We shall therefore adopt another approach; this will give us conditions involving the computation of only $n + 1$ determinants, the largest of order n. We shall see that the first n determinants are the same as in (10.2), while the last is obtained from Δ_n by replacing its first row by a vector whose components are computed by a certain rule from the coefficients a_{ij}.

As we saw in Section 2 (Remark 1), a necessary condition for the system (10.3) to be asymptotically stable in mean square is that the "non-random" system

$$\frac{dx_1}{dt} = x_2, \ \ldots, \ \frac{dx_{n-1}}{dt} = x_n, \quad \frac{dx_n}{dt} = - \sum_{i=1}^{n} b_i x_{n-i+1} \tag{10.4}$$

is asymptotically stable, i.e., the Routh-Hurwitz conditions (10.2) hold. It is known that under these assumptions there exists a positive definite quadratic form $V(x)$ such that

$$L_0 V = \sum_{i=1}^{n-1} x_{i+1} \frac{\partial V}{\partial x_i} - \sum_{i=1}^{n} b_i x_{n-i+1} \frac{\partial V}{\partial x_n} , \tag{10.5}$$

i.e. the total derivative along the trajectory of the system (10.4) is equal to a prescribed negative definite form $W(x)$.

We first assume that the quadratic form

$$a(x) = \sum_{i,j=1}^{n} a_{ij} x_{n-i+1} x_{n-j+1}$$

is positive definite. Then we have

LEMMA 10.1. *The trivial solution of the system* (10.3) *is asymptotically stable in mean square if and only if there exists a positive definite quadratic form*

$$V(x) = \sum_{i,j=1}^{n} d_{ij}x_i x_j$$

such that

$$L_0V = - a(x), \qquad d_{nn} < 1. \tag{10.6}$$

PROOF. Suppose that there exists a form $V(x) = \sum d_{ij}x_i x_j$ satisfying the conditions of the lemma. Then, by (10.5) and (10.6), we have

$$LV = L_0V + \frac{a(x)}{2}\frac{\partial^2 V}{\partial x_n^2} = (d_{nn} - 1)a(x) < 0.$$

Hence, by Theorem 3.2, it follows that the system (10.3) is asymptotically stable in mean square.

Conversely, if the system (10.3) is asymptotically stable, it follows from the same theorem that there exists a positive definite quadratic form

$$V_1(x) = \sum_{i,j=1}^{n} v_{ij}x_i x_j,$$

such that $LV_1 = -a(x)$, i.e.,

$$L_0V_1 = LV_1 - \frac{a(x)}{2}\frac{\partial^2 V_1}{\partial x_n^2} = - (v_{nn} + 1)a(x).$$

Thus $V(x) = V_1(x)/(v_{nn} + 1)$. Consequently $a_{nn} = v_{nn}/(v_{nn} + 1) < 1$, which we wished to show.

Thus in order to obtain the desired conditions we must express the coefficient d_{nn} of the form $V(x)$ defined by (10.6) in terms of the parameters b_i, a_{ij} of the system (10.3).

It follows from Theorem 3.2 that any function $V(x)$ satisfying (10.6) can be written as

$$V(x) = \int_0^\infty a(X^x(u))du.$$

This equality makes it possible to express the coefficients of $V(x)$ including d_{nn}, in terms of a fundamental system of solutions of equation (10.4). Next, as shown by Bedel'baev [1], we may express them in terms of the coefficients a_i, b_{ij}. Indeed, according to Bedel'baev [1],

$$d_{nn} = \frac{1}{2\Delta_n} \sum_{r=0}^{n-1} q_{nn}^{(r)} \Delta_{1,r+1}, \tag{10.7}$$

where $\Delta_{1,r+1}$ is the cofactor of the element in the first row and $(r+1)$-th column of the last Hurwitz determinant Δ_n, and the numbers $q_{nn}^{(r)}$ are related to the coefficients a_{ij} of the form $a(x)$ by

$$(-1)^{n-1} \sum_{i,j=1}^{n} a_{n-i+1,n-j+1} D_{ni}(\lambda) D_{nj}(-\lambda)$$

$$= \sum_{r=0}^{n-1} q_{nn}^{(r)} \lambda^{2(n-r-1)}. \tag{10.8}$$

Here $D_{nj}(\lambda)$ is the cofactor of the element in the n-th row and j-th column of the determinant

$$D(\lambda) = \begin{vmatrix} -\lambda & 1 & 0 & \dots & 0 & 0 \\ 0 & -\lambda & 1 & \dots & 0 & 0 \\ \cdot & \cdot & \cdot & \cdot & \cdot & \cdot \\ 0 & 0 & 0 & \dots & -\lambda & 1 \\ -a_n & -a_{n-1} & -a_{n-2} & \dots & -a_2 & -a_1 - \lambda \end{vmatrix}$$

of the system (10.3)

It is easy to see that

$$D_{ni}(\lambda) D_{nj}(-\lambda) = \lambda^{i+j-2}(-1)^{j-1}.$$

Hence, using (10.8), we infer that

$$\sum_{k=0}^{n-1} \lambda^{2k} \sum_{p+q=2(n-k)} a_{pq}(-1)^{q+1} = \sum_{k=0}^{n-1} q_{nn}^{(n-k-1)} \lambda^{2k}$$

and consequently

$$q_{nn}^{(n-k-1)} = \sum_{p+q=2(n-k)} a_{pq}(-1)^{q+1}. \tag{10.9}$$

It follows from Lemma 10.1, by (10.7) and (10.9), that if $a(x)$ is a positive definite quadratic form, then the system (10.3) is stable in mean square if and only if

$$\Delta_1 > 0; \quad \Delta_2 > 0; \quad \dots; \quad \Delta_n > 0; \quad \Delta_n > \Delta/2. \tag{10.10}$$

Here Δ is the determinant

$$\Delta = \begin{vmatrix} q_{nn}^{(0)} & q_{nn}^{(1)} & \dots & q_{nn}^{(n-1)} \\ 1 & b_2 & \dots & 0 \\ 0 & b_1 & \dots & 0 \\ \cdot & \cdot & \cdot & \cdot \\ 0 & 0 & \dots & b_n \end{vmatrix}, \tag{10.11}$$

which differs from the last Hurwitz determinant Δ_n only in its first row. The numbers $q_{nn}^{(r)}$ $(r = 0,\dots,n - 1)$ are related to the elements a_{ij} of the correlation matrix by formulas (10.9).

We shall now show that the assumption $a(x) > 0$ $(x \neq 0)$ is not essential. To do this we consider along with the system (10.3), another system

$$dX_1 = X_2dt; \qquad dX_2 = X_3dt; \quad \dots; \quad dX_{n-1} = X_ndt,$$

$$dX_n = - \sum_{i=1}^{n} b_iX_{n-i+1}dt - \sum_{i,j=1}^{n} X_{n-i+1}\sigma_{ij}d\xi_j(t)$$

$$+ \varepsilon \sum_{j=1}^{n} X_jd\eta_j(t). \tag{10.12}$$

Here $\eta_1(t),\dots,\eta_n(t)$ are assumed to be Wiener processes, independent of each other and of the processes $\xi_1(t),\dots,\xi_n(t)$, and $\varepsilon > 0$ is a small parameter.

It is easy to see that the differential generator associated with the system (10.12) is

$$L_\varepsilon = L + \tfrac{1}{2}\varepsilon^2|x|^2 \frac{\partial^2}{\partial x_n^2}.$$

Since the quadratic form

$$a_\varepsilon(x) = a(x) + \varepsilon^2|x|^2$$

is positive definite for any $\varepsilon > 0$, it follows that the system (10.12) is asymptotically stable in mean square if and only if we have that

$$\Delta_1 > 0, \qquad \Delta_2 > 0; \quad \dots \quad \Delta_n > 0;$$

$$\Delta_n > \frac{\Delta_\varepsilon}{2} = \frac{\Delta + \varepsilon^2\Delta_1}{2}, \tag{10.13}$$

$$\Delta_1 = \begin{vmatrix} 1 & -1 & 1 & \dots & (-1)^{n-1} \\ 1 & b_2 & b_1 & \dots & 0 \\ 0 & b_1 & b_3 & \dots & 0 \\ \cdot & \cdot & \cdot & \cdot & \cdot \\ 0 & 0 & 0 & \dots & b \end{vmatrix} .$$

If conditions (10.2) are satisfied, the determinant Δ_1 is positive. This follows from the fact that the coefficient $\tilde{d}_{nn}$ of x_n^2 in a positive definite form $\tilde{V}(x)$ satisfying the equation $L\tilde{V} = -|x|^2$ is

$$\tilde{d}_{nn} = \frac{\Delta_1}{2\Delta_n} .$$

Consequently

$$\Delta_n > \frac{\Delta}{2} . \tag{10.14}$$

We now assume that the system (10.3) is asymptotically stable in mean square. Then, by Remark 3 in Section 2, the system (10.12) is also stable for all sufficiently small $\varepsilon > 0$, whence (10.13) holds. From (10.13) and (10.14) we get (10.10).

Now suppose that the inequalities (10.10) are satisfied. Then there exists a sufficiently small $\varepsilon > 0$ for which the inequalities (10.13) hold, i.e., the system (10.12) is asymptotically stable in mean square for this ε. By Theorem 3.2, there exists a positive definite quadratic form $W(x) = \sum_{i,j=1}^{n} w_{ij}x_i x_j$ such that the form $L_\varepsilon W$ is negative definite. But then the function $LW = L_\varepsilon W - \varepsilon^2|x|^2 w_{nn}$ is also negative definite. Another application of Theorem 3.2 shows that the system (10.3) is stable in mean square.

We have thus proved

THEOREM 10.1. *The system* (10.3) *is asymptotically stable in mean square if and only if conditions* (10.10) *are satisfied, where the determinant* Δ *is given by* (10.11) *and the numbers* $q_{nn}^{(r)}$ $(r = 0,\dots,n - 1)$ *in the first row of* Δ *are expressed in terms of the coefficients* a_{ij} *by formulas* (10.9).

It is interesting to observe that the coefficients a_{ij} of the correlation matrix figuring in conditions (10.10) are those for which the sum $i + j$ is even. For example, for second- and third-order systems our necessary and sufficient conditions for asymptotic stability in mean square are

$$n = 2: \quad b_1 > 0; \qquad b_2 > 0; \qquad 2b_1b_2 > a_{11}b_2 + a_{22},$$

$$n = 3: \quad \begin{array}{l} b_1 > 0; \qquad b_3 > 0; \qquad b_1 b_2 > b_3; \\ 2(b_1 b_2 - b_3) b_3 > a_{11} b_2 b_3 + a_{33} b_1 + b_3(a_{22} - 2a_{13}). \end{array}$$

If the white noise processes $\dot{\eta}_1, \ldots, \dot{\eta}_n$ superimposed on the coefficients $_i$ of equation (10.1) are independent, i.e., $a_{ij} = 0$ for $i \neq j$, the determinant Δ assumes the particularly simple form

$$\Delta = \begin{vmatrix} a_{11} & -a_{22} & \ldots & (-1)^{n-1} a & a_{nn} \\ 1 & b_2 & \ldots & 0 & 0 \\ 0 & b_1 & \ldots & 0 & 0 \\ & & & & \\ 0 & 0 & \ldots & 0 & b_n \end{vmatrix} .$$

Conditions (10.9), (10.10) and (10.11) are sufficient for the system (10.3) to be asymptotically p-stable when $p \leq 2$. Let us now determine sufficient conditions for p-stability when $p > 2$. We first assume that the quadratic form is positive definite.

A necessary condition for asymptotic p-stability when $p = 2$, hence also when $p > 2$, is that there exists a positive definite quadratic form

$$V(x) = \sum_{i,j=1}^{n} d_{ij} x_i x_j ,$$

satisfying equation (10.6). We set

$$W(x) = [V(x)]^{p/2}.$$

It is readily seen that

$$LW = \frac{p}{2} V^{p/2-2} \left\{ V L_0 V + a(x) \left[d_{nn} V + (p-2) \left(\sum_{j=1}^{n} d_{nj} x_j \right)^2 \right] \right\}$$

$$= \frac{p}{2} V^{p/2-2} a(x) \left[(d_{nn} - 1) V + (p-2) \left(\sum_{j=1}^{n} d_{nj} x_j \right)^2 \right] . \tag{10.15}$$

By the well known inequality for positive definite self-adjoint matrices D (see, e.g., Gel'fand [1])

$$(Dx, y)^2 \leq (Dx, x)(Dy, y)$$

we get, by taking $y = (0, \ldots, 0, 1)$, that

$$\left(\sum_{j=1}^{n} d_{nj}x_j\right)^2 \leqslant d_{nn}V(x).$$

Using this relation, we conclude from (10.15) that

$$LW \leqslant \frac{p}{2} V^{p/2-1} a(x)[d_{nn}(p - 1) - 1].$$

If $d_{nn}(p - 1) > 1$, then it follows by Theorem 3.1 that the system (10.3) is asymptotically p-stable.

Thus, a sufficient condition for the system (10.3) to be p-stable ($p \geqslant 2$) is the occurrence of inequalities

$$\Delta_1 > 0; \quad \ldots; \quad \Delta_n > 0; \quad \Delta_n > \frac{p - 1}{2}. \tag{10.16}$$

The first n of these are also necessary.

That the condition $\Delta_n > \frac{p - 1}{2}$ is not necessary can be shown on examples. It is also readily seen that already conditions (10.16) are sufficient for p-stability ($p \geqslant 2$); there is then no need to assume that the quadratic form $a(x)$ is nonsingular.

11. Stochastic stability in the strong and weak senses

As before, let $X(t)$ be a solution of the linear system (7.6) with constant coefficients. We shall show that in a broad range of cases the, suitably normalized, limiting distribution of $\rho(t) = \ln|X(t)|$ as $t \to \infty$ is Gaussian. To do this we make use of the formula

$$\rho(t) = \rho_0 + \int_0^t Q(\Lambda^{\lambda_0}(s))ds + \sum_{r=1}^{k} (\sigma_r\Lambda^{\lambda_0}(s),\Lambda^{\lambda_0}(s))d\xi_r(s) \tag{11.1}$$

proved in Section 7, and we assume for simplicity that the non-degeneracy condition (7.7) (or (7.7'), p. 224, footnote) is satisfied. We then have

THEOREM 11.1. *If condition* (7.7) *or* (7.7') *is satisfied and* $D\rho(t) \to \infty$ *as* $t \to \infty$, *then*

$$\mathbf{P}\left\{\frac{\rho(t) - at}{\sqrt{D\rho(t)}} < x\right\}_{t\to\infty} \to \Phi(x) = \frac{1}{\sqrt{2\pi}}\int_{-\infty}^{x} e^{-y^2/2}dy.$$

Here $a = \displaystyle\int_{S_\ell} Q(\lambda)\nu(d\lambda)$.

PROOF. We have already seen (Section 7) that if condition (7.7) holds, then the transition probability of the process $\Lambda^{\lambda_0}(t)$ has a positive density. Hence, by the compactness of the phase space, the process $\Lambda^{\lambda_0}(t)$ satisfies Doeblin's condition. It is shown in Doob [3], Chapter V that for the transition function $P(\lambda,t,A)$ of the process $\Lambda(t)$ and any bounded measurable function $f(\lambda)$ on S_ℓ

$$|\mathbf{P}(\lambda,t,A) - \nu(A)| < ke^{-at}, \tag{11.2}$$

$$\left|\mathbf{E}\{f(\Lambda(t))/N_s\} - \int_{S_\ell} f(\lambda)\nu(d\lambda)\right| < k\|f\|e^{-a(t-s)}, \tag{11.3}$$

$$(\|f\| = \max_{S_\ell} |f(\lambda)|),$$

(where $k > 0$ and $\alpha > 0$ are constants). Using (11.1) and (11.2), we easily see that for certain constants $c_i > 0$

$$|\mathbf{E}\rho(t) - at| < |\rho_0| + c_1\int_0^t e^{-as}ds < c_2. \tag{11.4}$$

It follows from (11.4) and the assumptions of the theorem that

$$\frac{\rho(t) - \mathbf{E}\rho(t)}{\sqrt{D\rho(t)}} = \frac{\rho(t) - at}{\sqrt{D\rho(t)}} + o(1) \tag{11.5}$$

as $t \to \infty$. These relations, together with well known limit theorems for additive random functions (see Volkonskiĭ and Rozanov [1]), imply the assertion of the theorem.

REMARK. It is easy to give examples in which the condition $D\rho(t) \to \infty$ as $t \to \infty$ is not satisfied. However, it would be interesting to find effective sufficient conditions for this to hold.

In Section 1.5 we presented conditions for weak stochastic stability of systems whose right-hand side is perturbed by stochastic processes of a relatively general form. In fact, for equations with random right-hand side we studied conditions under which the solution $X^{s,x_0}(t)$ which satisfies the initial condition $X^{s,x_0}(t) = x_0$ fulfils the condition

$$\lim_{x_0\to 0}\ \sup_{t>s} \mathbf{P}\{|X^{s,x_0}(t)| > \varepsilon\} = 0. \tag{11.6}$$

It is clear that this definition is in general weaker than the definition

$$\lim_{x_0 \to 0} \mathbf{P}\{\sup_{t>s} |X^{s,x}(t)| > \varepsilon\} = 0 \tag{11.7}$$

we gave in Section 5.3. We claim that nevertheless a weakly stochastically stable linear system with constant coefficients satisfying the assumptions of Theorem 11.1 is also strongly stable. Indeed, if $a < 0$, the system is stable in both the strong and the weak sense. But if $a = 0$, it follows from Theorem 11.1 that the system is unstable in the weak sense, since the probability of the event $\{|X(t)| < 1\}$ converges by Theorem 11.1 to $\frac{1}{2}$ as $t \to \infty$.

These arguments would seem to justify the conjecture that, in general, strong and weak stochastic stability are equivalent for autonomous stochastic systems perturbed by white noise. However, the following example refutes this conjecture[11].

EXAMPLE. Let φ be the angle-coordinate of a point on the circle. We consider a process on the circle, governed by the Itô equation

$$d\varphi(t) = \left[-2\sin^2\frac{\varphi}{2} + \sin^3\frac{\varphi}{2}\cos\frac{\varphi}{2}\right]dt - 2\sin^2\frac{\varphi}{2}d\xi(t).$$

This system has a unique equilibrium position $\varphi = 0$. It is readily seen by means of Ito's formula (3.3.8) that the solution of this system satisfying the initial condition $\varphi(0) = \varphi_0$ is the function

$$\varphi(t) = 2\operatorname{arcctg}\left(t + \xi(t) + \operatorname{ctg}\frac{\varphi_0}{2}\right). \tag{11.8}$$

It follows from (11.8) that the point $t = 0$ is unstable in the sense of (11.7). Indeed, $t + \xi(t) \to \infty$ almost surely as $t \to \infty$, and hence any path for which $\varphi_0 < 0$ is sufficiently small in absolute value almost surely describes an almost complete circle.

On the other hand, we claim that for any $\varepsilon > 0$ there exist sufficiently large numbers $C(\varepsilon)$ and $R(\varepsilon)$ such that for all $t \geqslant 0$

$$\mathbf{P}\{|t + \xi(t) - C(\varepsilon)| < R(\varepsilon)\} < \varepsilon. \tag{11.9}$$

Indeed,

$$\mathbf{P}\{|t + \xi(t) - C| < R\} = \Phi\left(\frac{C + R - t}{\sqrt{t}}\right) - \Phi\left(\frac{C - R - t}{\sqrt{t}}\right) \tag{11.10}$$

(where Φ is the normal distribution function with parameters (0,1)). We set, for example, $C = T^3$. Then for $t \geqslant \frac{1}{2}R^3$ the difference $2R/\sqrt{t}$ between the arguments in (11.10) satisfies the

[11] The author's attention was drawn to the problem of the connection between strong and weak stochastic stability by N.N. Krasovskiĭ.

inequality

$$\frac{2R}{\sqrt{t}} \leqslant R^{-1/2} < \varepsilon$$

for sufficiently large R. But if $t < \frac{1}{2}R^3$, then

$$\frac{C - R - t}{\sqrt{t}} > \frac{\frac{1}{2}R^3 - R}{(\frac{1}{2}R^3)^{1/2}} \to \infty \quad \text{as} \quad R \to \infty.$$

These relations imply (11.9). From (11.8) and (11.9) it follows that the system is stable in the sense of definition (11.6); indeed, for sufficiently large $C(\varepsilon)$ we have the obvious inequality

$$\mathbf{P}\{|t + \xi(t) + C(\varepsilon)| < [C(\varepsilon)]^{1/3}\} < \varepsilon.$$

Thus a system which is weakly stable in the sense of (11.6) may be almost surely unstable. It would be interesting to construct an analogous example on the plane.

CHAPTER VII

SOME PROBLEMS IN THE THEORY OF STABILITY OF STOCHASTIC SYSTEMS

1. Stability in the first approximation[1]

Many problems concerning the stability of a nonlinear stochastic system can be reduced to problems about a linear system, obtained from the original system by dropping terms of higher than first order in x. Hence, the importance of systems of linear stochastic equations.

The first theorem on stability of stochastic systems in the linear approximation was proved in Kac and Krasovskiĭ [1], with the use of their method for studying randomly perturbed systems (see Section 5.1). They proved that the full system is stable in probability if the linearized system is exponentially stable in mean square. A similar result was established in Gihman [2,3] for the diffusion-type processes considered in Chapters III-VI. However, this method leaves unanswered the question of whether the linearization method is applicable to the stability theory of a broad range of systems; there exist linear systems which are almost surely asymptotically stable but not stable in mean square (see Section 6.3). This leads to the question of whether the full system is always stable whenever the corresponding linearized system has constant coefficients and is almost surely asymptotically stable. We intend to show that the answer to this question is positive.

We first consider the linear system

$$dX(t) = B(t)Xdt + \sum_{r=1}^{k} \sigma_r(t)Xd\xi_r(t) \tag{1.1}$$

with constant coefficients, i.e., $B, \sigma_1,\ldots,\sigma_k$ are constant matrices.

---

[1] See Nevel'son and Has'minskiĭ [2], Has'minskiĭ [12].

THEOREM 1.1. *If the linear system* (1.1) *with constant coefficients is asymptotically stable almost surely (or in probability), and the coefficients of the system*

$$dX(t) = b(t,X)dt + \sum_{r=1}^{k} \sigma_r(t,X)d\xi_r(t) \tag{1.2}$$

satisfy an inequality

$$|b(t,x) - Bx| + \sum_{r=1}^{k} |\sigma_r(t,x) - \sigma_r x| < \gamma |x| \tag{1.3}$$

in a sufficiently small neighbourhood of the point $x = 0$ *and with a sufficiently small constant* γ, *then the solution* $X = 0$ *of the system* (1.1) *is asymptotically stable in probability.*

PROOF. By Theorem 6.4.1, it will suffice to prove that if the system (1.1) is exponentially p-stable for some $p > 0$ and condition (1.3) is satisfied, then the system (1.2) is asymptotically stable in probability. We let

$$L_0 = \frac{\partial}{\partial t} + \left(Bx, \frac{\partial}{\partial x}\right) + \frac{1}{2}\sum_{r=1}^{k}\left(\sigma_r x, \frac{\partial}{\partial x}\right)^2,$$
$$L = \frac{\partial}{\partial t} + \left(b(t,x), \frac{\partial}{\partial x}\right) + \frac{1}{2}\sum_{r=1}^{k}\left(\sigma_r(t,x), \frac{\partial}{\partial x}\right)^2 \tag{1.4}$$

denote the differential generators of the systems (1.1) and (1.2), respectively. By Theorem 6.3.1, there exists a function $V(t,x)$ such that for some $k_i > 0$

$$k_1|x|^p \leqslant V(t,x) \leqslant k_2|x|^p; \qquad L_0V(t,x) \leqslant - k_3|x|^p,$$
$$\left|\frac{\partial V}{\partial x}\right| < k_4|x|^{p-1}, \qquad \left|\frac{\partial^2 V}{\partial x_i \partial x_j}\right| \leqslant k_4|x|^{p-2}. \tag{1.5}$$

By (1.3) , (1.4) and (1.5), we have

$$LV = L_0V + \left(b(t,x) - Bx, \frac{\partial}{\partial x}\right)V$$
$$+ \frac{1}{2}\sum_{r\,1}^{k}\left(\sigma_r(t,x) - \sigma_r x, \frac{\partial}{\partial x}\right)\left(\sigma_r(t,x) + \sigma_r x, \frac{\partial}{\partial x}\right)V$$
$$\leqslant - k_3|x|^p + \gamma k_4|x|^p + \gamma k_5|x|^p \tag{1.6}$$

in a sufficiently small neighbourhood of $x = 0$.

The constant k_5 in this inequality depends only on k_4 and on the supremum of absolute values of the coefficients in (1.1).

It follows from (1.6) that the function LV is negative definite in a sufficiently small neighbourhood of $x = 0$, provided $k_3/(k_4 + k_5)$. Moreover, according to (1.5), V is positive definite and has an infinitesimal upper limit. Applying Corollary 1 of Theorem 5.4.1 to the function V, we get the assertion of the theorem.

If the coefficients of the system (1.1) are time-dependent, we have the analogous

THEOREM 1.2. *Suppose that the coefficients of the system* (1.1) *are bounded functions of time, the trivial solution of* (1.1) *is stable in the large uniformly in t, and condition* (1.3) *is satisfied with a sufficiently small constant γ. Then the trivial solution of* (1.2) *is asymptotically stable in probability.*

By Theorem 6.5.1, it suffices to prove that if the system (1.1) is exponentially p-stable and condition (1.3) holds, then the assertion of the theorem is valid. This we did in the proof of Theorem 1.1.

REMARK 1. It is clear from the proofs of Theorems 1.1 and 1.2 that the constant γ in condition (1.3) depends only on k_3, k_4 and $\sup_{t>0} \|\sigma_r(t)\|$.

REMARK 2. It follows from Theorems 1.1 and 1.2 that the system (1.2) is asymptotically stable in probability if the linearized system

$$dX(t) = \frac{\partial b(t,0)}{\partial x} X dt + \sum_{r=1}^{k} \frac{\partial \sigma_r(t,0)}{\partial x} X d\xi_r(t)$$

is stable in the large uniformly in t, and the derivatives $\partial b/\partial x$ and $\partial \sigma_r/\partial x$ are bounded and uniformly (in t) continuous in x for $x = 0$.

2. Instability in the first approximation

We first recall some well known results for the deterministic case (see Ljapunov [1], Malkin [2]), confining ourselves to systems with constant coefficients.

LJAPUNOV'S THEOREM. *Suppose that at least one of the roots of the characteristic equation of the system*

$$\frac{dX}{dt} = BX \tag{2.1}$$

has positive real part. Let the vector $\varphi(x)$ *be such that* $|\varphi(x)| < A|x|^2$.

Then the solution $X \equiv 0$ *of the equation*

$$\frac{dX}{dt} = BX + \varphi(X) \tag{2.2}$$

is unstable.

Malkin observed in [2] that Ljapunov's arguments in fact yield a more general result:

MALKIN'S THEOREM. *If at least one of the roots of equation* (2.1) *has positive real part, then the trivial solution of the system* (2.2) *is unstable if* $|\varphi(x)| < \gamma|x|$, *where* γ *is a sufficiently small constant which depends only on the coefficients of a positive definite quadratic form satisfying the assumptions of Ljapunov's second instability theorem.*

It will be clear from the sequel that the situation is far more complicated in regard to stochastic systems. In particular, the examples given in Section 3 will show that the analog of Malkin's theorem fails to hold.

We shall first prove that the analog of Theorems 1.1 and 1.2 for instability is valid, provided the linearized system is unstable in a sufficiently strong sense. Theorem 6.3.3 implies that if the system (1.1) is exponentially q-stable, then there exists a function $V(t,x)$ such that

$$\left.\begin{array}{l} k_1|x|^{-q} \leqslant V(t,x) \leqslant k_2|x|^{-q}; \qquad L_0V \leqslant -k_3|x|^{-q}, \\ \left|\dfrac{\partial V}{\partial x}\right| \leqslant k_4|x|^{-q-1}; \qquad \left|\dfrac{\partial^2 V}{\partial x_i \partial x_j}\right| < k_4|x|^{-q-2} \\ \qquad\qquad (i,j = 1,\ldots,\ell). \end{array}\right\} \tag{2.3}$$

for certain constants $k_i > 0$.

THEOREM 2.1. *Let the coefficients of the linear system* (1.1) *be bounded functions of time, and let the trivial solution of this system be exponentially* q*-unstable for some* $q > 0$. *Moreover, suppose that inequality* (1.3) *holds with a sufficiently small constant* γ, *depending only on* $\sup_{t>0} \|\sigma_r(t)\|$ *and the constants* $k_1,\ldots,k_4$ *figuring in* (2.3). *Then the solution* $X(t) \equiv 0$ *of the system* (1.2) *is unstable in probability.*

PROOF. It follows from Theorem 6.3.3 that under the above assumptions there exists for the system (1.1) a function satisfying inequalities (2.3). Hence, as in the case of (1.6), we see that

$$LV \leqslant -k_3|x|^{-q} + \gamma k_4|x|^{-q} + \gamma k_5|x|^{-q},$$

in a sufficiently small neighborhood of $x = 0$, where $k_5 = (k_4, \sup_{t>0} \|\sigma_r(t)\|)$. It follows now that for sufficiently small γ, the function V satisfies all the assumptions of Theorem 5.4.2 (see the remark following that Theorem). This completes the proof.

From this theorem and Theorems 6.4.2 and 6.5.2, we get the following result.

THEOREM 2.2. *If for any* $x \neq 0$, $A > 0$, *the solutions of the linear system* (1.1) *satisfy the identity*

$$\lim_{T\to\infty} \sup_{s>0} \mathbf{P}\{ \inf_{u>s+T} |X^{s,x}(y)| < A\} \equiv 0, \tag{2.4}$$

and the elements of the matrices $B, \sigma_1, \ldots, \sigma_k$ *are bounded, then the solution* $X(t) \equiv 0$ *is unstable in probability for all systems of type* (1.2) *whose coefficients satisfy condition* (1.3) *with sufficiently small* γ.

THEOREM 2.3. *If the system* (1.1) *has constant coefficients, the assertion of Theorem* 2.2 *remains valid if assumption* (2.4) *is replaced by the requirement that for all* $x \neq 0$

$$\mathbf{P}\{|X^{s,x}(t)| \to \infty \text{ as } t \to \infty\} = 1. \tag{2.5}$$

A comparison of the theorems of Ljapunov and Malkin with Theorem 2.3 shows that for deterministic systems the latter furnishes a very poor result. Whereas according to the Ljapunov-Malkin theorems it is sufficient that at least one root of the characteristic equation of the linear system has a positive real part, Theorem 2.3 is valid for a deterministic linear system only when the real parts of *all* roots of the characteristic equation are positive (see Section 6.3). Nonetheless, if the system (1.1) with constant coefficients is nondegenerate in the sense that

$$\sum_{r=1}^{k} (\sigma_r x, \lambda)^2 > 0 \tag{2.6}$$

for all non-zero vectors x and λ, then Theorem 2.3 implies the

COROLLARY. *If inequality* (2.6) *is satisfied, then the assertion of Theorem* 2.3 *holds if conditions* (2.5) *are satisfied for at least one value of* x.

This corollary is obvious if we observe that, by Theorems 6.7.1 and 6.7.2, condition (2.5) holds for one x if and only if it holds for all $x \neq 0$.

There are two questions arising naturally in connection with the last theorem and its corollary.

1. Can we replace assumption (2.5) in Theorem 2.3 by the weaker assumption

$$\sup_{t>s} |X^{s,x}(t)| = \infty \quad \text{a.s.?}$$

2. Can the assertion of Theorem 2.3 be proved under the assumption that (2.5) holds for at least one $x \neq 0$, but without the nondegeneracy condition (2.6)?

We shall see in Section 3 that the answers to both these questions are in general negative.

3. Two examples

EXAMPLE 1. We again consider the one-dimensional system

$$dX(t) = b(t,X)dt + \sigma(t,X)d\xi(t), \tag{3.1}$$

such that the linearized system

$$dX(t) = b_0 X dt + \sigma_0 X d\xi(t) \tag{3.2}$$

has constant coefficients. If $b_0 < \sigma_0^2/2$, we can apply Theorem 1.1, and if $b_0 > \sigma_0^2/2$, we can use Theorem 2.3. If $b_0 = \sigma_0^2/2$, the linear system is unstable, but not asymptotically q-unstable for any $q > 0$ (see Section 6.1) and, moreover, we have

$$\mathbf{P}\{\sup_{t>s} |X^{s,x}(t)| = \infty\} = 1 \tag{3.3}$$

for $x \neq 0$. It follows also from the results of Section 6.1 that if $b_0 = \sigma_0^2/2$, then the system

$$dX(t) = (b_0 - \gamma)X dt + \sigma_0 X d\xi(t)$$

is asymptotically stable for any $\gamma > 0$.

This implies that the answer to the first of the questions posed at the end of Section 2 is negative. Thus the analog of Malkin's theorem is false here. However, the solution $X(t) \equiv 0$ of the system (3.1) is nevertheless unstable, if we assume that the differences $b(t,x) - b_0 x$ and $\sigma(t,x) - \sigma_0 x$ tend to zero sufficiently rapidly as $x \to 0$. Indeed, suppose that $b_0 = \sigma_0^2/2$ and that for some $k > 0$, $\alpha > 0$

$$|b(t,x) - b_0 x| + |\sigma(t,x) - \sigma_0 x| < k|x|^{1+\alpha}. \tag{3.4}$$

Consider the auxiliary function $V(x) = \ln\ln(1/|x|)$. The reader will easily verify that in this case $V \to \infty$ as $x \to 0$ and $\inf_{\varepsilon<|x|<\delta} LV < 0$ for sufficiently small fixed $\delta > 0$ and any $\varepsilon < \delta$.

The instability of (3.1) under the assumptions $b_0 = \sigma_0^2/2$ and (3.4) now follows from Theorem 5.4.2.

We have thus shown in the one-dimensional case that if the linear system with constant coefficients satisfies (3.3), and the full system is nearly linear in the sense of (3.4), then the latter is unstable in probability.

It would be interesting to know whether this remains true in the many-dimensional case.

EXAMPLE 2. Let $\Psi(z)$ denote a differentiable function of the real variable z, with compact support and bounded together with its first derivative. Suppose further that

$$\Psi(0) = 0; \qquad \Psi'(0) = -3; \qquad |\Psi(z)| < 1. \tag{3.5}$$

Using this function, we construct a Markov process on the plane, which is the solution of the system of Itô equations

$$\begin{aligned} dX_1(t) &= \left[X_1 + \varepsilon X_2 \Psi\left(\frac{X_1}{\varepsilon X_2}\right)\right]dt + \sigma X_1 d\xi_1(t), \\ dX_2(t) &= -X_2 dt + \delta X_1 d\xi_2(t). \end{aligned} \tag{3.6}$$

We first observe that for any $\varepsilon > 0$, $\delta > 0$ the coefficients of (3.6) have bounded derivatives with respect to x_1, x_2, and consequently they satisfy the existence conditions (Theorem 3.3.2). Further, for small ε and δ the coefficients of (3.6) are close to those of the deterministic system

$$\frac{dx_1}{dt} = x_1; \qquad \frac{dx_2}{dt} = -x_2 \tag{3.7}$$

in the sense of (1.3), where the constant γ in (1.3) may be assumed equal to $\min(\varepsilon,\delta)$. Finally, it is clear that the solutions of the system (3.7), except those for which $x_1(0) = 0$, have absolute values diverging to infinity as $t \to \infty$. Nevertheless, we can prove that for any $\varepsilon > 0$ and $\delta > 0$ the solution $X(t) \equiv 0$ of (3.6) is asymptotically stable in the large. This will furnish a negative anser to the second question at the end of Section 2[2].

We shall use in the proof the fact that all the coefficients of (3.6) are homogeneous functions of degree 1 and therefore the projection of the process $X(t)$ on the circle $|x| = 1$ is also a Markov process (see Section 6.7).

As in Section 6.8, we introduce the new variables

$$r(t) = \frac{1}{2}\ln(X_1^2(t) + X_2^2(t)) = \ln|X(t)|;$$

[2] In connection with this Nevel'son has proved that an unstable linear system perturbed by small linear random perturbations is also unstable.

$$\varphi(t) = \text{arctg}\,\frac{X_2(t)}{X_1(t)}$$

and apply Itô's formula (3.3.8). The result is

$$d\varphi(t) = -\left[2\sin\varphi\cos\varphi + \varepsilon\sin^2\varphi\Psi\left(\frac{\text{ctg}\varphi}{\varepsilon}\right)\right]$$

$$+\ \delta(\cos^2\varphi d\xi_2(t) - \sin\varphi\cos\varphi d\xi_1(t)), \tag{3.8}$$

$$dr(t) = \left[\cos^2\varphi - \sin^2\varphi + \varepsilon\sin\varphi\cos\varphi\Psi\left(\frac{\text{ctg}\varphi}{\varepsilon}\right)\right]$$

$$+\ \delta(\cos^2\varphi d\xi_1(t) + \sin\varphi\cos\varphi d\xi_2(t)). \tag{3.9}$$

The diffusion coefficient of the Markov process $\varphi(t)$ on the circle $0 \leqslant \varphi < 2\pi$ vanishes only at the points $\varphi_1 = \pi/2$, $\varphi_2 = 3\pi/2$. In view of (3.5), this means that $\varphi = \pi/2$ and $\varphi = 3\pi/2$ are solutions of equation (3.8). We claim that these solutions are stable. To prove this, we investigate the first-approximation equation in the neighborhood of the point $\varphi = \pi/2$. By (3.5), this equation is

$$d\left(\varphi - \frac{\pi}{2}\right) = -\left(\varphi - \frac{\pi}{2}\right)dt + \delta\left(\varphi - \frac{\pi}{2}\right)d\xi_1(t).$$

Since the first-approximation equation is asymptotically stable, it follows from Theorem 1.1 that the solution $\varphi = \pi/2$ of equation (3.8) is stable in probability. The stability of the solution $\varphi = 3\pi/2$ is proved in a similar fashion. The diffusion coefficient of the process $\varphi(t)$ is positive for $\varphi \neq \varphi_i$. Hence, reasoning as in the proof of Theorem 5.4.3, we see that $\varphi(t)$ has a limit as $t \to \infty$ for any initial condition. This limit is either $\pi/2$ or $3\pi/2$.

Thus, by equation (3.9) and Lemma 6.7.1, we see that

$$\lim_{t\to\infty}\frac{\ln|X(t)|}{t} = \lim_{t\to\infty}\frac{1}{t}\int_0^t\left[\cos^2\varphi(s) - \sin^2\varphi(s)\right.$$

$$\left. +\ \varepsilon\sin\varphi(s)\cos\varphi(s)\Psi\left(\frac{\text{ctg}\varphi(s)}{\varepsilon}\right)\right]ds = -1.$$

Thus, $\mathbf{P}\{|X(t)| \to 0 \text{ as } t \to \infty\} = 1$ and the solution of equation (3.6) is stable in the large, as required.

To conclude this section, we note that in this example condition (3.4) does not hold for any $\alpha > 0$. It is quite probable that the theorem on instability in the first approximation can be proved on the assumption that (2.5) holds for only one value of x and the full system is nearly linear in the sense of (3.4). This would be a natural generalization of the theorem of Ljapunov quoted at the beginning of Section 2.

4. Stability under damped random perturbations

Consider the one-dimensional system

$$dX(t) = -Xdt + \sigma(t)d\xi(t); \qquad X(0) = x_0. \tag{4.1}$$

It is easy to see that the solution of this system is

$$X(t) = x_0 e^{-t} + \int_0^t e^{s-t}\sigma(s)d\xi(s).$$

Hence

$$\mathbf{E}|X(t) - x_0 e^{-t}|^2 = \int_0^t e^{2(s-t)}\sigma^2(s)ds$$

$$\leqslant e^{-t}\int_0^{t/2}\sigma^2(s)ds + \int_{t/2}^t \sigma^2(s)ds.$$

It follows from this inequality that, although $X(t) \equiv 0$ is not a solution of the system (4.1), it is nonetheless true that any solution of (4.1) tends to zero as $t \to \infty$, provided that

$$\int_0^\infty \sigma^2(s)ds < \infty.$$

It is natural to expect a similar situation to obtain in a broader range of cases: Sufficiently rapidly damped persistent random perturbations do not affect the stability of an asymptotically stable system. Let us prove a result going in that direction.

THEOREM 4.1. *Consider the system*

$$\frac{dx}{dt} = F(t,x) \tag{4.2}$$

in E_ℓ. *Suppose that there exists for this system a positive definite and admitting an infinitesimally small upper limit function* $V(t,x)$ *satisfying*

$$\frac{d^0V}{dt} = \frac{\partial V}{\partial t} + \left(F, \frac{\partial V}{\partial x}\right) < -\alpha(t)\varphi(t,x) < 0, \tag{4.3}$$

$$\inf_{t>0} V(t,x) \to \infty \quad \text{as} \quad |x| \to \infty, \tag{4.4}$$

and such that for every $0 < \varepsilon < R < \infty$,

$$\inf_{\varepsilon<|x|<R} \varphi(t,x) > 0 \quad \text{for} \quad T > T_{\varepsilon,R} \tag{4.4'}$$

and

$$\int_0^\infty \alpha(t)dt = \infty \tag{4.5}$$

hold. Suppose further that the coefficients $\sigma_r(t,x)$ *of the stochastic equation*

$$dX(t) = F(t,X(t))dt + \sum_{r=1}^{k} \sigma_r(t,X(t))d\xi_r(t) \tag{4.6}$$

satisfy for a certain constant K_1 *and a certain positive and integrable on* $[0,\infty]$ *function* $g(t)$

$$\sum_{r=1}^{k} \sum_{i,j=1}^{\ell} \sigma_r^i \sigma_r^j \frac{\partial^2 V}{\partial x_i \partial x_j} = \sum_{r=1}^{k} \left(\sigma_r(t,x), \frac{\partial}{\partial x}\right)^2 V(t,x)$$

$$\leqslant (V(t,x) + K_1)g(t). \tag{4.7}$$

Then every solution of equation (4.6) *has almost surely limit* 0 *as* $t \to \infty$.

PROOF. We let the differential generator L of the process (4.6) act on the function

$$W(t,x) = (V(t,x) + K_1)\exp\left\{\int_t^\infty g(s)ds\right\}.$$

Evidently

$$LW = (LV - gV - K_1 g)\exp\left\{\int_t^\infty g(s)ds\right\}$$

$$\leqslant -\alpha(t)\varphi(t,x)\exp\left\{\int_t^\infty g(s)ds\right\}. \tag{4.8}$$

(4.4), (4.8) and the integrability of the function $g(t)$ imply that

$$LW \leqslant 0; \qquad \inf_{t\geqslant 0} W(t,x) \to \infty \quad \text{as} \quad |x| \to \infty.$$

Therefore (see Section 5.2 and Theorem 3.4.1) $W(t,X(t))$ is a bounded from below supermartingale. Theorem 5.2.1 implies now that almost surely there exists the limit

$$\eta = \lim_{t\to\infty} W(t,X(t)) = \lim_{t\to\infty} V(t,X(t)) + K_1. \tag{4.9}$$

Since $V(t,x)$ is positive definite and (4.9) holds, the theorem will be proved provided we show that $\eta = K_1$ holds almost surely.

For this purpose let us consider the domain $U_T(\varepsilon,R) = \{\varepsilon < |x| < R\} \times (t > T)$. Then (4.8) and the assumptions of the theorem imply that for a suitable choice of $T = T_{\varepsilon,R}$ the inequality $LW \leqslant -\alpha(t)\delta_{\varepsilon,R}$ holds in the domain $U_T(\varepsilon,R)$. Thus we may conclude by Theorem 3.7.1 that for every point $(s,x) \in U_T(\varepsilon,R)$ the moment $\tau(\varepsilon,R)$ at which the path of the process $X(t)$ exits from the domain $\varepsilon < |x| < R$ is almost surely finite. (More exactly, we should apply Theorem 3.7.1 only in the case when $\tau(\varepsilon,R) \geqslant T$, considering for $t \geqslant T$ the paths which leave the point $X(T)$ at the moment T.) Thus we may now apply Lemma 3.8.1 whose assumptions are satisfied for the function $W(t,x)$ in order to conclude that the process $X(t)$ is recurrent in the domain $|x| < \varepsilon$ for every $\varepsilon > 0$ and $x \in E_\ell$.

It is now easy to establish the equality $\mathbf{P}\{\liminf_{t\to\infty} |X(t)| = 0\} = 1$. Indeed, suppose that with positive probability we have $\liminf_{t\to\infty} |X^{s_0,x_0}(t)| > \delta$ for some $s_0 > 0$ and $x_0 \in E_\ell$. Then there exists a number $T > 0$ such that

$$\mathbf{P}\{\inf_{t>T} |X^{s\;,x}\;(t)| > \delta\} > p/2.$$

The last inequality contradicts the recurrence property of the process $X^{T,x}(t)$ as related to the domain $\{|x| < \delta\}$ and for all $|x| > \delta$.

Since $V(t,x)$ has an infinitesimally small upper limit, (4.9) implies that

$$\mathbf{P}\{\eta = K_1\} = 1.$$

As we have noted above, this suffices to complete the proof of the theorem.

REMARK 1. The assumption that there exists a function V

satisfying conditions (4.3), (4.4) and (4.5) is not very restrictive. For a wide class of stable in the large systems (4.2) it is possible to demonstrate the existence of a function V with the above properties (see Krasovskiĭ [1]). For instance (4.7) is satisfied if V grows not slower than $|x|^2$ as $|x| \to \infty$, it has bounded second derivatives with respect to the space variables, and moreover

$$\sum_{r=1}^{k} |\sigma_r(t,x)|^2 < (|x|^2 + 1)g(t). \tag{4.10}$$

REMARK 2. Theorem 4.1 throws some light upon the effect of damped random perturbations on a deterministic system which is stable in the large. If the system is dissipative in the sense of Section 1.2, one can prove in an analogous fashion that the constant R figuring in the definition of dissipativity satisfies the equality

$$P \overline{\lim_{t\to\infty}} |X(t,\omega)| < R\} = 1.$$

A more delicate analysis shows that in this case for sufficiently "well behaved" systems (4.2) the solution of equation (4.6) almost surely converges to one of the solutions of the deterministic system (4.2) as $t \to \infty$. We shall not go into the details here.

5. Application to stochastic approximation[3]

In 1951, Robbins and Monro [1] proposed an iterative procedure for the determination of the roots of the regression equation; they called this method stochastic approximation. Let us briefly describe it here.

Let $Y(x,\omega)$ be a family of random variables depending on a parameter x such that $\mathbf{E}Y(x,\omega) = R(x)$. Suppose that the distribution of $Y(x,\omega)$ is unknown to an "observer", who can only carry out certain "measurements" $y_n(x,\omega)$ of the variable for arbitrary values of the parameter. The problem is to determine a value of the parameter $x = x^0$ for which the function $R(x)$ assumes a prescribed value α.

The procedure proposed by Robbins and Monro to determine x^0 is as follows. Choose a sequence of positive numbers a_n such that

[3] A more detailed discussion of these problems is given in Chapters IV and V of the author's joint book with Nevel'son [4].

$$\sum_{n=1}^{\infty} a_n = \infty; \qquad \sum_{n=1}^{\infty} a_n^2 < \infty. \tag{5.1}$$

Fixing x_1 arbitrarily, define a sequence of numbers $x_2,\ldots,x_n,\ldots$ by the recurrence relation

$$x_{n+1} = x_n - a_n(y_n(x_n,\omega) - \alpha). \tag{5.2}$$

Under certain assumptions on the distribution of $Y(x,\omega)$, it can be shown that $x_n(\omega) \to x^0$ almost surely as $n \to \infty$. This result has since been generalized in various directions. A detailed bibliography of the literature up to 1965 can be found in Fabian [1].

A natural continuous analog of the procedure (5.2) is

$$\frac{dX}{dt} = - a(t)[Y(X(t),t,\omega) - \alpha], \tag{5.3}$$

where $Y(x,t,\omega)$ is a "stochastic process" with independent values and expectation $R(x)$. If we assume, as before, that the solution $X(t)$ is a continuous stochastic process[4], then this equation can be interpreted as the Itô equation

$$dX(t) = - a(t)[(R(X(t)) - \alpha)dt + \sigma(X(t))d\xi(t) \tag{5.4}$$

in E_ℓ.

We shall show that under certain assumptions concerning the functions $a(t)$, $R(t)$ and $\sigma(s)$ this process converges almost surely to a root of the equation $R(x) = \alpha$.

THEOREM 5.1. *Given two differentiable functions* $R(x)$ *and* $\sigma(x)$, *suppose that there exist a twice continuously differentiable function* $V(x)$ *and a constant* $k > 0$ *such that*

$$\left.\begin{array}{l} V(x) \to \infty \quad \text{as} \quad |x| \to \infty; \\ V(x^0) = 0; \qquad V(x) > 0 \quad \text{for} \quad x \neq x^0 \end{array}\right\} \tag{5.5}$$

and

$$(R(x) - \alpha)\frac{dV}{dx} > 0 \quad \text{for} \quad s \neq x^0,$$

$$\sigma^2(x)\frac{d^2V}{dx^2} < k(V(x) + 1).$$

Then the process (5.4) *satisfies the equality*

---

4 This restriction is not natural, and it can be eliminated by recourse to the stochastic differential equation of a Markov jump process (see, e.g., Skorohod [1]). However, jump processes are beyond the scope of this book.

$$P\{\lim_{t\to\infty} X(t) = x^0\} = 1$$

for all continuous positive functions $a(t)$ such that

$$\int_0^\infty a(t)dt = \infty; \qquad \int_0^\infty a^2(t)dt < \infty. \tag{5.6}$$

The proof follows at once from Theorem 4.1, if we set

$$(R(x) - \alpha)a(t) = - F(t,x - x^0).$$

COROLLARY. The assumptions of the theorem evidently hold for the function $V(x) = (x - x^0)^2$, if $R(x)$ and $\sigma(x)$ are such that

$$(x - x^0)(R(x) - \alpha) > 0 \quad \text{for} \quad x \neq x^0,$$

$$\sigma^2(x) < k(x^2 + 1).$$

For the discrete-time case analogous sufficient conditions for convergence of stochastic approximations were given in Gladyšev [1].

Theorem 4.1 also yields convergence conditions for stochastic approximations in many dimensions. We state the result.

THEOREM 5.2. *The many-dimensional stochastic approximation process*

$$dX(t) = - a(t)\left[(R(X(t)) - \alpha)dt + \sum_{r=1}^{k} \sigma_r(X(t))d\xi_r(t)\right] \tag{5.7}$$

converges almost surely to a solution of the equation

$$R(x) = \alpha, \tag{5.8}$$

if conditions (5.6) *are satisfied and there exists a function $V(x)$ satisfying condition* (5.5) *and the conditions*

$$\left(R(x) - \alpha, \frac{\partial V}{\partial x}\right) > 0 \quad \text{for} \quad x \neq x^0, \tag{5.9}$$

$$\sum_{r=1}^{k} \sum_{i,j=1}^{\ell} \sigma_r^i(x)\sigma_r^j(x) \frac{\partial^2 V}{\partial x_i \partial x_j} < k_1(V(x) + 1)$$

for a constant k_1.

REMARK. If condition (5.9) is satisfied only in a neighborhood of infinity, it can be shown that the process (5.7) converges to an invariant set of the system $dx/dt = -R(x) + \alpha$.

6. Stochastic approximations when the regression equation has several roots

Condition (5.9) guarantees the uniqueness of the solution of the equation $R(x) = \alpha$. However, it is interesting to study the properties of the Robbins-Monro procedure when the equation has several roots. We shall show that Ljapunov functions are also applicable in this case.

Not striving for maximal generality, we shall limit the discussion in this section to the case of a point x on the line $(-\infty < x < \infty)$. We shall assume that the coefficient $\sigma(x)$ of equation (5.4) satisfies

$$\sigma^2(x) < k(x^2 + 1) \tag{6.1}$$

for some constant $k > 0$.

THEOREM 6.1. *Suppose that the set* $A = \{x : R(x) = \alpha\}$ *consists of finitely many points* $x_1^{(0)}, \ldots, x_n^{(0)}$, *the derivative* $R'(x)$ *is continuous, and condition* (6.1) *is satisfied. Suppose moreover that*

$$(R(x) - \alpha)x > 0 \quad \text{for} \quad |x| > b \tag{6.2}$$

holds for some number $b > 0$.

Then for any function $a(t) > 0$ *satisfying the conditions* (5.6) *the Robbins-Monro process* $X(t)$ *defined by equation* (5.4) *converges almost surely to a point of the set* A.

PROOF. We set

$$R_1(x) \quad \begin{cases} R(x) - a & \text{for} \quad |x| \leqslant b, \\ \dfrac{R(b) - \alpha}{b}\, x & \text{for} \quad x > b, \\ \dfrac{R(-b) - \alpha}{-b}\, x & \text{for} \quad x < -b, \end{cases}$$

$$W(x) = \int_0^x R_1(y)dy;$$

$$V(t,x) = (W(x) + k_1)\exp\left\{\gamma\int_t^\infty a^2(s)ds\right\},$$

where the constants $k_1 > 0$ and $\gamma > 0$ will be determined later. The functions $R_1(x)$ and $R(x)$ are illustrated in Figure 3 and the function $W(x)$ in Figure 4.

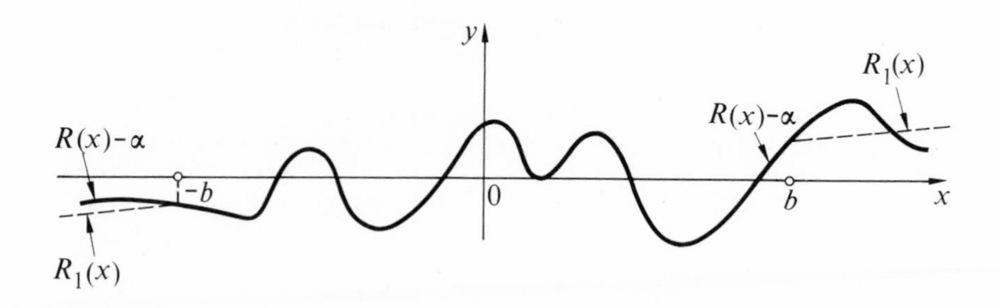

Figure 3

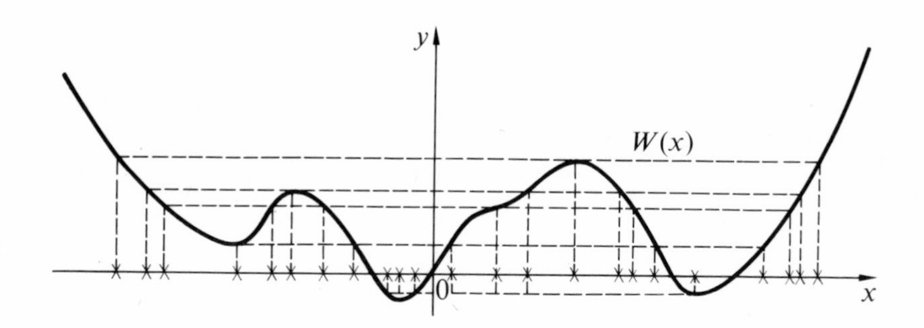

Figure 4

Applying to the function V the differential generator

$$L = \frac{\partial}{\partial t} - a(t)(R(x) - \alpha)\frac{\partial}{\partial x} + \frac{1}{2}a^2(t)\sigma^2(x)\frac{\partial^2}{\partial x^2}$$

of the process $X(t)$, we get

$$LV = -(W(x) + k_1)\gamma a^2(t)$$

$$- a(t)(R(x) - \alpha)R_1(x)\exp\left\{\gamma\int_t^\infty a^2(s)ds\right\}$$

$$+ \frac{1}{2}a^2(t)\sigma^2(x)R_1'(x)\exp\left\{\gamma\int_t^\infty a^2(s)ds\right\}.$$

By the above construction of the function $R_1(x)$, the derivative $R_1'(x)$ is bounded and the product $(R(x) - \alpha)R_1(x)$ is non-negative (see (6.2)). In addition, the function $W(x)$ increases like a parabola when $|x| \to \infty$. Hence, by (6.1), it follows that for a suitable choice of k_1 and γ we have $LV \leqslant 0$ for $t \geqslant t_0$; in fact, we have the even stronger condition

$$LV \leqslant - a(t)(R(x) - \alpha)R_1(x). \tag{6.3}$$

Thus the function $V(t,X(t))$ is a supermartingale for $t \geqslant t_0$ (see Section 5.2). Hence the limit

$$\lim_{t\to\infty} V(t,X(t)) = \xi \tag{6.4}$$

is almost surely finite. Moreover, it follows from (6.3) and (3.3.15) that

$$\int_{t_0}^{t} a(s)\mathbf{E}[(R(X(s)) - \alpha)R_1(X(s))]ds \leqslant \mathbf{E}V(t_0,X(t_0)) \tag{6.5}$$

for $t > t_0$. The integrand on the left of (6.5) is non-negative, and the function $a(s)$ is not integrable on $[t_0,\infty]$. Thus, there exists a sequence $t_n \to \infty$ such that

$$\lim_{n\to\infty} \mathbf{E}[(R(X(t_n)) - \alpha)R_1(X(t_n))] = 0.$$

It follows from this inequality and from Čebyšev's inequality that the sequence $R(X(t_n)) - \alpha$ converges to zero in probability. Then, as is well known, there exists a subsequence t_{n_k} such that

$$\lim_{k\to\infty} R(X(t_{n_k})) = \alpha \qquad \text{(a.s.)}.$$

Hence, it follows by (6.4) that the random variable ξ in (6.4) almost surely takes on only values from the finite set $W(x_1^{(0)}) + k_1, \ldots, W(x_n^{(0)}) + k_1$. Let A_1 be the set of points x such that $W(x) + k_1$ takes on one of these values. The assumptions of the theorem evidently imply that A_1 is a finite set (see Figure 4), in which the elements of A_1 are marked by crosses), and $A \subset A_1$. It follows from (6.4) that the process $X(t)$ converges to the set A_1 as $t \to \infty$. Moreover $X(t)$ almost surely converges to only one element of A_1, for otherwise it would follow from the continuity of $X(t)$ that (6.4) does not hold. Finally let us prove that the probability of the event $B = \{X(t) \to x_0 \text{ as } t \to \infty\}$ is zero if $x_0 \in A_1 \setminus A$. In fact, B implies the event

$$\lim_{t\to\infty} (R(X(t)) - \alpha)R_1(X(t)) = (R(x_0) - \alpha)^2 > 0.$$

Hence, by (5.6), we get

$$B \subset \left\{ \int_{t_0}^{\infty} a(s)(R(X(s)) - \alpha)R_1(X(s))ds = \infty \right\}.$$

This, together with (6.5), implies that $P(B) = 0$. Theorem 6.1 is proved.

Let A_2 denote the set of points $x \in A$ at which the function $R(x) - \alpha$ changes sign, from positive to negative. The next theorem shows that under certain additional assumptions the points of A_2 cannot be limits of the stochastic approximation process (5.4). For the discrete-time case this was conjectured by Fabian (2.3). Similar questions have been considered by Krasulina [1].

THEOREM 6.2. *Under the assumptions of Theorem* 6.1, *let* x^* *be a point of the set* A_2 *such that* $R'(x^*) < 0$ *and* $\sigma^2(x^*) > 0$. *Assume moreover that the function* $a(t)$ *satisfies* $|a'(t)| < ka^2(t)$ *(*$k > 0$ *constant). Then* x^* *cannot be a limit point of the stochastic approximation process* (5.4).

For the proof we need the following two lemmas.

LEMMA 6.1. *Let* $X(t)$ *be a diffusion process with differential generator* L *and let* D *be a bounded domain. Then* $X(t)$ *almost surely exits from* D *in a finite time, if there exists a function* $V(t,x)$ *such that in* $I \times D$

$$LV \leqslant 0, \tag{6.6}$$

$$\inf_{x \in D} V(t,x) \to \infty \quad \text{as} \quad t \to \infty. \tag{6.7}$$

PROOF. Let τ denote the first exit time from D of the path of the process and let $\tau(t) = \min(\tau,t)$. As mentioned in Section 5.2, the process $V(\tau(t),X(\tau(t)))$ is a supermartingale. By Theorem 5.2.1

$$\lim_{t\to\infty} V(\tau(t),X(\tau(t))) = \xi$$

exists and is finite. The finiteness of ξ and relation (6.7) imply the assertion.

LEMMA 6.2. *The function*

$$W(z) = \int_0^z dv \int_0^v \frac{e^{u-v}}{\sqrt{uv}}\, du \tag{6.8}$$

is a solution of the equation

$$zw'' + \left(z + \frac{1}{2}\right)w' = 1. \tag{6.9}$$

Moreover, W satisfies

$$W(z) = \ln z + O(1) \qquad (z \to \infty), \tag{6.10}$$

$$0 < zW'(z) < c \quad \text{for} \quad z \to 0. \tag{6.11}$$

PROOF. The relation (6.9) is verified directly, while (6.10) and (6.11) are consequences of the estimates

$$W'(z) = \frac{e^{-z}}{\sqrt{u}} \int^{z} \frac{e^{u}}{\sqrt{u}}\, du = \frac{1}{z} + \frac{1}{2z^2} + O\left(\frac{1}{z^3}\right) \qquad (z \to \infty),$$

$$W'(z) = 2 + O(z) \qquad (z \to 0),$$

which are proved by integration by parts.

PROOF OF THEOREM 6.2. (1) Without loss of generality, we may assume that $x^* = 0$. Let us prove that, if $\varepsilon > 0$ is sufficiently small, then the process $X(t)$, beginning at an arbitrary time $t > 0$ at almost any point of the interval $(-\varepsilon, \varepsilon)$, reaches the ends of this interval in a finite time. It follows from Theorem 6.1 and the strong Markov property of the process $X(t)$ that this assertion is equivalent to Theorem 6.2.

Using Lemma 6.1, we can thus reduce the proof of Theorem 6.2 to the construction of a function $V(t,x)$ in $\{t > 0\} \times (-\varepsilon, \varepsilon)$ satisfying conditions (6.6), (6.7). Applying the usual methods of stability theory, we shall first construct a function $V(t,x)$ satisfying the conditions (6.6), (6.7) for

$$\tilde{L} = \frac{\partial}{\partial t} + a(t)\beta x \frac{\partial}{\partial x} + \frac{1}{2} a^2(t)\sigma^2 \frac{\partial^2}{\partial x^2},$$

where $\beta = -R'(0)$, $\sigma_0^2 = \sigma^2(0)$. It will then be an easy matter to show that after some modification the function will satisfy these conditions for the "full" operator

$$L = \frac{\partial}{\partial t} - a(t)(R(x) - \alpha) \frac{\partial}{\partial x} + \frac{1}{2} a^2(t)\sigma^2(x) \frac{\partial^2}{\partial x^2}. \tag{6.12}$$

We shall look for the function $V(t,x)$ in the form

$$V(t,x) = \Phi(t) - W(z), \qquad z = \frac{x^2}{\varphi(t)},$$

where W is the function defined by (6.8) and the functions $\varphi(t)$ will be specified below. Simple computations lead to the equality

$$\tilde{L}V(t,x) = \Phi'(t) + zW'(z)\left(\frac{\varphi'}{\varphi} - 2\beta a(t)\right) - \frac{a^2(t)\sigma_0^2}{\varphi(t)}[2zW'' + W'].$$

Since $2zW'' + W' = 2 - 2zW'$ by (6.9), it follows now that

$$\tilde{L}V(t,x) = \Phi'(t) - \frac{2a^2(t)\sigma_0^2}{\varphi(t)} + zW'(z)\left[\frac{\varphi'}{\varphi} + \frac{2\sigma_0^2 a^2(t)}{\varphi} - 2\beta a(t)\right].$$

We now define $\varphi(t)$ by

$$\varphi(t) = 2\sigma^2 \exp\left\{2\beta\int_0^t a(s)ds\right\}\int_t a^2(s)\exp\left\{-2\beta\int_0^s a(u)du\right\}ds$$

(the convergence of the integral follows from (5.6)). Thus, we obtain the equality

$$LV(t,x) = \Phi'(t) - \frac{2a^2(t)\sigma_0^2}{\varphi(t)}.$$

Setting

$$\tilde{\Phi}(t) = \ln f(t);$$

$$f(t) = \left[\int_t^\infty a^2(s)\exp\left\{-2\beta\int_0^s a(u)du\right\}ds\right]^{-1},$$

we see now that for this choice of $\varphi(t)$ and $\Phi(t)$ the function $V(t,x)$ satisfies the condition $\tilde{L}V(t,x) = 0$. Since $W(z)$ is an increasing function for $z > 0$ (see (6.11)), it follows that $V(t,x) \geq V(t,\varepsilon)$ for $|x| \leq \varepsilon$ and therefore (6.7) will follow if we can show that $V(t,\varepsilon) \to \infty$ as $t \to \infty$. When $t \to \infty$, it follows from (6.10) that

$$V(t,\varepsilon) = \ln f(t) - \ln\left[\frac{\varepsilon^2}{2\sigma_0^2}\exp\left\{-2\beta\int_0^t a(s)ds\right\}f(t)\right] + O(1)$$

$$\geq \beta\int_0^t a(s)ds + O(1). \tag{6.13}$$

We have thus verified (6.6) and (6.7) for the function $V(x,t)$ and the operator $\tilde{L}$.

(2) We now prove that if $\gamma < 2\beta$, then the function

$$V_1(t,x) = V(t,x) - \gamma\int_0^t a(s)ds = \Phi(t) - W\left(\frac{x^2}{\varphi(t)}\right) - \gamma\int_0^t a(s)ds$$

will satisfy conditions (6.6) and (6.7) in the domain $(t > 0) \times (-\varepsilon,\varepsilon)$ for the operator (6.12), provided $\varepsilon > 0$ is sufficiently small.

(6.7) follows for $V_1(t,x)$ from (6.13). We now apply the operator L to the function V_1. Since $\tilde{L}V = 0$, it follows that

$$LV_1 = LV - \gamma a(t) = (L - \tilde{L})V - \gamma a(t)$$

$$= - a(t)(R(x) - \alpha - R'(0)x) \frac{\partial V}{\partial x}$$

$$+ \frac{1}{2} a^2(t)(\sigma^2(x) - \sigma_0^2) \frac{\partial^2 V}{\partial x^2} - \gamma a(t)$$

$$\leqslant \delta\left(a(t)\left|x \frac{\partial V}{\partial x}\right| + a^2(t)\left|\frac{\partial^2 V}{\partial x^2}\right|\right) - \gamma a(t) \tag{6.14}$$

for $x \in (-\varepsilon,\varepsilon)$. The constant $\delta > 0$ in this inequality can be made arbitrarily small by a suitable choice of ε. It is clear from the form of the function V and from (6.9), (6.11) that

$$\left.\begin{aligned} \left|x \frac{\partial V}{\partial x}\right| &= |zW'(z)| < k_1, \\ \left|\frac{\partial^2 V}{\partial x^2}\right| &= \frac{4}{\varphi(t)}\,|1 - zW'(z)| < \frac{k_1}{\varphi(t)} \end{aligned}\right\} \tag{6.15}$$

for some constant k_1. Now, applying de l'Hôpital's rule and using the inequality $|a'(t)| < ka^2(t)$, we easily see that

$$a(t)f(t)\exp\left\{-2\beta\int_0^t a(s)ds\right\} = \frac{2\sigma_0^2 a(t)}{\varphi(t)}$$

is bounded. Hence, for a suitable constant k_2,

$$\frac{a^2(t)}{\varphi(t)} < k_2 a(t). \tag{6.16}$$

The relations (6.14), (6.15) and (6.16) now imply the required inequality $LV_1 \leqslant 0$. This we wished to show.

7. Some generalizations

In this section we shall survey a few stability problems not yet discussed in this book. The discussion will inevitably be quite sketchy. In some cases we shall only make reference to the literature, not touching upon the problem proper.

1. *Stability and excessive functions*[4]

Let $X = (X(t), \mathbf{P}_x)$ be a temporally homogeneous right-continuous strong Markov process in a Banach space E. Here $\mathbf{P}_x$ is the measure generated by the "initial condition" $X(0) = x$ (for more details, see Dynkin [3]). We denote by $\|x\|$ the norm of an element x, and by $\mathfrak{B}$ the σ-algebra of measurable sets in E.

An extremely useful tool for investigating the properties of Markov processes is provided by the excessive functions (see Dynkin [3]). An *excessive function* is a function $V(x)$ with the properties

$$0 \leqslant T_t V(x) = \int_E P(x,t,dy)V(y) \leqslant V(x), \qquad (t \geqslant 0,\ x \in E),$$

$$T_t V(x) \to V(x) \quad \text{as} \quad t \downarrow 0.$$

It is known (Dynkin [3], p. 499) that an excessive function V satisfies the inequality

$$\mathbf{E}_x V(X(\zeta)) \leqslant V(x) \tag{7.1}$$

for any Markov time ζ. The function V is said to be excessive for the process X in an open set U if inequality (7.1) is satisfied for all $\zeta \leqslant \tau_U$, where τ_U is the first exit time of a path of the process from U.

Recall that a nonempty set $D \in \mathfrak{B}$ is said to be invariant for the process X if $P(x,t,D) = 1$ for $x \in D$, $t \geqslant 0$.

DEFINITION. *An invariant point $x_0 \in E$ for the process X is said to be stable in probability for the process X if*

$$\inf_{\|y-x_0\|\to 0} \mathbf{P}_y\{\sup_{t>0} \|X(t) - x_0\| > \varepsilon\} = 0.$$

LEMMA 7.1. *A sufficient condition for a point x_0 to be stable in probability for the process X is that there exists a function V, which is excessive for the process X in a neighborhood of x_0, and which satisfies $V(x_0) = 0$ and $\inf_{\|x-x_0\|>\varepsilon} V(x) = V_\varepsilon > 0$ for $\varepsilon > 0$.*

PROOF. The proof follows from (7.1) and Čebyšev's inequality, since

$$V_\varepsilon \mathbf{P}_x\{\sup_{t>0} \|X(t) - x_0\| > \varepsilon\} \leqslant \mathbf{E}_x V(X(\tau_{U_\varepsilon(x_0)}(t))) \leqslant V(x).$$

[4] See Bucy [1], Has'minskiĭ [6].

It is evident that the above stability condition is too general to be of much interest. In certain special cases, however, one can derive more specific conditions. For example, the proof of Theorem 5.3.1 essentially reduces to verifying that a non-negative function V which is defined in some neighbourhood U of the origin is of class $\mathbf{C}_2^0(U)$ and satisfies the inequality $LV \leqslant 0$, is excessive in U.

For right-continuous strong Markov processes, the fact that non-negative functions for which the weak differential generator is non-positive, are excessive, can be established by means of a well-known theorem of Dynkin [3, p. 191]. Together with Lemma 7.1, this enables one to derive stability conditions for such processes (see Kushner [2,3]). A generalization of Theorem 5.3.1 to jump processes has been considered by Gihman and Dorogovčev [1].

2. *Stability of the invariant set*

Let $\rho(x,U) = \inf_{x_0 \in U} \|x - x_0\|$ denote the distance of a point x from the set U. An invariant set U of the process X is said to be stable in probability if

$$\lim_{\rho(y,U)\to 0} \mathbf{P}_y\{\sup_{t>0} \rho(X(t),U) > \varepsilon\} = 0.$$

The proof of the following lemma is similar to that of Lemma 6.1.

LEMMA 7.2. *An invariant set D is stable in probability for the process X if there exists a function $V(x)$, excessive for X in a neighborhood of the set D, such that:*

1. $V(x) = 0$ *for* $x \in D$;
2. $\inf_{\rho(x,D)>\varepsilon} V(x) = V_\varepsilon > 0$ *for* $\varepsilon > 0$.

The following theorem is the analog of Theorem 5.3.1 for stability of the invariant set. We prove it by using Lemma 7.2 and proceeding as in the proof of Theorem 5.3.1.

THEOREM 7.1. *Let $X(t)$ be the process described by the operator*

$$L = \frac{\partial}{\partial t} + \left(b(t,x), \frac{\partial}{\partial x}\right) + \frac{1}{2}\sum_{r=1}^{k}\left(\sigma_r(t,x), \frac{\partial}{\partial x}\right)^2. \tag{7.2}$$

Assume that there exists a function $V(t,x)$, twice continuously differentiable with respect to x and continuously differentiable with respect to t, vanishing for $x \in D$ and satisfying the conditions

$$LV \leqslant 0; \qquad \inf_{\rho(x,D)>\varepsilon;t>0} V(t,x) = V_\varepsilon > 0 \quad \text{for} \quad \varepsilon > 0$$

in a neighborhood of the set D.

Then the invariant set D of the process X is stable in probability.

REMARK. If the set D is inaccessible (see Section 5.2) to the process, the smoothness conditions imposed on V can be weakened. To be precise, the function V need not be smooth at the points x of the set $\Gamma = \{x:\rho(x,D) = 0\}$. Moreover, using Lemma 7.2 one can establish analogous stability conditions for the invariant set of a jump process.

3. *Equations whose coefficients are Markov processes*

Several authors (Kac and Krasovskiĭ [1], Frisch [1] and others) have considered the properties of systems described by equations of type

$$\frac{dY}{dt} = F(Y,t,X(t)), \tag{7.3}$$

where Y,F are vectors in E_m and $X(t)$ is a Markov process with values in E_ℓ. If the process $X(t)$ is governed by the operator (7.2), one can investigate the stability of the system (7.3) with the use of Theorem 7.1. In fact, it is clear that the pair $(X(t), Y(t))$ is also a Markov process, whose differential generator is defined on sufficiently smooth functions by

$$L_1V = \frac{\partial V}{\partial t} + \left(F(y,t,x), \frac{\partial V}{\partial y}\right) + \left(b(t,x), \frac{\partial V}{\partial x}\right)$$

$$+ \frac{1}{2}\sum_{r=1}^{k}\left(\sigma_r(t,x), \frac{\partial}{\partial x}\right)^2 V.$$

Thus, investigation of the stability of the path $Y(t) \equiv 0$ for the process (7.3) (on the assumption that $F(0,t,x) \equiv 0$) reduces to investigation of the stability of the m-dimensional hyperplane $y = 0$ for the $(\ell + m)$-dimensional Markov process $(X(t),Y(t))$. Hence, using Theorem 7.1, we get

THEOREM 7.2. *Suppose that for some* $\varepsilon_0 > 0$ *and all* $t > 0$, $|y| < \varepsilon_0$, *there exists a function* $V(t,x,y)$, *continuously differentiable with respect to* $t, y \in E_m$, *twice continuously differentiable with respect to* $x \in E_\ell$ *everywhere except perhaps for the set* $y = \{0\}$, *and such that*

$$L_1V < 0; \qquad V(t,x,0) = 0; \qquad \inf_{t>0,|y|>\varepsilon} V(t,x,y) = V_\varepsilon > 0$$

for $0 < \varepsilon < \varepsilon_0$.

Then the solution $Y \equiv 0$ *of the system* (7.3) *is stable in probability, in the sense that*

$$\lim_{|y|\to 0} \mathbf{P}\{\sup_{t>0} |Y(t)| > \varepsilon; X(0) = x; Y(0) = y\} = 0.$$

Many of the problems considered hitherto in simpler situations are of interest for equations of type (7.3). For example, we might study stability in probability of a linear system

$$\frac{dY}{dt} = F(X(t))Y. \tag{7.4}$$

This problem has an extremely simple solution if $X(t)$ is a temporally homogeneous ergodic process and $m = 1$. Then

$$Y(t) = Y(0)\exp\left\{\int_0^t F(X(s))ds\right\}. \tag{7.5}$$

Hence, by the strong law of large numbers (Section 4.5), it follows at once that if

$$\overline{F} = \int F(x)\mu(dx) < 0$$

holds, where μ is the stationary distribution of the process $X(t)$, then the process $Y(t)$ is asymptotically stable. The same arguments show that the system (7.4) is unstable if $\overline{F} > 0$.

Conditions for p-stability of systems of type (7.4) are quite complicated, even in the case $m = 1$. In fact, it follows from (7.5) that

$$\mathbf{E}\left\{\frac{|Y(t)|^p}{X(0)} = x; Y(0) = y\right\} = |y|^p \mathbf{E}_x \exp\left\{p\int_0^t F(X(s))ds\right\}.$$

If $X(t)$ is, say a temporally homogeneous diffusion process in E_ℓ with local characteristics $b(x)$ and $\sigma_1(x),\ldots,\sigma_k(x)$, then, as is known (Dynkin [3]), the function

$$u(t,x) = \mathbf{E}_x \exp\left\{p\int_0^t F(X(s))ds\right\}$$

is a solution of the equation

$$\frac{\partial u}{\partial t} = \left(b(x), \frac{\partial u}{\partial x}\right) + \frac{1}{2}\left(\sigma\ (x), \frac{\partial}{\partial x}\right)^2 u + pF(x)u, \tag{7.6}$$

which satisfies the initial condition

$$u(0,x) = 1. \tag{7.7}$$

Thus the problem of p-stability of the system (7.4) for $m = 1$ is reduced to investigation of the limiting behavior of the solution of problem (7.6),(7.7) as $t \to \infty$.

If $m > 1$, complications arise. Frisch [1] introduces a linear system of partial differential equations; solving this system, one can use quadratures to determine the moments of a process $Y(t)$ which satisfies (7.4) for arbitrary m.

The case of a temporally homogeneous Markov process $X(t)$ with finitely many states has been studied by Kac and Krasovskiĭ [1]. In particular, their paper presents algebraic criteria for the stability of systems in the mean square.

4. *Stability under persistent perturbation by white noise*

We have already studied the problem of stability under persistent random perturbations (Section 1.6). Our attention was then centered upon random perturbations with finite expectation. It is to be expected that if we narrow down the class of admissible perturbations, we shall be able to derive broader stability criteria. The author has considered in [7] the stability of deterministic systems under perturbation by white noise. We now briefly present some results of that paper.

We consider the equation

$$\frac{dx}{dt} = F(t,x) \qquad (F(t,0) \equiv 0) \tag{7.8}$$

in E_ℓ.

We call the solution $x = 0$ of equation (7.8) *stable under persistent perturbation by white noise* if the solutions $X(t)$ of the Itô equation

$$dX(t) = F(t,X)dt + \sum_{r=1}^{k} \sigma_r(t,X)d\xi_r(t) + b(t,X)dt \tag{7.9}$$

have the following property: For any $\varepsilon > 0$, there exists a $\gamma > 0$ such that for all processes $X(t)$ satisfying equation (7.9), with initial condition $|X(0)| < \gamma$ and coefficients σ_r and b such that

$$\sum_{r=1}^{k} |\sigma_r(t,x)| + |b(t,x)| < \gamma, \tag{7.10}$$

the inequality

$$P\{|X(t)| > \varepsilon\} < \varepsilon$$

holds for all $t > 0$.

THEOREM 7.3 (Has'minskiĭ [7]). *The solution* $x = 0$ *is stable under persistent perturbation by white noise if there exists a positive definite function* $V(t,x) \in \mathbf{C}_2$ *in the domain* $(t > 0) \times E_\ell$ *which has an infinitesimal upper limit and satisfies the conditions:*

1. $\inf_{t>0} V(t,x) \to \infty$ *as* $|x| \to \infty$.

2. *For every* $\varepsilon > 0$, *there exist positive constants* $\alpha_1(\varepsilon)$, $\alpha_1(\varepsilon)$ *and* $\gamma(\varepsilon)$ *such that for all* $\sigma_r(t,x)$ *and* $b(t,x)$ *satisfying condition* (7.10) *with* $\gamma = \gamma(\varepsilon)$ *the inequality*

$$\frac{\partial V}{\partial t} + \frac{1}{2}\sum_{r=1}^{k}\left(\sigma_r(t,x), \frac{\partial}{\partial x}\right)^2 V + \left(F(t,x) + b(t,x), \frac{\partial V}{\partial x}\right)$$

$$\leqslant - \alpha_1(\varepsilon) - \alpha_2(\varepsilon)\frac{\left|\frac{\partial V}{\partial x}\right|^2}{V} \tag{7.11}$$

holds in the domain $|x| > \varepsilon$.

One can easily infer from Theorem 7.3 simpler stability conditions for persistent perturbations. For example, in Has'minskiĭ [7] the theorem is applied to investigate the effect of random perturbation by white noise on the performance of absolutely stable controlled systems with a single final control element. It is shown that the system is stable under such perturbations if the response curve of the final control element has a bounded derivative.

It is also readily inferred from Theorem 7.3 that an exponentially p-stable stochastic system is stable under persistent perturbation by white noise (see Nevel'son and Has'minskiĭ [2]).

5. *Boundedness in probability of the output process of a nonlinear stochastic process*

Suppose that the solution $X(t) \equiv 0$ of the system of Itô equations

$$dX(t) = b(t,X(t))dt + \sum_{r=1}^{k}\sigma_r(t,X(t))d\xi_r(t) \tag{7.12}$$

in E_ℓ is exponentially p-stable for $p = 1$. Assume that the system (7.12) is "driven" by a continuous stochastic process $\zeta(t)$, whose absolute value has bounded expectation, and which is independent of the Wiener processes $\xi_1(t),\ldots,\xi_k(t)$. The "output process" $Y(t)$ of the resulting system is described by the equation

$$dY(t) = b(t,Y(t))dt + \sum_{r=1}^{k} \sigma_r(t,Y(t))d\xi_r(t) + \zeta(t)dt. \quad (7.13)$$

It is shown in Nevel'son and Has'minskiĭ [2] that the expectation of the process $|Y(t)|$ is bounded uniformly in t. It follows that the "output process" $Y(t)$ is bounded in probability.

The case of a system described by a linear n-th order equation driven by white noise has been considered in detail by Dym [1]. See also Nevel'son [4].

CHAPTER VIII

STABILIZATION OF CONTROLLED STOCHASTIC SYSTEMS[1]

1. Preliminary remarks

As mentioned in the preface, the stability theory of stochastic differential equations was developed mainly to meet the needs of stabilization of moving systems subjected to random perturbations. In this chapter we shall consider some problems concerning the stabilization of controlled stochastic systems. The results achieved to date in this field are rather sparse, despite the fact that the basic formulations of the problems and the fundamental equations have been known for some time (Kalman [1], Krasovskiĭ and Lidskiĭ [1], Wonham [3] etc.). The only results of any significance are those pertaining to linear systems and employing quadratic control criteria. We devote to them the exposition which now follows, based on the material of Chapters V through VII.

In conformity with the assumptions adopted hitherto, we shall consider controlled systems of the type

$$dX(t) = b(t,X,u) + \sum_{r=1}^{k} \sigma_r(t,X,u)d\xi_r(t). \tag{1.1}$$

Here $b(t,x,u)$ and $\sigma_r(t,x,u)$ are vector-valued functions, jointly continuous in all arguments, u is a scalar control parameter and $\xi_r(t)$, $r = 1,\dots,k$, are independent Wiener processes such that $\mathbf{E}\xi_r(t) = 0$, $\mathbf{E}\xi_r^2(t) = t$. We shall assume that the control u in the system (1.1) is a function of t and $X(t)$, $u = u(t,X(t))$. Then the process described by equation (1.1) is Markovian[2]. A

---

[1] This chapter was written jointly with M.B. Nevel'son.

[2] In the literature this type of control is known as Markov control, or control employing the feedback principle.

function $u = u(t,x)$ is said to be admissible if the coefficients $b(t,x,u(t,x))$, $\sigma_r(t,x,u(t,x))$ are continuous, have continuous derivatives with respect to x which are bounded uniformly in $t > 0$, and moreover $u(t,0) = 0$. The class of admissible controls is denoted by $\mathfrak{U}$. Each function (admissible control) $u \in \mathfrak{U}$ can be associated with a Markov process $X_u^{s,x}(t)$, which is the solution of equation (1.1) with initial condition $X_u^{s,x}(s) = x$.

By analogy with the deterministic case (see Krasovskiĭ [2]), we shall consider the following two stabilization problems:

I. Asymptotic (exponential) p-stabilization: To determine an admissible control $u = u_0(t,x)$ such that the system (1.1) with $u = u_0(t,x)$ is asymptotically (exponentially) p-stable (see Section 5.7).

II. Optimal stabilization to minimize a given cost: To determine a control $u = u_0(t,x)$ for which the functional

$$J^{s,x_0}(u) = \int_s^{\infty} \mathbf{E}K(t,X_u^{s,x_0}(t),u(t,X_u^{s,x_0}(t)))dt$$

(i.e. the cost) attains a minimum. Here (s,x_0) is a fixed initial point, and $K(t,x,u) \geqslant 0$ for $t \geqslant 0$, $x \in E_\ell$, $u \in (-\infty,\infty)$.

REMARK 1.1. Under the assumptions of the theorem proved in the next section, the function $u_0(t,x)$ solving the optimal stabilization problem turns out to be the same for all initial points (s,x_0).

REMARK 1.2. Problem II is in general not a stabilization problem. For example, if $K(t,x,u) = 0$ for $|x| > R$, the optimal strategy may be to force the path of the process $X_u^{s,x}(t)$ out of the R-neighborhood of $x = 0$. In the sequel, however, we shall confine ourselves to costs satisfying the condition that for any $u \in (-\infty,\infty)$ and certain constants $p > 0$, $c > 0$,

$$K(t,x,u) > c|x|^p \tag{1.2}$$

holds. Under this restriction Problems I and II prove to be intimately connected. In fact, suppose that the control $u_0(t,x)$ is a solution of Problem II for a function $K(t,x,u)$ satisfying condition (1.2). It then follows from Lemma 5.7.1 that

$$\lim_{t\to\infty} \mathbf{E}|X_{u_0}^{s,x}(t)|^p = 0 \quad \text{as} \quad t \to \infty. \tag{1.3}$$

Under certain additional assumptions, this implies that the system (1.1) is asymptotically and exponentially stable.

2. Bellman's principle

In this section we wish to prove a theorem which is a modification of Bellman's principle regarding problems of optimal stabilization of stochastic systems (see Krasovskiĭ [1], Krasovskiĭ and Lidskiĭ [1], Wonham [3], Fleming [1]).

Let V be a function of class $\mathbf{C}_2^{(0)}(E)$. Substituting $u = u(t,x)$ in

$$L_u = \frac{\partial}{\partial t} + \left(b(t,x,u),\ \frac{\partial}{\partial x}\right) + \frac{1}{2}\sum_{r=1}^{k}\left(\sigma_r(t,x,u),\ \frac{\partial}{\partial x}\right)^2,$$

we get the differential generator of a certain Markov process $X_u^{s,x}(t)$.

THEOREM 2.1. *Suppose that there exist functions* $V_0(t,x) \in \mathbf{C}_2^{(0)}(E)$, $u_0(t,x) \in \mathfrak{U}$, *satisfying for all* $t \geqslant 0$, $x \in E_\ell$, $u \in (-\infty,\infty)$ *and certain positive constants* p, n, k_1, k_2 *the conditions*

$$V_0(t,x) \leqslant k_1|x|^p, \qquad \left|\frac{\partial V_0}{\partial x_i}\right| \leqslant k_1(|x|^n + 1), \tag{2.1}$$

$$L_{u_0}V_0(t,x) + K(t,x,u_0(t,x)) \equiv 0, \tag{2.2}$$

$$L_uV_0(t,x) + K(t,x,u) \geqslant 0, \tag{2.3}$$

$$K(t,x,u) \geqslant k_2|x|^p. \tag{2.4}$$

Then the function $u_0(t,x)$ *is a solution of the optimal stabilization problem for the system* (1.1) *in the sense of minimizing the cost* $J^{2,x_0}(u)$ *and moreover*

$$J^{s,x_0}(u_0) = \min_{u \in \mathfrak{U}} J^{s,x_0}(u) = V_0(s,x_0). \tag{2.5}$$

Further, the control $u_0(t,x)$ *makes the system* (1.1) *exponentially p-stable*

PROOF. Let $u = u(t,x)$ be any admissible control. Applying Itô's formula (3.3.8) to the function $V(t,X^{s,x_0}(t))$ and noting that by the second of the inequalities (2.1) and Lemma 5.2.2 all the stochastic integrals appearing in Itô's formula have zero expectation, we get

$$\mathbf{E}V_0(t,X_u^{s,x_0}(t)) - V_0(s,x) = \mathbf{E}\int_s^t L_uV_0(v,X_u^{s,x_0}(v))dv. \tag{2.6}$$

Setting $u = u_0(t,x)$ in this equality and applying (2.2), we obtain

$$\mathbf{E}\int_s^t K(v,X_{u_0}^{s,x_0}(v),u_0(v,X_{u_0}^{s,x_0}(v)))dv$$

$$= V_0(s,x) - \mathbf{E}V_0(t,X_{u_0}^{s,x_0}(t)). \tag{2.7}$$

Letting $t \to \infty$, we get $J^{s,x_0}(u_0) < \infty$. Using this inequality, (2.4) and Remark 1.2, we see that the process $X_{u_0}^{s,x_0}(t)$ satisfies (1.3). From (1.3) and (2.1) we get

$$\mathbf{E}V_0(t,X_{u_0}^{s,x_0}(t)) \leqslant k_1\mathbf{E}|X_{u_0}^{s,x_0}(t)|^p \to 0 \quad \text{as} \quad t \to \infty.$$

Hence, letting $t \to \infty$ in (2.7), we have

$$J^{s,x_0}(u_0) = V_0(s,x_0).$$

Now, if $u(t,x)$ is any admissible control such that $J^{s,x_0}(u) < \infty$, then as before, we readily see that

$$\lim_{t\to\infty} \mathbf{E}V_0(t,X_u^{s,x_0}(t)) = 0.$$

Using this equality and the relation

$$\mathbf{E}V_0(t,X_u^{s,x_0}(t))$$

$$= V_0(s,x_0) + \mathbf{E}\int_s^t K(v,X_u^{s,x_0}(v),u(v,X^{s,x_0}(v)))dv$$

$$\geqslant V_0(s,x_0),$$

which follows from (2.6), (2.3), we finally see by letting $t \to \infty$ that

$$\min_{u\in\mathfrak{U}} J^{s,x_0}(u) \geqslant V_0(s,x_0).$$

It remains to prove that the system (1.1) is exponentially p-stable when $u = u_0(t,x)$. By Theorem 5.7.1, it will suffice to prove that

$$V_0(t,x) \geqslant k_3 |x|^p$$

for some constant $k_3 > 0$. From (2.4) and (2.5) we get

$$V_0(s,x) = J^{s,x_0}(u_0) \geqslant k_2 \int_s^\infty \mathbf{E} |X_{u_0}^{s,x_0}(v)|^p dv.$$

Thus there exists for any $x, s \geqslant 0$ a $T = T(s,x)$ such that

$$\mathbf{E} |X_{u_0}^{s,x}(T)|^p < \frac{1}{2} |x|^p.$$

In view of the inequality $L_{u_0}(|x|^p) \geqslant -k_4 |x|^p$ (see (5.7.7)), the above and Itô's formula imply that

$$\begin{aligned} V_0(s,x) &\geqslant k_2 \int_s^\infty \mathbf{E} |X_{u_0}^{s,x}(v)|^p dv \\ &\geqslant -k_5 \int_s^T \mathbf{E} L_{u_0}(|X_{u_0}^{s,x}(v)|^p) dv \\ &= k_5 (|x|^p - \mathbf{E} |X_{u_0}^{s,x}(T)|^p) \\ &\geqslant \frac{k_5}{2} |x|^p = k_3 |x|^p. \end{aligned}$$

This completes the proof.

For subsequent work, it is useful to combine conditions (2.2) and (2.3) into one equation

$$\min_{u \in (-\infty,\infty)} [L_u V_0(s,x) + K(s,x,u)] = 0 \tag{2.8}$$

(Bellman's equation).

REMARK 2.1. Condition (2.4), which imposes a restriction on the function $K(t,s,u)$ for all u, seems at first sight rather stringent. One might expect that (2.4) could be replaced by the weaker condition $K(t,x,u_0(t,x)) > k_2 |x|^p$. However, the following example will show that if condition (2.4) is thus weakened the assertion of Theorem 2 is no longer true.

Consider the optimal stabilization problem for the deterministic system

$$\frac{dx_1}{dt} = -x_1, \qquad \frac{dx_2}{dt} = x_2 + u$$

with $K(t,x_1,x_2,u) = x_1^2 + u^2$. In this case

$$L_u = - x_1 \frac{\partial}{\partial x_1} + (x_2 + u) \frac{\partial}{\partial x_2}.$$

Conditions (2.1), (2.2) and (2.3) are easily seen to hold for the function

$$V_0 = \frac{x_1^2}{2} + 2x_2^2, \qquad u_0 = - 2x_2.$$

It is also clear that condition (2.4) holds, say, with $u = c_1x_1 + c_2x_2$, when $c_2 \neq 0$. Nevertheless, the control u_0 is not optimal in our sense, since the optimal control is evidently $u = 0$.

REMARK 2.2. We have proved that the control u_0 is optimal for all controls of class $\mathfrak{U}$. One might expect a "higher-quality" control to exist in the class of controls which allow for the past history of the process $X(t)$ from the initial time s to the present t. However, it can be proved that u_0 is also optimal in this broader class of admissible controls (see Nevel'son and Has'minskiĭ (3)).

REMARK 2.3. We may consider simultaneously with the problem II of optimal stabilization also the problem of minimizing the functional

$$\mathcal{J}_T^{s,x_0}(u) = \int_s^T \mathbf{E}K(t,X_u(t),u(t,X_u^{s,x_0}(t)))dt, \tag{2.9}$$

where T is a constant larger than s. By repeating almost word for word the considerations in the proof of Theorem 2.1, we may easily show that if there exist functions $V_0^T(t,x) \in C_2^{(0)}(E)$, $u_{0T}(t,x) \in \mathfrak{U}$ which satisfy for all $T \geqslant t \geqslant s$, $x \in E_\ell$, $u \in (-\infty,\infty)$ and certain positive constants p,n_1,k_1,k_2 the conditions (2.1), (2.3) and (2.4) and the identity

$$V_0^T(T,x) \equiv 0, \tag{2.10}$$

then

$$\min_{u \in \mathfrak{U}} \mathcal{J}_T^{s,x}(u) = \mathcal{J}_T^{s,x}(u_0) = V_0^T(s,x).$$

3. Linear systems

Let us apply Theorem 2.1 to the investigation of the system

$$dX = \left[B(t)dt + \sum_{r=1}^{k} \sigma_r(t)\xi_r(t)\right]X$$

$$+ \left[h(t)dt + \sum_{r=1}^{k} \varphi_r(t)d\xi_r(t)\right]u, \tag{3.1}$$

which is linear in x and u. Here $B(t)$ and $\sigma_r(t)$ are $\ell \times \ell$ matrices, $h(t)$ and $\varphi_r(t)$ vectors in E_ℓ. The elements of the matrices $B(t)$, $\sigma_r(t)$ and the components of the vectors $h(t)$, $\varphi_r(t)$ are assumed to be continuous and bounded functions of time.

We consider the optimal stabilization problem for the system (3.1), with the kernel of the cost defined by

$$K(t,x,u) = (\alpha(t)x,x) + \lambda u^2. \tag{3.2}$$

Here $\alpha(t)$ is a bounded symmetric $\ell \times \ell$ matrix, which is positive definite uniformly in $t \geqslant s$; $\lambda > 0$.

We wish to find an optimal Ljapunov function $V_0(t,x)$ satisfying the assumptions of Theorem 2.1 and given by a negative definite quadratic form

$$V_0(t,x) = (C(t)x,x), \tag{3.3}$$

where $C(t)$ is a symmetric $\ell \times \ell$ matrix.

Clearly

$$L_u = \frac{\partial}{\partial t} + \left(B(t)x + h(t)u, \frac{\partial}{\partial x}\right)$$

$$+ \frac{1}{2}\sum_{r=1}^{k} \left(\sigma_r(t)x + \varphi_r(t)u, \frac{\partial}{\partial x}\right)^2$$

is the operator for the system (3.1). Equation (2.8), relating the optimal Ljapunov function $V_0(t,x)$ and the optimal control $u_0(t,x)$, has the form

$$\frac{\partial V_0}{\partial t} + \left(B(t)x, \frac{\partial}{\partial x}\right)V_0 + \frac{1}{2}\sum_{r=1}^{k} \left(\sigma_r(t)x, \frac{\partial}{\partial x}\right)^2 V_0 + (\alpha(t)x,x)$$

$$= - \min_{u \in U} \left\{u\left[\left(h(t), \frac{\partial}{\partial x}\right)V_0 + \sum_{r=1}^{k} \left(\sigma_r(t)x, \frac{\partial}{\partial x}\right)\left(\varphi_r(t), \frac{\partial}{\partial x}\right)V_0\right] + \right.$$

(Contd)

(Contd)
$$+ u^2\left[\frac{1}{2}\sum_{r=1}^{k}\left(\varphi_r(t), \frac{\partial}{\partial x}\right)^2 V_0 + \lambda\right]\Bigg\}$$

$$= - u_0\left[\left(h(t), \frac{\partial}{\partial x}\right)V_0 + \sum_{r=1}^{k}\left(\sigma_r(t)x, \frac{\partial}{\partial x}\right)\left(\varphi_r(t), \frac{\partial}{\partial x}\right)V_0\right]$$

$$- u_0^2\left[\frac{1}{2}\sum_{r=1}^{k}\left(\varphi_r(t), \frac{\partial}{\partial x}\right)^2 V_0 + \lambda\right] . \quad (3.4)$$

The function $u_0(t,x)$ in (3.4) obviously has the form

$$u_0(t,x) = - \frac{\left(h(t), \frac{\partial}{\partial x}\right)V_0 + \sum_{r=1}^{k}\left(\sigma_r(t)x, \frac{\partial}{\partial x}\right)\left(\varphi_r(t), \frac{\partial}{\partial x}\right)V_0}{2\lambda + \sum_{r=1}^{k}\left(\varphi_r(t), \frac{\partial}{\partial x}\right)^2 V_0} . \quad (3.5)$$

Substituting (3.3) into (3.5), we get

$$u_0(t,x) = - \frac{(h(t),C(t)x) + \sum_{r=1}^{k}(C(t)\varphi_r(t),\sigma_r(t)x)}{\lambda + \sum_{r=1}^{k}(C(t)\varphi_r(t),\varphi_r(t))} . \quad (3.6)$$

This implies that when the optimal Ljapunov function is defined by (3.3), the optimal control is linear in x.

Using (3.4) and (3.6), we get the following equation for $V_0(t,x)$:

$$\frac{\partial V_0}{\partial t} + \left(B(t)x, \frac{\partial}{\partial x}\right)V_0 + \frac{1}{2}\sum_{r=1}^{k}\left(\sigma_r(t)x, \frac{\partial}{\partial x}\right)^2 V_0 + (\alpha(t)x,x)$$

$$= \frac{\left[\left(h(t), \frac{\partial}{\partial x}\right)V_0 + \sum_{r=1}^{k}\left(\sigma_r(t)x, \frac{\partial}{\partial x}\right)\left(\varphi_r(t), \frac{\partial}{\partial x}\right)V_0\right]^2}{4\left[\lambda + \frac{1}{2}\sum_{r=1}^{k}\left(\varphi_r(t), \frac{\partial}{\partial x}\right)^2 V_0\right]} .$$

Since the matrix $C(t)$ is symmetric, this equation is equivalent to

$$\frac{dC}{dt} + CB(t) + B^*(t)C + \sum_{r=1}^{k} \sigma_r^*(t)C\sigma_r(t) + \alpha(t)$$

$$= \frac{\left[Ch(t) + \sum_{r=1}^{k} \sigma_r^*(t)C\varphi_r(t)\right]\left[h^*(t)C + \sum_{r=1}^{k} \varphi_r^*(t)C\sigma_r(t)\right]}{\lambda + \sum_{r=1}^{k} (C\varphi_r(t),\varphi_r(t))}. \qquad (3.7)$$

From Theorem 2.1 we now infer

LEMMA 3.1. *If equation* (3.7) *has a solution* $C(t)$ *which is bounded and positive definite for all* $t \geq s$, *then equation* (3.6) *minimizes the functional*

$$J^{s,x}(u) = \int_s^\infty \mathrm{E}[(\alpha(t)X_u^{s,x}(t),X_u^{s,x}(t)) + \lambda u^2(t,X_u^{s,x}(t))]dt.$$

In conclusion we note that the above remains valid in the deterministic case $\sigma_r(t) \equiv 0$, $\varphi_r(t) \equiv 0$. In particular, equation (3.7) is then simply a matrix Riccati equation:

$$\frac{dC}{dt} + CB(t) + B^*(t)C - \frac{Ch(t)h^*(t)C}{\lambda} + \alpha = 0. \qquad (3.8)$$

4. Method of successive approximations

In the last section we set up a nonlinear differential equation for the matrix $C(t)$ of the optimal Ljapunov function $V_0(t,x)$ associated with a linear control system

$$dX = (B(t)X + h(t)u)dt + \sum_{r=1}^{k} (\sigma_r(t)X + \varphi_r(t)u)d\xi_r(t) \qquad (4.1)$$

possessing the cost function

$$K(t,x,u) = (\alpha(t)x,x) + \lambda u^2.$$

This equation has a fairly complicated form. Even in the deterministic case ($\sigma_r \equiv 0$, $\varphi_r \equiv 0$), when it becomes the well-known Riccati equation, it is not easy to investigate. We would therefore like to have easily checkable conditions for the existence of a bounded positive definite solution to the above equation, i.e., existence conditions for an optimal linear control. Such conditions can be given in terms of Ljapunov functions. While deriving them, we shall also describe a convenient method for practical computation of the optimal control: The method of successive approximations whose basic idea is due to Bellman [2]. This method has been applied to other problems of optimal control by Wonham [3,4], Fleming [1] and others.

THEOREM 4.1. *Suppose that there exists an admissible equation which stabilizes the system* (4.1) *so that exponential stability in the mean square is obtained. Let further* $\lambda > 0$ *be arbitrary and let* $\alpha(t)$ *be a positive definite uniformly with respect to* $t \geq 0$ *symmetric matrix with continuous bounded coefficients. Then there exists a linear control* $u_0(t,x) \in \mathcal{U}$ *which is optimal in the sense of the quality criterion* $J^{s,u}(u)$. *Moreover*

$$\mathcal{J}^{s,x}(u_0) = \min_{u \in \mathcal{U}} \mathcal{J}^{s,x}(u) = V_0(s,x_0) = (C_0(s)x,x), \tag{4.2}$$

where $C_0(s)$ *is the unique bounded positive definite solution of equation* (3.7).

PROOF. Let us consider the minimization problem for the functional (2.9), where $K(t,x,u)$ is given by (3.2) and T is a constant. Remark 2.3 implies that to solve this problem it is enough to find functions $V_0^T(t,x)$, $u_0^T(t,x)$ satisfying the conditions (2.1), (2.3), (2.4) and (2.10).

Let $u_{1T}(t,x)$ be an arbitrary control that is admissible and linear in x, for instance $u_{1T}(t,x) \equiv 0$. We define the function $V_1^T(s,x)$ by the formula

$$V_1^T(s,x) = \int_s^T \mathrm{E}K(v,X_{u_{1T}}^{s,x}(v),u_{1T}(v,X_{u_{1T}}^{s,x}(v)))dv.$$

Lemma 5.6.2 implies that this function is twice continuously differentiable with respect to x and once with respect to s. It is evident from the Markov property that

$$V_1^T(s,x) = \int_s^{s+\Delta} \mathrm{E}K(v,X_{u_{1T}}^{s,x}(v),u_{1T}(v,X_{u_{1T}}^{s,x}(v)))dv$$

$$+ \mathrm{E}V_1^T(s+\Delta,X_{u_{1T}}^{s,x}(s+\Delta)).$$

This and Itô's formula imply

$$L_{u_{1T}}V_1^T(s,x) + (\alpha(s)x,x) + \lambda u_{1T}^2(s,x) = 0,$$

$$V_1^T(T,x) \equiv 0. \tag{4.3}$$

On the other hand the linearity of $u_{1T}(t,x)$ implies that the process $X_{u_{1T}}^{s,x}(t)$ is described by a system of linear stochastical equations. Therefore (see Chapter VI) $V_1^T(s,x)$ is a quadratic form in x

$$V_1^T(s,x) = (C_1^T(s)x,x).$$

Let us define now the function $u_{2T}(s,x)$, i.e. the second approximation to the optimal control, by the equation

$$\min_{u\in(-\infty,\infty)} [L_u V_1^T(s,x) + (\alpha(s)x,x) + \lambda u^2]$$

$$= L_{u_{2T}} V_1^T(s,x) + (\alpha(s)x,x) + \lambda u_{2T}^2. \tag{4.4}$$

(4.3) and (4.4) imply

$$L_{u_{2T}} V_1^T(s,x) + (\alpha(s)x,x) + \lambda u_{2T}^2 \leqslant 0. \tag{4.5}$$

Moreover we get from (4.4) that

$$u_2(s,x)$$

$$= -\frac{\left(h(s), \dfrac{\partial V_1^T(s,x)}{\partial x}\right) + \sum\limits_{r=1}^{k} \left(\sigma_r(s)x, \dfrac{\partial}{\partial x}\right)\left(\varphi_r(s), \dfrac{\partial}{\partial x}\right) V_1^T(s,x)}{2\lambda + \sum\limits_{r=1}^{k} \left(\varphi_r(s), \dfrac{\partial}{\partial x}\right)^2 V_1^T(s,x)}$$

$$= -\frac{(h(s),C_1^T(s)x) + \sum\limits_{r=1}^{k} (\sigma_r(s)x, C_1^T(s)\varphi_r(s))}{\lambda + \sum\limits_{r=1}^{k} (C_1^T(s)\varphi_r(s),\varphi_r(s))},$$

which means that $u_{2T}(s,x)$ is a linear function. Suppose now that the function $V_2^T(s,x) = (C_2^T(s)x,x)$ is given by

$$V_2^T(s,x) = \int_s^T \mathbf{E}K(v, X_{u_{2T}}^{s,x}(v), u_{2T}(v, X_{u_{2T}}^{s,x}(v)))dv.$$

In a similar way as (4.3), we obtain

$$L_{u_{2T}} V_2^T(s,x) + (\alpha(s)x,x) + \lambda u_{2T}^2 = 0,$$

$$V_2^T(T,x) = 0. \tag{4.6}$$

This and (4.5) imply that the difference $U^T(s,x) = V_1^T(s,x) - V_2^T(s,x)$ satisfies

$$L_{u_{2T}} U^T(s,x) \leqslant 0, \qquad U^T(T,x) = 0.$$

These estimates and the equality

$$\mathbf{E} U^T(T, X_{u_{2T}}^{s,x}(T)) - U^T(s,x) = \int_s^T \mathbf{E} L_{u_{2T}} U^T(v, X_{u_{2T}}^{s,x}(v))dv,$$

which is a consequence of Itô's formula, imply easily that

$$- U^T(s,x) \leqslant \mathbf{E} U^T(T, X_{u_{2T}}^{s,x}(T)) - U^T(s,x) \leqslant 0.$$

Therefore $U^T(s,x) \geqslant 0$ and we have shown that

$$V_1^T(s,x) \geqslant V_2^T(s,x). \tag{4.7}$$

Proceeding further in the same fashion we find the functions $u_{3T}(s,x), u_{4T}(s,x), \ldots$ from the identities

$$\min_{u \in (-\infty,\infty)} [L_u V_{n-1}^T(s,x) + (\alpha(s)x,x) + \lambda u^2]$$

$$= L_{u_{nT}} V_{n-1}^T(s,x) + (\alpha(s)x,x) + \lambda u_{nT}^2, \tag{4.8}$$

where

$$V_n^T(s,x) = \int_s^T \mathbf{E} K(v, X_{u_{nT}}^{s,x}(v), u_{nT}(v, X_{u_{nT}}^{s,x}(v)))dv.$$

Evidently the equalities

$u_{nT}(s,x)$

$$= - \frac{\left(h(s), \frac{\partial V_{n-1}^{T}(s,x)}{\partial x}\right) + \sum_{r=1}^{k} \left(\sigma_r(s)x, \frac{\partial}{\partial x}\right)\left(\varphi_r(s), \frac{\partial}{\partial x}\right) V_{n-1}^{T}(s,x)}{2\lambda + \sum_{r=1}^{k} \left(\varphi_r(s), \frac{\partial}{\partial x}\right)^2 V_{n-1}^{T}(s,x)}, \tag{4.9}$$

$$L_{u_{nT}} V_n^T(s,x) + (\alpha(s)x,x) + \lambda u_{nT}^2(s,x) = 0 \tag{4.10}$$

hold. Also precisely as above we show that $V_n^T(s,x) \geq V_{n+1}^T(s,x)$ for every $n = 1,2,\ldots$. The function $V_n^T(s,x) = (C_n^T(s)x,x)$ is a non-negative definite quadratic form. It is well known that a monotone decreasing sequence of non-negative definite quadratic forms converges to a quadratic form. Let in our case this limit be $V_0^T(s,x) = (C_0^T(s)x,x)$. This and (4.9) imply the existence of the limit

$$\lim_{n\to\infty} u_{nT}(s,x) = u_{0T}(s,x)$$

$$= \nu_{1T}(s)x_1 + \ldots + \nu_{\ell T}(s)x_\ell. \tag{4.11}$$

Finally, (4.8), (4.10) and (4.11) imply that the functions $V_0^T(s,x), u_{0T}(s,x)$ satisfy

$$\min_{u\in(-\infty,\infty)} [L_u V_0^T(s,x) + (\alpha(s)x,x) + \lambda u^2]$$

$$= L_{u_{0T}} V_0^T(s,x) + (\alpha(s)x,x) + \lambda u_{0T}^2, \tag{4.12}$$

and moreover $V_0^T(T,x) = 0$, and

$$u_{0T}(s,x)$$

$$= - \frac{\left(h(s), \frac{\partial V_0^T(s,x)}{\partial x}\right) + \sum_{r=1}^{k} \left(\sigma_r(s)x, \frac{\partial}{\partial x}\right)\left(\varphi_r(s), \frac{\partial}{\partial x}\right) V_0^T(s,x)}{2\lambda + \sum_{r=1}^{k} \left(\varphi_r(s), \frac{\partial}{\partial x}\right)^2 V_0^T(s,x)}. \tag{4.13}$$

This, together with Remark 2.3 implies that

$$\min_{u\in\mathcal{U}} \mathcal{J}_T^{s,x}(u) = \mathcal{J}_T^{s,x}(u_{0T}) = (C_0^T(s)x,x).$$

Thus the existence of a control which stabilizes the system

(4.1) so that exponential stability in mean square is obtained, implies that

$$(C_0^T(s)x,x) \leqslant \min_{u\in\mathcal{U}} \mathcal{J}^{s,x}(u) \leqslant k|x|^2 \tag{4.14}$$

holds for a certain positive constant k. We may deduce now from (4.14) and (4.13) the existence of the limits

$$\lim_{T\to\infty} V_0^T(s,x) = V_0(s,x), \tag{4.15}$$

$$\lim_{T\to\infty} u_{0T}(s,x) = u_0(s,x), \tag{4.16}$$

where $V_0(s,x) = (C_0(s)x,x)$ is a quadratic form and $u_0(s,x)$ is an admissible control.

The expression in (4.12) which is preceded by the min sign is a parabola with respect to u and (4.15), (4.16) imply that the coefficients of this parabola have limits as $T \to \infty$, for any fixed s,x. Moreover the coefficient $A(s,x,t)$ at u^2 satisfies the inequality $A(s,x,T) \geqslant \lambda$. It follows that the functions $V_0(s,x)$, $u_0(s,x)$ are also related to each other by the Bellman equation

$$\min_{u\in(-\infty,\infty)} [L_u V_0(s,x) + (\alpha(s)x,x) + \lambda u^2]$$

$$= L_{u_0} V_0(s,x) + (\alpha(s)x,x) + \lambda u_0^2 = 0.$$

The assertion which we wished to prove follows now by Theorem 2.1.

5. The case of an *n*-th order equation

We shall show now that the results of Sections 3 and 4, combined with certain ideas from Section 6.10, can be applied to controlled systems governed by n-th order equations:

$$y^{(n)} + [b_1 + \dot{\eta}_1(t)]y^{(n-1)} + \ldots + [b_n + \dot{\eta}_n(t)]y$$

$$= [h + \sigma\xi(t)]u. \tag{5.1}$$

Here b_i, $i = 1,..,n$, $h \neq 0$ and σ are assumed to be constants and $\dot{\eta}_i(t)$, $i = 1,..,n$, $\xi(t)$ are Gaussian white noise processes with zero expectation such that

$$\mathbf{E}\dot{\eta}_i(t)\dot{\eta}_j(s) = a_{ij}\delta(t - s), \qquad \mathbf{E}\dot{\xi}(t)\dot{\xi}(s) = \delta(t - s).$$

Also, the process $\dot{\xi}(t)$ is assumed to be independent of $\dot{\eta}_1(t),\ldots,$ $\dot{\eta}_n(t)$. As before (see Section 6.10), the substitution

$$y = X_1, \qquad y' = X_2, \quad \ldots, \quad y^{(n-1)} = X_n$$

converts (5.1) into a system of Itô equations

$$\left.\begin{aligned} dX_1 &= X_2 dt, \quad \ldots, \quad dX_{n-1} = X_n dt, \\ dX_n &= \left[- \sum_{i=1}^{n} b_i X_{n-i+1} + hu \right] dt \\ &\quad - \sum_{i,j=1}^{n} \alpha_{ij} X_{n-i+1} d\xi_j + \sigma u d\xi, \end{aligned}\right\} \tag{5.2}$$

where $\xi_1,\ldots,\xi_n$, are independent Wiener processes and the matrix $((\alpha_{ij}))$ is defined by $((\alpha_{ij}))((\alpha_{ji})) = ((a_{ij}))$.

The operator L_u associated with the system (5.2) is

$$L_u = \sum_{i=1}^{n} x_{i+1} \frac{\partial}{\partial x_i} + \left(hu - \sum_{i=1}^{n} b_i x_{n-i+1}\right)\frac{\partial}{\partial x_n} + \frac{1}{2}\left(\sigma^2 u^2 + \sum_{i,j=1}^{n} a_{n-i+1\,n-j+1} x_i x_j\right)\frac{\partial^2}{\partial x_n^2} . \tag{5.3}$$

We shall solve the optimal stabilization problem for the system (5.2) in the sense of minimizing the cost $J^x(u)$ with kernel

$$K(x,u) = (\alpha x,x) + u^2 \tag{5.4}$$

(α is a positive definite matrix with constant elements). In other words, we shall minimize the functional

$$J^x(u) = \mathbf{E}\int_0^\infty [(\alpha X_u^x(t),X_u^x(t)) + u^2(X^x(t))]\,dt \tag{5.5}$$

on the set of solutions of the system (5.2).

By Remark 2.1, an optimal control $u = u_0(x)$ solving this problem will make the system (5.2) asymptotically stable in mean square.

Equation (3.7) for the optimal Ljapunov function $V_0(x) = (Cx,x)$ is in this case

$$CB + B^*C - \frac{Chh^*C}{1 + \sigma^2 c_{nn}} + c_{nn}A + \alpha = 0, \tag{5.6}$$

where $h^* = (0,\ldots,0,h)$, $A = ((\alpha_{n-i+1n-j+1}))$,

$$B = \left(\left(\begin{matrix} 0 & 1 & 0 & \ldots & 0 \\ 0 & 0 & 1 & \ldots & 0 \\ \cdot & \cdot & \cdot & \cdot & \cdot \\ 0 & 0 & 0 & \ldots & 1 \\ -b_n & -b_{n-1} & -b_{n-2} & \ldots & -b_1 \end{matrix}\right)\right)$$

If there exists a positive definite matrix $C = ((c_{ij}))$ satisfying equation (5.6), then, by (3.6), the optimal control is defined by

$$u_0(x) = -\frac{h\sum_{i=1}^{n} c_{in}x_i}{1 + \sigma^2 c_{nn}}.$$

We first assume that $\sigma \neq 0$ and that the matrix A is positive definite. We set $\alpha = A/\sigma^2$ in equation (5.6), thus getting

$$CB + B^*C - \frac{Chh^*C}{1 + \sigma^2 c_{nn}} + \left(\frac{1}{\sigma^2} + c_{nn}\right)A = 0. \tag{5.7}$$

This equation can obviously be written as

$$DB + B^*D - Dh_1h_1^*D^* + A = 0, \tag{5.8}$$

where

$$D = \frac{C}{1/\sigma^2 + c_{nn}}, \qquad h_1^* = \frac{h^*}{\sigma} = (0,\ldots,0,h_1). \tag{5.9}$$

We now consider the auxiliary problem: Among the solutions of the deterministic controlled system

$$y^{(n)} + b_1y^{(n-1)} + \ldots + b_ny = h_1u \tag{5.10}$$

find one that minimizes the functional

$$J_1(u) = \int_0^\infty \left[\sum_{i,j=1}^{n} a_{n-i+1,n-j+1}X_iX_j + u^2\right]dt. \tag{5.11}$$

Here, as before, we denote $X_i = y^{(i-1)}$, $i = 1,\ldots,n$. Setting $u = \nu_1x_1 + \ldots + \nu_nx_n$ and suitably choosing the numbers $\nu_1,\ldots,\nu_n$, we can guarantee that the characteristic equation of the system (5.10) has any preassigned coefficients. Therefore Theorem 4.1 implies that the functional $J_1(u)$ can be minimized. Consequently equation (3.8), which in this case is the same as (5.8), has a unique positive definite solution $D_0 = ((d_{ij}^{(0)}))$. Hence we see by (5.9) that the matrix

$$C_0 = ((c_{ij}^{(0)})) = \frac{D_0}{\sigma^2(1 - d_{nn}^{(0)})}$$

satisfies equation (5.7).

Assume that

$$d_{nn}^{(0)} < 1.$$

Then the matrix C_0 is positive definite and hence (see Section 3) there exists a linear control

$$u_0(x) = - \frac{h \sum_{i=1}^{n} c_{in}^{(0)} x_i}{(1 + \sigma^2 c_{nn}^{(0)})},$$

which minimizes the functional (5.5).

Conversely, suppose that the system (5.2) has an admissible control $u = u(x)$ such that $J^x(u) < \infty$. Then (see Theorem 4.1) equation (5.7) has a positive definite solution C_0. It is also obvious that

$$D_0 = \frac{C_0}{1/\sigma^2 + c_{nn}^{(0)}}.$$

Thus,

$$d_{nn}^{(0)} = \frac{c_{nn}^{(0)}}{1/\sigma^2 + c_{nn}^{(0)}} < 1.$$

It follows from the results of Section 2 that the optimal control

$$u = \hat{u}_0(x) = \hat{v}_1 x_1 + \ldots + \hat{v}_n x_n = (\hat{v}, x),$$

solving problem (5.10), (5.11) is related to the coefficients $d_{ij}^{(0)}$ of the optimal Ljapunov function $\hat{V}_0(x) = (D_0 x, x)$ by

$$\hat{u}_0(x) = - h_1 \sum_{i=1}^{n} d_{in}^{(0)} x_i.$$

Hence it follows that

$$d_{nn}^{(0)} = -\frac{\hat{v}_n}{h_1},$$

and thus, in order to determine the coefficient $d_{nn}^{(0)}$ it suffices to find the coefficient $\hat{v}_n$ of the optimal control for problem (5.10), (5.11).

We have thus proved

LEMMA 5.1. *If the matrix A is positive definite and $\sigma \neq 0$, then the system* (5.2) *can be stabilized for a quadratic cost if and only if the coefficient $\hat{v}_n$ of the optimal control $\hat{u}_0(x) = \hat{v}_1 x_1 + \ldots + \hat{v}_n x_n$ of problem* (5.10), (5.11) *satisfies the inequality*

$$-\hat{v}_n/h_1 < 1.$$

To compute the coefficient $\hat{v}_n$, we shall use the following well-known fact from the theory of deterministic controlled systems (see Lur'e [1]); this is a corollary of the Pontrjagin maximum principle (see Pontrjagin, Boltjanskiĭ, Gamkrelidze, Miščenko [1]).

LEMMA 5.2. *Let $\tilde{u}_0 = \tilde{v}_1 x_1 + \ldots + \tilde{v}_n x_n = (\tilde{v}, x)$ be an optimal control minimizing the functional*

$$J^x(u) = \int_0^\infty [(\alpha X^x, X^x) + u^2]dt$$

on the set of solutions of the deterministic system

$$\dot{X} = PX + qu \tag{5.12}$$

(P is a constant $n \times n$ matrix, q a column-vector in E_n, u a scalar control). Further, let $\Delta(\lambda)$ and $\Delta_{\tilde{v}}(\cdot)$ be the characteristic polynomials of the system (5.12) *for $u = 0$ and $u = \tilde{u}_0$, respectively, and $H_k(\lambda)$ the determinant obtained from $\Delta(\lambda)$ when its k-th column is replaced by q. Let the matrix $\alpha = ((\alpha_{ij}))$ be positive definite. Then the equation*

$$\Delta(\lambda)\Delta(-\lambda) + \sum_{i,j=1}^{n} \alpha_{ij} H_i(\lambda) H_j(-\lambda) = 0$$

has exactly n roots $\tilde{\lambda}_1, \ldots, \tilde{\lambda}_n$ with positive real part, and moreover for all λ

$$(\lambda + \tilde{\lambda}_1)(\lambda + \tilde{\lambda}_2)\ldots(\lambda + \tilde{\lambda}_n) = (-1)^n \Delta_{\tilde{v}}(\lambda).$$

We apply this lemma in the special case of the deterministic system (5.10) and the cost (5.11), assuming as before that the matrix A is positive definite. It is obvious that the characteristic polynomial of the system (5.10) for $u = 0$ is

$$D(\lambda) = \lambda^n + b_1\lambda^{n-1} + \ldots + b_n,$$

and for the optimal control $\hat{u}_0 = \hat{v}_1 x_1 + \ldots + \hat{v}_n x_n$ it is

$$\Delta_{\hat{v}}(\lambda) = \lambda^n + (b_1 - \hat{v}_n h_1)\lambda^{n-1} + \ldots + (b_n - \hat{v}_1 h_1).$$

Thus, letting $\lambda_1, \ldots, \lambda_n$ denote those roots of the equation

$$H(\lambda) = D(\lambda)D(-\lambda) + h_1^2 \sum_{i,j=1}^{n} (-1)^{i-1} a_{n-i+1\,n-j+1}\lambda^{i+j-2} = 0,$$

which have positive real parts, we see from Lemma 5.2 that

$$(\lambda + \lambda_1)(\lambda + \lambda_2)\ldots(\lambda + \lambda_n) = \Delta_{\hat{v}}(\lambda)(-1)^n. \tag{5.13}$$

Equating the coefficients at λ^{n-1} in (5.13), we get

$$\hat{v}_n = \frac{b_1 - \alpha}{h_1}, \tag{5.14}$$

where

$$\alpha = \sum_{i=1}^{n} \lambda_i.$$

Lemma 5.1 and (5.14) imply the following proposition: If $\sigma \neq 0$ and the matrix A is positive definite, then a linear control which makes the system (5.2) stable in mean square exists if and only if

$$\alpha - b_1 < \left(\frac{h}{\sigma}\right)^2. \tag{5.15}$$

Now suppose that the matrix A is only positive semi-definite (in this case the equation $H(\lambda) = 0$ may have roots with zero real part). Let us consider along with (5.2) another linear system (see also Section 10.6):

$$y^{(n)} + [b_1 + \dot{\eta}_1(t) + \dot{\zeta}_1(t)]y^{(n-1)} \tag{5.16}$$

$$+ \ldots + [b_n + \dot{\eta}_n(t) + \dot{\zeta}_n(t)]y = (h + \sigma\dot{\xi}(t))u,$$

where $\dot{\zeta}_i(t)$ are white noise processes independent of $\dot{\eta}_1, \ldots, \dot{\eta}_n, \dot{\xi}$, such that

$$\mathbf{E}\dot{\zeta}_i(t) = 0, \qquad \mathbf{E}\dot{\zeta}_i(t)\dot{\zeta}_j(s) = \varepsilon^2\delta(t - s)\delta_{ij},$$

ε being a small parameter.

Since the matrix $A_\varepsilon = A + \varepsilon^2 J$ (where J is the $n \times n$ unit matrix) is positive definite, it follows from what was proved above that a [linear] optimal control exists for the system (5.16) if and only if

$$\alpha_\varepsilon - b_1 < \left(\frac{h}{\sigma}\right)^2, \tag{5.17}$$

where α_ε is the sum of the roots of the equation

$$H_\varepsilon(\lambda) = H(\lambda) + h_1^2\varepsilon^2(1 - \lambda^2 + \ldots + (-1)^{n-1}\lambda^{2n-2}) = 0,$$

which have positive real part.

Now suppose that inequality (5.15) is satisfied. It follows that if we take ε sufficiently small, then (5.17) is satisfied. Consequently, there is a linear control $u = u_\varepsilon(x)$ which makes the system (5.16), hence also the system (5.2), asymptotically stable in mean square, i.e., $J^x(u_\varepsilon) < \infty$. By Theorem 4.1, the system (5.2) has a [linear] optimal control in the sense of (5.5).

Conversely, suppose that the optimal control problem for the system (5.2) is solvable. Then also the system (5.16) has an optimal control for all sufficiently small $\varepsilon < \varepsilon_0$, and thus inequality (5.17) holds for $\varepsilon < \varepsilon_0$. Letting $\varepsilon \to 0$ in (5.17), we get

$$\alpha - b_1 \leqslant \left(\frac{h}{\sigma}\right)^2. \tag{5.18}$$

It can be shown that in fact the inequality in (5.18) may be assumed to be strong. This is easily verified for the cases $n = 1$, $n = 2$.

We have thus proved the following theorem (Nevel'son [5]).

THEOREM 5.1. *Let* $\sigma \neq 0$ *and let* $\lambda_1, \ldots, \lambda_k$ $(k \leqslant n)$ *be those roots of the equation* $H(\lambda) = 0$ *which have positive real part. Then, assuming that*

$$\sum_{i=1}^{k} \lambda_i - b_1 < \left(\frac{h}{\sigma}\right)^2, \tag{5.19}$$

we get that the optimal stabilization problem for the system (5.2) *in the sense of minimizing the cost* (5.5) *has a solution in the class of linear controls. But if we assume*

$$\sum_{i=1}^{k} \lambda_i - b_1 > \left(\frac{h}{\sigma}\right)^2,$$

then no such solution exists. If the matrix A is positive definite ($k = n$), *then inequality* (5.19) *is not only a sufficient but also a necessary condition for the existence of a linear optimal control stabilizing the system* (5.2).

We shall now show that if $\sigma = 0$, then the system (5.2) is always stabilizable in the sense of (5.5).

For this we need the fact that the absolute values of the roots of the equation

$$f_\varepsilon(y) = \varepsilon y^n + c_1(\varepsilon)y^{(n-1)} + \ldots + c_n(\varepsilon), \tag{5.20}$$

where $c_i(\varepsilon)$ are bounded for $\varepsilon \in [0,1]$, increase no faster than ε^{-1} as $\varepsilon \to 0$. In fact, if there exists a sequence $\varepsilon_n \to 0$ for which equation (5.20) has a root $y_k(\varepsilon_n)$ such that $\varepsilon_n|y_k(\varepsilon_n)| \to \infty$, then

$$f_{\varepsilon_n}(y_k(\varepsilon_n)) = [y_k(\varepsilon_n)]^{n-1}(\varepsilon_n y_k(\varepsilon_n) + O(1))$$

as $n \to \infty$, and thus $y_k(\varepsilon_n)$ cannot be a root of equation (5.20).

Let $\lambda_1(\sigma),\ldots,\lambda_{2n}(\sigma)$ denote the roots of the equation

$$\sigma^2 H(\lambda) = \sigma^2 D(\lambda)D(-\lambda) + h^2 \sum_{i,j=1}^{n} (-1)^{i-1} a_{n-i+1n-j+1}\lambda^{i+j-2}$$

$$= 0.$$

This equation clearly involves only even powers of λ. Therefore, regarding it as an equation in λ^2 and using the same argument as before, we see that

$$|\lambda_i(\sigma)|^2 = O\left(\frac{1}{\sigma^2}\right), \qquad i = 1,\ldots,2n.$$

Consequently the sum α of the positive roots of the equation $H(\lambda) = 0$ increases no faster than σ^{-1} as $\sigma \to 0$. Thus inequality (5.19) holds for sufficiently small $\sigma^* > 0$. Applying Theorem 5.1, we conclude now that there exists a linear control $u = u^*(x)$ for which the system

$$y^n + (b_1 + \dot{\eta}_1)y^{n-1} + \ldots + (b_n + \dot{\eta}_n)y = u(b + \sigma^*\xi) \tag{5.21}$$

is asymptotically stable in mean square.

In the sequel it will be convenient to let $L_{u,\sigma}$ denote the differential operator (5.3), thus explicitly indicating the dependence on σ.

Since the system (5.21) is stable, there exists a positive definite quadratic form $V(x)$ such that $L_{u,\sigma^*}V < 0$ for $x \neq 0$. It follows from

$$L_{u,\sigma^*}V = L_{u,0}V + \tfrac{1}{2}(\sigma^*)^2 u^2 \frac{\partial^2 V}{\partial x_n^2}$$

that also $L_{u,0}V < 0$ for $x \neq 0$, i.e., for $u = u^*(x)$, $\sigma = 0$, the system (5.2) is asymptotically stable in mean square.

We have thus proved the following

THEOREM 5.2. *If* $\sigma = 0$, *the optimal stabilization problem for the system* (5.1) *in the sense of* (5.5) *has a solution for any values of the parameters figuring therein, provided* $h \neq 0$.

In particular, consider the stochastic system

$$y^{(n)} + b_1 y^{(n-1)} + \ldots + b_n y = (h + \sigma\dot{\xi})u. \tag{5.22}$$

In this case

$$H(\lambda) = D(\lambda)D(-\lambda).$$

Let $\mu_1,\ldots,\mu_r$ and $\nu_1,\ldots,\nu_m$, $r + m \leqslant n$, be the roots of the polynomial $D(\lambda)$ with positive and negative real parts, respectively. Obviously,

$$\alpha = \mu_1 + \ldots + \mu_r - \nu_1 - \ldots - \nu_m.$$

Since moreover $\mu_1 + \ldots + \mu_r + \nu_1 + \ldots + \nu_m = -b$, it follows from Theorem 5.1 that the system (5.22) is stabilizable in mean square if and only if

$$2\sigma^2 \sum_{i=1}^{r} \mu_i < h^2.$$

We note in conclusion that Theorem 5.2 yields the following necessary and sufficient conditions for the system (5.2) to be stabilizable in the cases $n = 1$, $n = 2$:

$$\sqrt{b_1^2 + \left(\frac{h}{\sigma}\right)^2 a_{11}} - b_1 < \left(\frac{h}{\sigma}\right)^2 \quad \text{for } n = 1,$$

$$\sqrt{b_1^2 + \left(\frac{h}{\sigma}\right)^2 a_{22} - 2b_2 + 2\sqrt{b_2^2 + \left(\frac{h}{\sigma}\right)^2 a_{11}}} - b_1 < \left(\frac{h}{\sigma}\right)^2$$

$$\text{for } n = 2.$$

APPENDIX TO THE ENGLISH EDITION

We shall be concerned here with some results about stability of stochastic systems which were obtained since the time of the Russian edition of this book. We shall give detailed proofs only for a few of the results. Some theorems will be only stated and others merely mentioned. This varying degree of detail in our exposition is certainly not motivated by our feelings concerning the importance of the material. Rather, we have given throughout the priority to those results which are essentially connected with the main part of the book.

More information concerning other new interesting results can be found in the books of Morozan [7] and Bunke [4], in the paper entitled "Stability of stochastic dynamical systems" and printed in the Lecture Notes in Mathematics, Vol. 294 (1972), in the survey papers of Vonem [1] and Kozin [1] and possibly other recent works.

1. Moment stability and almost sure stability for linear systems of equations whose coefficients are Markov processes

1. Consider the equation

$$\dot{Y}(t) = F(X(t))Y(t), \tag{1.1}$$

where $Y(t)$ is an m-dimensional vector, $F(x)$ is an $m \times m$ matrix and $X(t)$ is a random Markov process in E_ℓ with a corresponding differential generator L (see Dynkin [3]).

Let us describe a general method due to Benderskiĭ [2] of obtaining equations for the moments of positive integral degree of the process $Y(t)$. It is evident that $(X(t),Y(t))$ is a Markov process in $E_{m+\ell}$. Its generating differential operator L acts on sufficiently smooth functions $V(x,y)$, $x \in E_m$, $y \in E_\ell$ by

$$LV(x,y) = LV(x,y) + \left(F(x,y), \frac{\partial V(x,y)}{\partial y}\right) .$$

We denote by $Y^{x,y}(t)$ the solution of equation (1.1) satisfying the initial conditions $X(0) = x$, $Y(0) = y$. The expectation of $Y^{x,y}(t)$ will be denoted by $u(x,y,t)$. Under quite general assumptions the coordinates $u_i = u_i(x,y,t)$ of the vector $u(x,y,t)$ satisfy the equation (see Dynkin [3])

$$\frac{\partial u_i}{\partial t} = Lu_i \tag{1.2}$$

and the initial conditions $u_i(x,y,0) = y_i$. In the particular case when $X(t)$ is a diffusion process, the above is a consequence of Lemma 3.6.

Moreover (1.1) implies that

$$Y^{x,y}(t) = A(x,t,\omega)y,$$

where $A(x,t\omega)$ is the fundamental matrix of the system (1.1) corresponding to the initial condition $X(0) = x$ for the process $X(t)$. Consequently the function

$$u(x,y,t) = \mathbf{E}A(x,t,\omega)y = B_1(x,t)y \tag{1.3}$$

depends linearly on y. Substituting (1.3) in (1.2) we obtain the equation

$$\frac{\partial B_1(x,t)}{\partial t} = LB_1(x,t) + B_1(x,t)F(x), \qquad B_1(x,0) = J \tag{1.4}$$

for the matrix $B_1(x,t)$. (Here J stands for the $m \times m$ identity matrix.) After having found $B_1(x,t)$ form (1.4), we may calculate the first moment of $Y(t)$ from (1.3).

The above method allows us to find moments of arbitrary degree of the process $Y(t)$. Indeed, we may derive from (1.1) the equation

$$\frac{d(Y(t) \times Y(t))}{dt} = [F(X(t)) \times J + J \times F(X(t))](Y(t) \times Y(t)).$$

This is an equation of the form (1.1) for the product $Y(t) \times Y(t)$ (see p. 259). Let $B_2(x,t)$ be the matrix which satisfies

$$\mathbf{E}[Y^{x,y}(t) \times Y^{x,y}(t)] = B_2(x,t)(y \times y).$$

Then, by the above, this matrix satisfies also

$$\frac{\partial B_2(x,t)}{\partial t} = LB_2(x,t) + B_2(x,t)[F(X(t)) \times J + J \times F(X(t))],$$
$$B_2(x,0) = J \times J. \tag{1.5}$$

Also in the same way one can obtain equations for the moments of arbitrary high degree of the process $Y(t)$. Let us consider the following particular case.

Suppose that $X(t)$ is a diffusion process with differential generator

$$L = \left(b(x), \frac{\partial}{\partial x}\right) + \frac{1}{2}\sum_{r=1}^{k}\left(\sigma_r(x), \frac{\partial}{\partial x}\right)^2,$$

where $b(x)$ and $\sigma_r(x)$ are vectors in E_ℓ.

Then (1.4) and (1.5) are second order partial differential equations for the matrices $B_1(x,t)$ and $B_2(x,t)$. An analogous method for obtaining in this particular case equations for the moments was proposed by Frisch [1].

Suppose now that $X(t)$ is a stationary Markov process with finitely many states $x_1,\ldots,x_p$. Let us assume that the probability of transition from the i-th state to the j-th state during a time interval Δt is given for $\Delta t \to 0$ by $P_{ij}(\Delta t) = a_{ij}\Delta t + O(\Delta t)$, $i,j = 1,\ldots,p$, where a_{ij} are some constants. The action of the differential operator L of the process $X(t)$ on a function $V(x)$ defined at the points $x_1,\ldots,x_p$ is given by

$$LV(x) = \sum_{k=1}^{p} a_{jk}[V(x_k) - V(x_j)].$$

Hence we get for the matrices $B_1(x_j,t)$, $B_2(x_j,t)$ the following system of linear equations

$$\frac{dB_1(x_j,t)}{dt} = \sum_{k=1}^{p} a_{jk}[B_1(x_j,t) - B_1(x_j,t)] + B_1(x_j,t)F(x_j),$$

$$B_1(x_j,0) = J,$$

$$\frac{dB_2(x_j,t)}{dt} = \sum_{k=1}^{p} a_{jk}[B_2(x_k,t) - B_2(x_j,t)]$$
$$+ B_2(x_j,t)[F(x_j) \times J + J \times F(x_j)],$$

$$B_2(x_j,0) = J \times J, \qquad j = 1,\ldots,p.$$

We see thus that for a time-homogeneous Markov process $X(t)$ with

p states the stability problem for the solution of equation (1.1) is reduced to the investigation of the stability of the solution of a system of ordinary differential equations. Benderskiĭ [1], [2], McKenna and Morrison [1] and others have discussed various partial cases of the above. For a system with discrete time analogous results were obtained by Benderskiĭ in [3]. Darkovskiĭ and Leĭbovič [1] considered the system (1.1) in the "mixed" case when the process $X(t)$ undergoes changes of value only at discrete moments of time.

2. Benderskiĭ and Pastur [1], [2] have also investigated almost sure stability of the system (1.1). They applied ideas which are contained in Sections 6, 7 and 8 of Chapter VI above. Let us give a short presentation of their approach.

Just as in Section 6.7, let us introduce new variables defined by

$$\lambda = \frac{x}{|x|}, \qquad \rho = \ln|x|.$$

Then we obtain from (1.1) for every Markov process

$$\frac{d\lambda}{dt} = (F(X(t)) - (F(X(t))\lambda,\lambda)\mathcal{I})\lambda. \tag{1.6}$$

This is a system of differential equations for the process $\lambda(t)$ on the sphere $|\lambda| = 1$ in E_n. Here

$$\frac{d\rho}{dt} = (F(X(t))\lambda(t),\lambda(t)).$$

This implies the following formula, analogous to (6.7.10)

$$\frac{\rho(T) - \rho(0)}{T} = \frac{1}{T}\int_0^T (F(X(t))\lambda(t),\lambda(t))dt. \tag{1.7}$$

We conclude that the stability problem, and also the problem of determining the exact rate of growth of the solution of (1.1) reduce to calculating the limit of the right side of (1.7) as $T \to \infty$. Let us note here that if $X(t)$ is a stationary process, then Theorem 2.2.1 implies that the equation (1.6) possesses a stationary and a stationarily related to $X(t)$ solution $\lambda_0(t)$.

For certain concrete cases Benderskiĭ and Pastur gave conditions under which every solution of equation (1.6) tends to a stationary one and the pair $X(t),\lambda_0(t)$ is an ergodic stationary process. In this situation there exists a non-probabilistic limit

$$\lim_{T\to\infty} \frac{\rho(T) - \rho(0)}{T} = \int (F(x)\ell,\ell)dP(x,\ell),$$

where $P(x,\ell)$ is the joint distribution of $X(t),\lambda_0(t)$. It is evident that in this case we have exponential almost sure stability if and only if the constant

$$a = \int (F(x)\ell,\ell)dP(x,\ell)$$

is negative.

The above program has been realized to the fullest in case of second order equations (see the papers of Benderskiĭ and Pastur [1], [2])

$$y'' + X(t)y = 0. \tag{1.8}$$

If we introduce the notation $y(t) = y_1$, $y'(t) = y_2$ we obtain a system of type (1.1) with

$$F(x,t) = \begin{Vmatrix} 0 & 1 \\ -X(t) & 0 \end{Vmatrix}.$$

It is convenient in this case, just as in Section 6.8 to replace here the vector $\lambda(t)$ by the variable $\varphi(t) = \arctan(y_1(t)/y_2(t))$ defined on the circle. Evidently we have

$$\frac{d\varphi(t)}{dt} = -(\sin^2\varphi(t) + X(t)\cos^2\varphi(t)).$$

Moreover (1.7) implies that

$$\lim_{T\to\infty} \frac{\rho(T) - \rho(0)}{T} = \lim_{T\to\infty} \frac{1}{2T}\int_0^T [1 - X(t)]\sin(2\varphi(t))dt,$$

provided the last limit exists. In case of a stationary ergodic process $X(t)$ Benderskiĭ and Pastur [1] have given quite general conditions under which the above limit exists. If $X(t)$ is a Markov process, then the pair $X(t),\varphi(t)$ is also a Markov process and we can apply the methods of Chapter VI to calculate its joint stationary distribution P. In particular, if $X(t)$ is a Markov process which takes on only two values, then P can be calculated in an open form (see Benderskiĭ and Pastur [2]). In conclusion let us remark that equation (1.8) is important because it makes its appearance also in physics; the behaviour of the solutions of this equation is closely connected with the behaviour of the solutions of Schroedinger's equation with a random potential (see Benderskiĭ and Pastur [2]).

2. Almost sure stability of the paths of one-dimensional diffusion processes

We shall be concerned in this section with the one-dimensional stochastic Itô equation whose coefficients are independent of time. Thus it is rather a particular model to which we have devoted much consideration in Chapter V. However, in contrast to what we did in the Chapters V-VII, we shall not investigate here the stability of the trivial solution, but the stability of an arbitrary path of the corresponding Markov process.

The idea of the basic (and somewhat unexpected) result can be described as follows: If $X(t)$ is a recurrent process in E_ℓ which satisfies the equation

$$dx(t) = b(X(t))dt + \sigma(X(t))d\xi(t), \tag{2.1}$$

then, except for the trivial special case when the coefficients b and σ are periodic, the process $X(t)$ is an almost surely stable in the large solution of equation (2.1). Thus the solutions of equation (2.1) with various initial conditions are getting asymptotically close to each other as $t \to \infty$, as we might have expected. This is in general not the situation in case of deterministic systems.

To present all this more precisely, let us make the following assumptions:

1. The functions $b(x)$ and $\sigma(x)$ satisfy a Lipschitz condition on every compact set $K \subset E_\ell$ and moreover $\sigma(x)$ never vanishes.

2. If we define

$$Q(x) = \int_0^x \exp\left\{-2\int_0^y \frac{b(z)}{\sigma^2(z)}\,dz\right\}dy,$$

then

$$Q(\pm\infty) = \pm\infty. \tag{2.2}$$

Only Assumption 2 seems to be somewhat restrictive. We have shown in Section 3.8 (Example 2) that Assumption 2 is equivalent to the recurrence property of the Markov process described by equation (2.1). We have observed in Section 4.2 that a recurrent process $X(t)$ spends an infinite amount of time in every neighborhood of every point x. We shall prove now the stronger result that for every $x, x_0 \in E_\ell$, $T > 0$ and $\alpha > 0$

$$\mathbf{P}_x\{\bigcup_{\tau > T} \sup_{\tau < t < \tau+1} |X(t) - x_0| < \alpha\} = 1. \tag{2.3}$$

Let τ_α denote the instant at which the path of the process $X(t)$ reaches for the first time the set $|x - x_0| = \alpha$. Then we conclude by Lemma 3.6.2 that the function $\mathbf{P}_x\{\tau_\alpha > s\} = u(s,x)$ satisfies the equation

$$\frac{\partial u}{\partial s} = b(x)\,\frac{\partial u}{\partial x} + \frac{1}{2}\,\sigma^2(x)\,\frac{\partial^2 u}{\partial x^2}$$

and the initial conditions

$$u(0,x) = 1, \qquad |x - x_0| < \alpha,$$

$$u(s,x_0 \pm \alpha) = 0.$$

The strong maximum principle for parabolic equations implies that

$$\inf_{|x-x_0| \leqslant \alpha/2} \mathbf{P}_x\{\tau_\alpha > 1\} = \inf_{|x-x_0| \leqslant \alpha/2} u(1,x) = \beta > 0.$$

Just as in the proofs of Lemmas 4.2.1 and 4.3.1, let us consider now the cycles (parts of paths of the process $X(t)$) contained between the sets $\Gamma = \{|x - x_0| = \alpha/2\}$ and $\Gamma_1 = \{|x - x_0| = \alpha\}$ with the time parameter exceeding T (there are infinitely many such cycles due to the recurrence property of the process). Let A_i denote the event that the second half of the i-th cycle lasts for a time longer than 1. Then

$$\mathbf{P}\{\bigcup_{\tau>T}\{\sup_{\tau<t<\tau+1} |X(t) - x_0| < \alpha\}\} \geqslant \mathbf{P}\{\bigcup_{i=1}^{\infty} A_i\}.$$

Since

$$\mathbf{P}\{A_i/\overline{A}_1,\ldots,\overline{A}_{i-1}\} \geqslant \inf_{|x-x_0| \leqslant \alpha/2} \mathbf{P}_x\{\tau_\alpha > 1\} \quad \beta > 0,$$

we have

$$\mathbf{P}\{\bigcup_{i=1}^{\infty} A_i\} = 1 - \mathbf{P}\{\overline{A}_1\}\mathbf{P}\{\overline{A}_2/\overline{A}_1\}\mathbf{P}\{\overline{A}_3/\overline{A}_1\overline{A}_2\}\ldots = 1$$

and therefore (2.3) is proved.

REMARK 2.1. Using the method of the above proof, we can establish a more general result: Given an arbitrary recurrent diffusion process in E_ℓ with a non-degenerate diffusion matrix, one can assemble the segments of its paths into a set which is everywhere dense in $C[0,h]$. Here $C[0,h]$ denotes the space of E_ℓ-valued continuous functions defined on the interval $[0,h]$ with the metric topology of uniform convergence.

Let us consider now the stochastic process

$$Y(t) = Q(X(t)).$$

(This is equivalent to considering the process $X(t)$ in the canonical scale, see Dynkin [3]). By (2.2), this transformation maps E_ℓ onto the whole real line. Moreover Itô's formula implies

$$dY(t) = \sigma_1(Y(t))d\xi(t), \tag{2.4}$$

where

$$\sigma_1(y) = \sigma(Q^{-1}(y))q(Q^{-1}(y))$$

$$q(y) = Q'(y).$$

It follows easily from Assumptions 1 and 2 that $Y(t)$ is a regular and recurrent process. Let us show, for instance that it is regular. The differential generator of the process (2.4) is given by

$$L = \frac{\partial}{\partial t} + \frac{1}{2}\sigma_1^2(y)\frac{\partial^2}{\partial y^2}.$$

It follows from this without difficulty that the auxiliary function

$$V(y) = |y|$$

satisfies for $|y| > 1$ the conditions (3.4.2) and (3.4.3). This and Theorem 3.4.1 (see also the Remark following it) imply the regularity of the process $Y(t)$.

LEMMA 2.1. *Let $Y_i(t)$, $i = 1,2$ denote the solutions of equation* (2.4) *which satisfy the initial conditions $Y_i(0) = y_i$, $y_1 < y_2$. Then there exists a non-negative random variable $\zeta < \infty$ such that*

$$\lim_{t\to\infty} [Y_2(t) - Y_1(t)] = \zeta.$$

PROOF. Let us first observe that $Y_2(t) \geqslant Y_1(t)$ holds for all $t \geqslant 0$. This intuitively obvious fact follows from the "comparison theorem" of Skorohod given in Section 3 of Chapter V. (One can prove this also in another way by observing that the solution of equation (2.4) with given initial condition is unique in the case when the initial condition is specified at a random and not dependent on the past time moment τ, and next selecting $\tau = \inf\{t: Y_1(t) = Y_2(t)\}$. Let $Z(t) = Y_2(t) - Y_1(t)$. The above implies that $Z(t) \geqslant 0$. Let τ_R denote the moment of the first exit from the circle of radius τ_R of the two-dimensional Markov process $(Y_1(t),Y_2(t))$. Properties of the stochastic integral imply now that the process

$$Z(\tau_R \wedge t) = y_2 - y_1 + \int_0^{\tau_R \wedge t} [\sigma_1(Y_2(s)) - \sigma_1(Y_1(s))]d\xi(s)$$

is a positive supermartingale, that is

$$\mathbf{E}(Z(\tau_R \wedge t)/N_s) = Z(\tau_R \wedge s), \qquad t > s. \tag{2.5}$$

Here N_s denotes the σ-algebra of events generated by the run of the process $\xi(t)$ along the time interval $[0,s]$. The regularity of this process implies that $\lim_{R\to\infty} \tau_R = \infty$, whence, passing to the limit in (2.5) as $R \to \infty$, and applying Fatou's lemma, we get

$$\mathbf{E}(Z(t)/N_s) \leqslant Z(s). \tag{2.6}$$

This inequality means that $Z(t)$ is a positive supermartingale. To complete the proof of the lemma, it suffices to apply Theorem 5.2.1.

COROLLARY 2.1. *For every initial* $y \in E_\ell$, *the solution of equation* (2.4) *is stable in the mean and in probability*

Indeed, (2.6) implies that for all $t \geqslant 0$

$$\mathbf{E}[Y_2(t) - Y_1(t)] \leqslant y_2 - y_1,$$

whence

$$\lim_{y_2 - y_1 \to 0} \sup_{t \geqslant 0} \mathbf{E}[Y_2(t) - Y_1(t)] = 0.$$

The inequality for supermartingales implies also

$$\mathbf{P}\{\sup_{t>0} [Y_2(t) - Y_1(t)] \geqslant \varepsilon\} \to \frac{y_2 - y_1}{\varepsilon} \to 0 \quad \text{as} \quad y_2 - y_1 \to 0$$

for every positive ε.

The following lemma gives stability conditions in the large for any solution of equation (2.4).

LEMMA 2.2. *Let* $Y_i(t)$, $i = 1,2$ *denote the solutions of equation* (2.4) *which satisfy the initial conditions* $Y_i(0) = y_i$, $y_1 < y_2$. *Then the following assertions hold true.*

(1) *If the function* $\sigma_1(y)$ *is not periodic, then* $\zeta \equiv 0$.

(2) *If the function* $\sigma_1(y)$ *is periodic with period* ϑ *and* $k = (y_2 - y_1)/\vartheta$ *is an integer (thus, in particular if* $\sigma_1(y)$ *does not*

depend on y), then $Y_2(t) - Y_1(t) = y_2 - y_1$.

If the function $\sigma_1(y)$ *is not identically constant and is periodic with period* ϑ, *and* $k = (y_2 - y_1)/\vartheta$ *is not an integer, then the distribution of* ξ *is concentrated at the two points* $\vartheta[k]$ *and* $\vartheta([k] + 1)$.

PROOF. Let us show that almost surely

$$\sigma_1(y + \zeta) = \sigma_1(y) \tag{2.7}$$

for all $y \in E_\ell$. Suppose that this is not the case. Then the continuity of $\sigma_1(y)$ implies that

$$\mathbf{P}\{\inf_{\substack{|y-y_0-\zeta|<\delta \\ |z-y_0|<\delta}} |\sigma_1(y) - \sigma_1(z)| > \delta\} \geqslant p_1 \tag{2.8}$$

for some y_0 and some positive δ and p_1. Lemma 2.1 implies that the stochastic integral

$$\int_0^t [\sigma_1(Y_2(s)) - \sigma_1(Y_1(s))]d\xi(s)$$

has a finite limit as $t \to \infty$. Using this and the inequality

$$\mathbf{P}\{\int^\infty [\sigma_1(Y_2(s)) - \sigma_1(Y_1(s))]^2 ds\}$$

$$\leqslant 1 - 2\mathbf{P}\{\xi(\delta^2) > a\}$$

$$+ \mathbf{P}\{\sup_{T\leqslant t<\infty} \int_T^t [\sigma_1(Y_2(s)) - \sigma_1(Y_1(s))]d\xi(s) > a\}$$

which is valid for every $a > 0$ (see Gihman and Skorohod [2], p. 32), we get that for every $\varepsilon > 0$, there is a $T_1(\varepsilon)$ such that

$$\mathbf{P}\{\int_T^\infty [\sigma_1(Y_2(s)) - \sigma_1(Y_1(t))]^2 ds > \delta^2\} < \varepsilon \tag{2.9}$$

for all $T \geqslant T_1(\varepsilon)$.

Let us select $T_2(\varepsilon)$ so that

$$\mathbf{P}\{\sup_{t>T_2(\varepsilon)} |Y_2(t) - Y_1(t) - \zeta| > \delta/2\} < \varepsilon \tag{2.10}$$

holds. (2.3) and (2.10) imply that there exists a random variable $\tau > \max(T_1(\varepsilon),T_2(\varepsilon))$ such that

$$\sup_{\tau<s<\tau+1} |Y_1(s) - y_0| < \delta/2,$$

$$\mathbf{P}\{\sup_{\tau<s<\tau+1} |Y_2(s) - y_0 - \zeta| > \delta\} < \varepsilon.$$

We deduce from (2.8), (2.9) and the last two inequalities that

$$\varepsilon > \mathbf{P}\{\int_\tau [\sigma_1(Y_2(s)) - \sigma_1(Y_1(s))]^2 ds > \delta^2\}$$

$$\geqslant \mathbf{P}\{\int_\tau^{\tau+1} [\sigma_1(Y_2(s)) - \sigma_1(Y_1(s))]^2 ds > \delta^2\} \geqslant p_1 - \varepsilon.$$

Since ε is here arbitrary, the last inequality is in contradiction with $p_1 > 0$. Thus equality (2.7) is proved.

(2.7) implies immediately the first assertion of the theorem.

Suppose now that the function $\sigma_1(y)$ is periodic, with period ϑ and $k = (y_2 - y_1)/\vartheta$ is an integer. Then the function $\tilde{Y}_1(t) = Y_1(t) + y_2 - y_1 = Y_1(t) + k\vartheta$ satisifes equation (2.4) and the initial condition $\tilde{Y}_1(0) = y_2$. The second assertion of the theorem follows now from the uniqueness of the solution of equation (2.4).

Finally, assume that $\sigma_1(y)$ is a not identically constant periodic function with period ϑ and $k = (y_2 - y_1)/\vartheta$ is not an integer. Let us consider the system

$$\begin{aligned} dY(t) &= \sigma_1(Y(t))d\xi(t), \\ dZ(t) &= (\sigma_1(Y(t) + Z(t)) - \sigma_1(Y(t)))d\xi(t). \end{aligned} \tag{2.11}$$

This system has got two solutions: $Z_1(t) = Y_2(t) - Y_1(t)$ and $\tilde{Y}_1(t) = Y_1(t)$, $\tilde{Y}_2(t) = Y_\vartheta(t)$, $Z_2(t) = Y_\vartheta(t) - Y_0(t)$, where $Y_\vartheta(t)$ and $Y_0(t)$ are the solutions of the first equation in (2.11) defined by the initial conditions $Y_\vartheta(0) = \vartheta([k] + 1)$ and $Y_0(0) = 0$. From the uniqueness of the solutions of this system and from $y_2 - y_1 = Z_1(0) < Z_2(0) = \vartheta([k] + 1)$ follows $Z_1(t) \leqslant Z_2(t)$ for all t. Moreover the second assertion of the theorem, proved above, implies that $Z_2(t) \equiv \vartheta([k] + 1)$. Thus $Y_2(t) - Y_1(t) \leqslant \vartheta([k] + 1)$. We show in the same way that $\vartheta[k] \leqslant Y_2(t) - Y_1(t)$ holds for all t. These inequalities and (2.7) imply the third assertion of the theorem.

Let us return now to the original equation (2.1). We put $r(x_1,x_2) = |Q(x_2) - Q(x_1)|$. It is evident that $r(x_1,x_2)$ defines a distance in E_ℓ. Lemmas 2.1 and 2.2 yield now the following theorem which gives criteria for stability in the large of an arbitrary solution of equation (2.1) which describes a recurrent Markov process.

THEOREM 2.2. *Suppose that Assumptions* 1 *and* 2 *are satisfied. Let* $X_1(t), X_2(t)$ *be solutions of equation* (1.1) *which satisfy the initial conditions* $X_1(0) = x_1$, $X_2(0) = x_2$. *Then there exists a finite limit*

$$\lim_{t\to\infty} r(X_1(t), X_2(t)) = \zeta.$$

The identity $\zeta \equiv 0$ *holds for every* x_1, x_2 *if and only if the function* $\sigma_1(y) = \sigma(Q^{-1}(y))q(Q^{-1}(y))$ *is not periodic. In the case when* $\sigma_1(y)$ *is a periodic function with period* ϑ, *and the number* $k = |Q(x_2) - Q(x_1)|$ *is not an integer, the distribution of* ξ *is concentrated in the two points* $[k]\vartheta$ *and* $([k] + 1)\vartheta$, *and if* k *is an integer, then* $r(X_1(t), X_2(t)) \equiv |Q(x_2) - Q(x_1)|$. *The last equality holds also in the case when* $\sigma_1(y)$ *does not depend on* y.

COROLLARY 2.2. *Let* $X(t)$ *be a positive Markov process. Then any two solutions* $X_1(t)$, $X_2(t)$ *of equation* (2.1) *with initial conditions* $X_1(0) = x_1$, $X_2(0) = x_2$ *satisfy*

$$\lim_{r\to\infty} r(X_1(t), X_2(t)) = 0.$$

Indeed, since (2.1) describes a positive Markov process, we have (see Dynkin [3]) that

$$\int_{-\infty}^{\infty} \frac{dx}{\sigma^2(x)q(x)} < \infty.$$

Thus

$$\int_{-\infty}^{\infty} \frac{dy}{\sigma_1(y)} = \int_{-\infty}^{\infty} \frac{dQ(x)}{(\sigma(x)q(x))^2} = \int_{-\infty}^{\infty} \frac{dx}{\sigma^2(x)q(x)} < \infty.$$

It follows from this that the function $\sigma_1(y)$ is not periodic.

The theorem which now follows says that every solution of Itô's homogeneous stochastic differential equation in E_ℓ describing a recurrent Markov process is stable in the mean and in probability with respect to the metric $r(x_1, x_2)$.

THEOREM 2.3. *Suppose that Assumptions* 1 *and* 2 *above are satisfied. Then any two solutions* $X_1(t)$, $X_2(t)$ *of equation with initial conditions* $X_1(0) = x_1$, $X_2(0) = x_2$ *satisfy*

$$\lim_{x_2 - x_1 \to 0} \sup_{t \geqslant 0} r(X_1(t), X_2(t)) = 0$$

$$\lim_{x_2 - x_1 \to 0} \mathbf{P}\{\sup_{t \geqslant 0} r(X_1(t), X_2(t)) = 0,$$

for every positive ε.

The theorem is a consequence of Corollary 2.1 and of the recurrence property of the process $X(t)$.

3. Reduction principle

There is known in the stability theory of deterministic systems the so-called reduction principle which is basic for the investigation of critical stability situations. This principle permits us to reduce the investigation of the stability of an $(\ell + m)$-dimensional system $X(t), Y(t)$ to investigating the stability of two systems: the ℓ-dimensional system of the first approximation of the vector $X(t)$ (the coefficients of this approximation are assumed independent of y), and the m-dimensional system obtained by substituting $X = 0$ in the equations for Y (see Malkin [1], p. 383 and Remark on p. 529). Here the reduction principle will be used for the simpler case when both ramified systems are by linear approximation uniformly stable in the large. Although the result cannot be directly applied to the investigation of critical cases, it nevertheless offers a possibility of simplifying stability investigations in some cases of practical importance.

Thus let there be given an $(\ell + m)$-dimensional Markov process $X(t), Y(t)$ described by the system of stochastic differential equations

$$\begin{aligned} dX(t) &= b(t,X(t),Y(t))dt + \sum_{r=1}^{k} \sigma_r(t,X(t),Y(t))d\xi_r(t) \\ dY(t) &= \tilde{b}(t,X(t),Y(t))dt + \sum_{r=1}^{k} \tilde{\sigma}_r(t,X(t),Y(t))d\xi_r(t), \end{aligned} \tag{3.1}$$

where the vectors X, b, σ_r are ℓ-dimensional and Y, b, $\tilde{\sigma}_r$ are m-dimensional. As usually, let us assume that the coefficients b, σ_r, $\tilde{b}$, $\tilde{\sigma}_r$ satisfy conditions (5.1.2) and (5.1.3) so that, in particular, the system (3.1) has the trivial solution

$$X(t) \equiv 0, \qquad Y(t) \equiv 0.$$

Moreover let us assume that the derivatives with respect to x,y of the coefficients of the system (3.1) are uniformly continuous with respect to t, and

$$\frac{\partial b(t,0,0)}{\partial y} \equiv 0, \qquad \frac{\partial \sigma_r(t,0,0)}{\partial y} \equiv 0.$$

Thus in the system of equations for the first approximation

$$dX(t) = \frac{\partial b(t,0,0)}{\partial x} X(t)dt + \sum_{r=1}^{k} \frac{\partial \sigma_r(t,0,0)}{\partial x} X(t)d\xi_r(t) \quad (3.2)$$

$$dY(t) = \left[\frac{\partial b(t,0,0)}{\partial x} X(t) + \frac{\partial b(t,0,0)}{\partial y} Y(t)\right]dt \quad (3.3)$$

$$+ \sum_{r=1}^{k} \left[\frac{\partial \sigma_r(t,0,0)}{\partial x} X(t) + \frac{\partial \sigma_r(t,0,0)}{\partial y} Y(t)\right]d\xi_r(t)$$

the component of $X(t)$ is also a Markov process. Theorem 7.1.2 (and the subsequent Remark 2) imply that if the trivial solution is uniformly stable in the large for the system of equations (3.2),(3.3), then it is asymptotically stable in probability for the system (3.1). The theorem which now follows allows us to say somewhat more.

For a one-dimensional process $X(t)$ the theorem was proved by Pinsky [1] who used another method.

THEOREM 3.1. *Suppose the above assumptions about the coefficients of the system* (3.1) *are verified. Assume further that the trivial solution is uniformly stable in the large for the system* (3.2) *and for the system*

$$dY(t) = \frac{\partial \tilde{b}(t,0,0)}{\partial y} Y(t)dt + \sum_{r=1}^{k} \frac{\partial \tilde{\sigma}_r(t,0,0)}{\partial y} Y(t)d\xi_r(t) \quad (3.4)$$

Then the trivial solution is asymptotically stable in probability for the system (3.1).

PROOF. The assumptions of the theorem and the Theorems 6.5.1 and 6.3.1 imply that there exist for a sufficiently small $p > 0$ two homogeneous functions $V_1(t,x)$, $V_2(y,x)$ of homogeneity degree p such that

$$\left.\begin{array}{l} k_1|x|^p \leqslant V_1(t,x) \leqslant k_2|x|^p, \\[2ex] \left|\dfrac{\partial V_1(t,x)}{\partial x_i}\right| \leqslant k_3|x|^{p-1}, \qquad \left|\dfrac{\partial^2 V_1(t,x)}{\partial x_i \partial x_j}\right| \leqslant k_3|x|^{p-2}, \\[2ex] \qquad\qquad i,j = 1,\dots,\ell, \\[1ex] L_1V_1(t,x) \leqslant -|x|^p, \end{array}\right\} \quad (3.5)$$

$$\left.\begin{array}{l} k_1|y|^p \leqslant V_2(t,y) \leqslant k_2|y|^p, \\[1ex] \left|\dfrac{\partial V_2(t,y)}{\partial y_i}\right| \leqslant k_3|y|^{p-1}, \qquad \left|\dfrac{\partial^2 V_2(t,y)}{\partial y_i \partial y_j}\right| \leqslant k_3|y|^{p-2}, \\[2ex] \qquad\qquad i,j = 1,\dots,m, \\[1ex] L_2V_2(t,y) \leqslant -|y|^p, \end{array}\right\} \quad (3.6)$$

where

$$L_1 = \frac{\partial}{\partial t} + \left(\frac{\partial b(t,0,0)}{\partial x} x, \frac{\partial}{\partial x}\right) + \frac{1}{2}\sum_{r=1}^{k}\left(\frac{\partial \sigma_r(t,0,0)}{\partial x} x, \frac{\partial}{\partial x}\right)^2,$$

$$L_2 = \frac{\partial}{\partial t} + \left(\frac{\partial \tilde{b}(t,0,0)}{\partial y} y, \frac{\partial}{\partial y}\right) + \frac{1}{2}\sum_{r=1}^{k}\left(\frac{\partial \tilde{\sigma}_r(t,0,0)}{\partial y} y, \frac{\partial}{\partial y}\right)^2$$

are the differential generators of the systems (3.2) and (3.4) respectively. (Here and below we adopt the same notation $(\cdot,\cdot)$ for the inner product in E_ℓ as well as in E_m). The differential generator of the system (3.2),(3.3) is

$$L = \frac{\partial}{\partial t} + \left(\frac{\partial b(t,0,0)}{\partial x} x, \frac{\partial}{\partial x}\right) + \frac{1}{2}\sum_{r=1}^{k}\left(\frac{\partial \sigma_r(t,0,0)}{\partial x} x, \frac{\partial}{\partial x}\right)^2$$

$$+ \left(\frac{\partial \tilde{b}(t,0,0)}{\partial x} x + \frac{\partial \tilde{b}(t,0,0)}{\partial y} y, \frac{\partial}{\partial y}\right)$$

$$+ \frac{1}{2}\sum_{r=1}^{k}\left(\frac{\partial \tilde{\sigma}_r(t,0,0)}{\partial x} x + \frac{\partial \tilde{\sigma}_r(t,0,0)}{\partial y} y, \frac{\partial}{\partial y}\right)^2$$

$$+ \sum_{r=1}^{k}\left(\frac{\partial \sigma_r(t,0,0)}{\partial x} x, \frac{\partial}{\partial x}\right)$$

$$\times \left(\frac{\partial \tilde{\sigma}_r(t,0,0)}{\partial x} x + \frac{\partial \tilde{\sigma}_r(t,0,0)}{\partial y} y, \frac{\partial}{\partial y}\right).$$

Let us consider now the auxiliary function

$$W(t,x,y) = [V_1^{2/p}(t,x) + \varepsilon V_2^{2/p}(t,y)]^{p/2} + AV_1(t,x), \qquad (3.7)$$

where V_1, V_2 are functions satisfying the condition (3.5),(3.6), and the values of the constants $\varepsilon > 0$, $A > 0$ will be specified below. (3.5) and (3.6) evidently imply that

$$k_4(|x|^p + |y|^p) \leqslant W(t,x,y) \leqslant k_5(|x|^p + |y|^p) \qquad (3.8)$$

holds for certain $k_4 > 0$, $k_5 > 0$ and arbitrary $\varepsilon > 0$, $A > 0$. Moreover it is evident that $x = 0$ is an unattainable invariant set for the process $X(t), Y(t)$. Hence in the case when $X(0) \neq 0$, we can apply Itô's formula and consequently also Theorem 5.7.1

to any function W which is not differentiable on the hyperplane $x = 0$. Considering this and (3.8), we see that it will suffice to prove that outside the set $x = 0$ the function $W(t,x,y)$, defined by (3.7) satisfies

$$LW(t,x,y) \leqslant - k_6(|x|^p + |y|^p), \qquad k_6 > 0 \tag{3.9}$$

for some ε and A.

Indeed, by the remark we made above, (3.8), (3.9) and Theorem 5.7.1 imply that the system (3.2),(3.3) is exponentially p-stable. The assertion of our theorem follows now by Theorem 7.1.2.

Let $W_i = V_i^{2/p}$. Then evidently

$$L_1V_1(t,x) = \frac{p}{2} W_1^{p/2-1} L_1W_1$$
$$+ \frac{1}{8} p(p - 2)W_1^{p/2-2} \sum_{r=1}^{k} \left(\frac{\partial\sigma_r(t,0,0)}{\partial x} x, \frac{\partial W_1}{\partial x}\right)^2 .$$

From this and from (2.5) follows easily

$$W_1L_1W_1 + \frac{p - 2}{4} \sum_{r=}^{k} \left(\frac{\partial\sigma_r(t,0,0)}{\partial x} x, \frac{\partial W_1}{\partial x}\right)^2 \leqslant - k_7|x|^4$$

for some constant $k_7 > 0$. In a similar way we obtain

$$W_2L_2W_2 + \frac{p - 2}{4} \sum_{r\ 1}^{k} \left(\frac{\partial\tilde{\sigma}_r(t,0,0)}{\partial y} y, \frac{\partial W_2}{\partial y}\right)^2 \leqslant - k_8|y|^4,$$

$$k_8 > 0.$$

These inequalities, together with (3.5),(3.6) imply that for some constants $k_9, k_{10} > 0$ which do not depend on ε

$$L[(W_1 + \varepsilon W_2)]^{p/2}$$
$$= \frac{p}{2} (W_1 + \varepsilon W_2)^{p/2-2}$$
$$\times\left\{(W_1 + \varepsilon W_2)L_1W_1 + \frac{p - 2}{4} \sum_{r=1}^{k} \left(\frac{\partial\sigma_r(t,0,0)}{\partial x} x, \frac{\partial W_1}{\partial x}\right)^2\right.$$
$$+ \varepsilon(W_1 + \varepsilon W_2)L_2W_2$$
$$+ \frac{p - 2}{4} \varepsilon^2 \sum_{r=1}^{k} \left(\frac{\partial\tilde{\sigma}_r(t,0,0)}{\partial x} x + \frac{\partial\tilde{\sigma}_r(t,0,0)}{\partial y} y, \frac{\partial W_2}{\partial y}\right)^2 +$$

(Contd)

$$\text{(Contd)} \quad + \varepsilon(W_1 + \varepsilon W_2)\left(\frac{\partial \tilde{b}(t,0,0)}{\partial x}\, x,\ \frac{\partial W_2}{\partial y}\right)$$

$$+ \varepsilon(W_1 + \varepsilon W_2)\sum_{r=1}^{k}\left(\frac{\partial \tilde{\sigma}_r(t,0,0)}{\partial x}\, x,\ \frac{\partial}{\partial y}\right)^2 W_2$$

$$+ \varepsilon(W_1 + \varepsilon W_2)\sum_{r=1}^{k}\left(\frac{\partial \tilde{\sigma}_r(t,0,0)}{\partial x}\, x,\ \frac{\partial}{\partial y}\right)$$

$$\times \left(\frac{\partial \tilde{\sigma}_r(t,0,0)}{\partial y}\, y,\ \frac{\partial}{\partial y}\right) W_2$$

$$+ \frac{\varepsilon(p-2)}{4}\left(\frac{\partial \tilde{\sigma}_r(t,0,0)}{\partial x}\, x + \frac{\partial \tilde{\sigma}_r(t,0,0)}{\partial y}\, y,\ \frac{\partial W_2}{\partial y}\right)$$

$$\times \left(\frac{\partial \sigma_r(t,0,0)}{\partial x}\, x,\ \frac{\partial W_1}{\partial x}\right)\Bigg\}$$

$$\leqslant k_9(W_1 + \varepsilon W_2)^{p/2-2}$$

$$\times [-k_{10}(|x|^4 + \varepsilon^2|y|^4) + \varepsilon|x|^2|y|^2$$

$$+ \varepsilon|y||x|^3 + \varepsilon^2|y|^3|x|]. \qquad (3.10)$$

Moreover $V_1 = L_1V_1 \leqslant -|x|^p$ by (3.5). From this and from the obvious inequalities

$$|x|^p \geqslant \frac{|x|^p W_1^{2-p/2}}{(W_1 + \varepsilon W_2)^{2-p/2}} \geqslant k_{11}(W_1 + \varepsilon W_2)^{p/2-2}|x|^4$$

we find that

$$LV_1 \leqslant -k_{11}(W_1 + \varepsilon W_2)^{p/2-2}|x|^4. \qquad (3.11)$$

The inequalities (3.10) and (3.11) yield an estimate

$$LW \leqslant k_9(W_1 + \varepsilon W_2)^{p/2-2}$$

$$\times [-Ak_{12}|x|^4 - k_{10}\varepsilon^2|y|^4 + \varepsilon|x|^2|y|^2$$

$$+ \varepsilon|y||x|^3 + \varepsilon^2|y|^3|x|],$$

with independent of ε and A constants k_i.

It is evident that if ε is sufficiently small and A is sufficiently large, then the expression between square brackets is a negative definite form of degree four (it is easiest to see this by putting $y\sqrt{\varepsilon} = z$). This implies (3.9) and hence also the assertion of the theorem.

4. Some further results

1. We have shown in Chapter VI that it is possible to give necessary and sufficient conditions for the asymptotical stability in the mean of a quadratical linear stochastic system (1.2) in E_ℓ with constant coefficients in terms of the eigenvalues of a certain $\ell(\ell + 1)/2 \times \ell(\ell + 1)/2$ matrix $\tilde{B}$. If ℓ is large, then the determination of this condition requires, in general, to calculate large order determinants. In Section 6.10 we considered a partial case of (1.2). This was described by an equation of degree ℓ, and for that case we gave other, more effective criteria which could be verified by calculating $\ell + 1$ determinants up to order ℓ inclusive. (Another special case in which it is possible to lower the degree was considered in Section 3.) Jakubovič and Levit [1] have shown that it is possible to obtain analogous conditions for certain linear systems of special types arising in a series of examples. To begin with, let us show one of their results concerning a special case.

Let us consider Itô's equation in E_ℓ

$$\dot{X}(t) = AX(t) + \mu b\sigma(X(t))\dot{\xi}(t), \qquad \sigma(X) = c^*X = (c,X), \quad (4.1)$$

where A is a constant $\ell \times \ell$ matrix, b and c are constant ℓ-dimensional column vectors and $\xi(t)$ is a standard Wiener process.

THEOREM 4.1. *The quadratic system* (4.1) *is asymptotically stable in the mean if and only if the matrix A is stable and*

$$\mu^2\int_0^\infty (c^*e^{At}b)^2 dt < 1 \quad (4.2)$$

is satisfied.

PROOF. We show first the necessity. Since the system (4.1) is stable in the mean square, Theorem 6.3.2 implies that there exists a positive definite quadratic form $\tilde{V}(x)$ which satisfies $L\tilde{V}(x) \leqslant -|x|^2$, where

$$L = L_0 + \frac{\mu^2}{2}\left(b\sigma(x), \frac{\partial}{\partial x}\right)^2, \qquad L_0 = \left(Ax, \frac{\partial}{\partial x}\right)$$

is the differential generator of the process (4.1). Since

$L_0\tilde{V}(x) \leqslant L\tilde{V}(x)$, we conclude from the above that the deterministic system $\dot{X} = AX$ is stable, i.e. the matrix A is stable.

Further, by the same Theorem 6.3.2, there exists non-negative definite quadratic form $V(x) = (Hx,x)$ such that

$$LV(x) = -\sigma^2(x). \tag{4.3}$$

This implies

$$L_0V(x) + \mu^2(Hb,b)\sigma^2(x) = -\sigma^2(x)$$

or

$$L_0V(x) = -\sigma^2(x)(1 + \mu^2V(b)). \tag{4.4}$$

Let $V_0(x) = (H_0x,x)$ be a non-negative definite quadratic form which satisfies the equation

$$L_0V_0(x) = -\sigma^2(x). \tag{4.5}$$

(A form like this exists due to the stability of the matrix A). Then, by (4.4) we have $V_0(x) = (1 + \mu^2V(b))^{-1}V(x)$ and therefore

$$V_0(b) < \frac{1}{\mu^2}. \tag{4.6}$$

Since

$$V_0(x) = \int_0^\infty c^*e^{At}x\ {}^2 dt \tag{4.7}$$

holds, we see now that (4.2) follows from (4.6).

We prove now the sufficiency part. Thus let A be a stable matrix satisfying (4.2). Following Morozan [8], let us consider two quadratic forms $V_0(x) = (H_0x,x)$ and $V_1(x) = (H_1x,x)$ defined by the equalities (4.5) and $L_0V_1(x) = -\ x\ ^2$ respectively. We put $\gamma_0 = V_0(b)$, $\gamma_1 = V_1(b)$. By (4.7), $\mu^2\gamma_0 < 1$. Thus it is possible to find δ such that

$$\frac{2}{\mu} - \delta\gamma_1 > 0, \qquad \left(\frac{2}{\mu} - \delta\gamma_1\right)^2 > 4\gamma_0. \tag{4.8}$$

The inequalities (4.8) imply that the quadratic equation $\lambda^2 - (2/\mu - \delta\gamma_1)\lambda + \gamma_0 = 0$ has a positive root; we shall denote it by λ_0. Let

$$V(x) = \frac{V_0(x)}{\lambda_0} + \delta V_1(x). \tag{4.9}$$

It is easy to verify that then

$$\frac{2}{\mu} - V(b) = \lambda_0, \tag{4.10}$$

$$L_0 V(x) = -\frac{\sigma^2(x)}{\lambda_0} - \delta|x|^2. \tag{4.11}$$

From these two equalities we obtain

$$\begin{aligned} LV(x) &= L_0 V(x) + \mu^2 V(b)\sigma^2(x) \\ &= -\frac{\sigma^2(x)}{\lambda_0} - \delta_0|x|^2 + \mu^2\left(\frac{2}{\mu} - \lambda_0\right)\sigma^2(x) \\ &= -\sigma^2(x)\left[\frac{1}{\sqrt{\lambda_0}} - \mu\sqrt{\lambda_0}\right]^2 - \delta_0|x|^2 \\ &\leqslant -\delta_0|x|^2. \end{aligned}$$

This, together with Theorem 6.3.2 implies that the system is asymptotically stable in mean square.

The result just obtained can be extended to linear stochastic systems of the form

$$\dot{X}(t) = AX(t) + \sum_{r=1}^{k} \mu_r b_r \sigma_r(X(t))\dot{\xi}_r(t), \quad \sigma_r(X) = c_r^* X \tag{4.12}$$

where b_r and c_r are constant column vectors in E_ℓ, μ_r are certain constants and $(\xi_1(t),\ldots,\xi_k(t))$ is a k-dimensional standard Wiener process. More exactly, we have the following theorem whose proof can be found in the paper of Jakubovič and Levit quoted above.

THEOREM 4.2. *The system* (4.12) *is asymptotically stable in mean square if and only if the matrix A is stable and the eigenvalues of the matrix* $((\int_0^\infty (\mu_i c_j^* e^{At} b_i)^2 dt$, $i,j = 1,\ldots,k$ *have absolute value less than one.*

Willems [1] has proved an analog of Theorem 4.2 for the case of discrete time.

2. A series of papers is devoted to the question of absolute stability of stochastic systems which are obtained from a deterministic system

$$\dot{X}(t) = AX(t) + b\varphi(\sigma), \qquad \sigma = c^*X = (c,X) \tag{4.13}$$

by introducing in some or other way noise into its equations. Here A is a constant $\ell \times \ell$ matrix, b and c are constant vectors in E_ℓ, and $\varphi(0)$ is a real-valued function of σ satisfying $\varphi(0) = 0$.

(4.13) is the system of automatic control with one executive.

There exist well-known conditions, due to Popov [1] which guarantee absolute stability of the system (4.13), that is, the stability of its trivial solution, uniformly over all non-linearizations from a certain class Φ. Lewit [1] established analogous conditions for absolute stability for various non-linearity classes Φ in case of the equation

$$X(t) = AX(t) + (b\varphi(\sigma) + \varkappa'\xi(t)), \qquad \varkappa' = r'X.$$

Here r is a constant $\ell \times k$ matrix, and $\xi(t)$ is a k-dimensional Gaussian "white noise". Let us further mention that this problem is also dealt with in the book of Morozan [7] and in the papers of Willems [2], Brockett and Willems [1], where more references can be found, and in other works.

Morozan [8] considers the equation

$$\dot{X}(t) = AX(t) + b\varphi(\sigma)\dot{\xi}(t), \tag{4.14}$$

where $\xi(t)$ is a standard Wiener process. He proves in this paper the following elegant stability criterion for the trivial solution in the non-linearity class $\varphi_\mu = \{\varphi(\sigma):\sigma\varphi(\sigma) \leqslant \mu\sigma^2, \varphi(\sigma)$ satisfies a local Lipschitz condition$\}$.

THEOREM 4.3. *The trivial solution of the system* (4.13) *is exponentially stable in mean square, uniformly with respect to all non-linearities* $\varphi(\sigma)$ *in the class* Φ_μ *if and only if the matrix A is stable and condition* (4.2) *holds.*

PROOF. Suppose that the system (4.14) is exponentially stable in mean square for every function $\varphi(\sigma)$ in the class Φ_μ. Since the function $\varphi(\sigma) = \mu\sigma$ is an element of this class, we have, in particular, that the linear system

$$\dot{X}(t) = (A + \mu bc'\dot{\xi}(t))X(t)$$

is asymptotically stable in mean square. Then Theorem 4.1 implies that A is a stable matrix and that inequality (4.2) holds.

Suppose now that A is a stable matrix and that inequality (4.2) holds. Let L_φ denote the differential generator of the process (4.14). It is evident that

$$L_\varphi = L_0 + \frac{1}{2}\left(b\varphi(\sigma), \frac{\partial}{\partial x}\right)^2, \qquad L_0 = \left(Ax, \frac{\partial}{\partial x}\right).$$

Let us define the function $V(x)$ as in the proof of Theorem 4.1. Then we get, using also (4.10), (4.11), and the inequality $\sigma\varphi(\sigma) \leqslant \mu\sigma^2$ that

$$L_{\varphi}V(x) = L_0V(x) + V(b)\varphi^2(\sigma)$$

$$\leqslant -\frac{\sigma^2(x)}{\lambda_0} - \delta_0|x|^2 + V(b)\varphi^2(\sigma)$$

$$= -\frac{\sigma^2(x)}{\lambda_0} - \delta_0|x|^2 - \varphi^2(\sigma)\left[-V(b) + \frac{2}{\mu}\right]$$

$$+ 2\sigma\varphi(\sigma) - 2\left[\sigma\varphi(\sigma) - \frac{1}{\mu}\varphi^2(\sigma)\right]$$

$$\leqslant -\left(\frac{\sigma(x)}{\sqrt{\lambda_0}} - \sqrt{\lambda_0}\varphi(\sigma)\right)^2 - \delta_0|x|^2$$

$$\leqslant -\delta_0|x|^2.$$

This, together with Theorem 6.3.2 implies that the solution $X(t)$ of equation (4.14) satisfies for every initial condition $X(0) = x$ the inequality

$$M|X(t)|^2 \leqslant A|x|^2e^{-\alpha t}$$

where A and α do not depend on the function $\varphi \in \Phi_\mu$. The theorem is proved.

3. Let us conclude this section by mentioning some further investigations related to the questions considered in this book.

To begin with, there is a series of works by Friedman [1], Friedman and Pinsky [1-3] and Pinsky [1]. In these papers there are given stability conditions for a point and for an invariant set of a similar sort as we gave here in Chapter V and in Subsection 2 of Section 7.7. Moreover, special considerations are devoted to the angle component $\varphi(t)$ for processes in the plane. An interesting generalization of this invariance theorem of La Salle to stochastic systems was given by Kushner [6], [7]. Morozan [8] clarified the idea of boundedness in various probabilistic meanings for solutions of stochastic dynamical systems of a more general type than the ones considered in the present book.

Much attention has been given to the investigation of stability problems for stochastic systems with lag. Kolmanovskiĭ [2], Kushner [9] and others proved general theorems of the Lyapunov type. Šaĭhet [1] considered stability in the first approximation, and Car'kov [1] gave stability criteria in mean square for linear systems.

BIBLIOGRAPHY

Ahmetkaliev, T.

[1] *On the connection between the stability of stochastic difference and differential systems*, Differencial'nye Uravnenija **1** (1965), 1016-1026 = Differential Equations **1** (1965), 790-798. MR **32** #9112.

[2] *Stability of difference stochastic systems under constantly acting perturbations*, Differencial'nye Uravnenija **2** (1966), 1161-1169 = Differential Equations **2** (1966), 600-604. MR **34** #4634.

Ariaratnam, S. and Graefe, P.

[1] *Linear systems with stochastic coefficients. I,II,III*, Internat. J. Control (1) **1** (1965), 239-248; ibid. (1) **2** (1965), 161-169; 205-211.

Aström, K.J.

[1] *On a first-order stochastic differential equation*, Internat. J. Control (1) **1** (1965), 301-326. MR **33** #1170.

Barbašin, E.A. and Krasovskiĭ, N.N.

[1] *On stability of motion in the large*, Dokl. Akad. Nauk USSR **86** (1952), 453-456. (Russian) MR **14**, 646.

Bedel'baev, A.K.

[1] *On the structure of Ljapunov functions which are quadratic forms and their application to the stability of control systems*, Avtomatika (Kiev) **1** (1958), 37-43. (Ukrainian) RŽMat. **1960** #6472.

Bellman, R.E.

[1] *Stability of differential equations*, McGraw-Hill, New York, 1953; Russian transl., IL, Moscow, 1954. MR **15**#794; **17** #734.

[2] *Dynamic programming*, Princeton Univ. Press, Princeton, N.J., 1957; Russian transl., IL, Moscow, 1960. MR **19** #820; **22** #4898.

[3] *Stochastic transformation and functional equations*, Proc. Sympos. Appl. Math., vol. 16, Amer. Math. Soc., Providence, R.I., 1964, pp. 171-177. MR **28** #4269

Benderskiĭ, M.M.

[1]* *Determination of the region of stability of a second order linear differential equation with random coefficients of a special form.* (Russian) Differential'nye Uravnienija **5** (1969), 1885-1888. MR **41** # 597.

[2]* *Determination of the moments of the solutions of a system of linear differential equations with random Markov coefficients.* (Russian) Trudy FTINT, Matematičeskaja Fizika i Funkcional'nyĭ Analiz, (Har'kov) **1** (1969), 64-71.

[3]* *On the asymptotic behaviour of the moments of the solutions of a linear system with random coefficients.* (Russian) Trudy FTINT, Matematičeskaja Fizika i Funkcional'nyĭ Analiz, (Har'kov) **3** (1972), 15-21.

Benderskiĭ, M.M. and Pastur, L.A.

[1]* *Asymptotical behaviour of the solutions of a second order equation with random coefficients.* (Russian) Teorija Funkciĭ i Funkcional'nyĭ Analiz, (Har'kov) **22** (1973), 3-14.

[2]* *The spectrum of the one-dimensional Schrodinger equation with random potential.* (Russian) Mat. Sb. (N.S.) **82** (**124**) (1970), 273-284. MR **41** # 7228.

Bertram, J.E. and Sarachik, P.E.

[1] *Stability of circuits with randomly time-varying parameters*, Proc. Internat. Sympos. on Circuit and Information Theory (Los Angeles, Calif.), IRE Trans. **CT-6** (1959), 809-823.

Bharucha-Redi, A.T.

[1] *On the theory of random equations*, Proc. Sympos. Appl. Math., vol. 16, Amer. Math. Soc., Providence, R.I., 1964, pp. 40-69. MR **32** #6498.

Blagoveščenskiĭ, Ju. N. and Freĭdlin, M.I.

[1] *Certain properties of diffusion processes depending on a parameter*, Dokl. Akad. Nau, USSR **138** (1961), 508-511 = Soviet Math. Dokl. **2** (1961), 633-636. MR **25** #2632.

Bogdanoff, J.L. and Kozin, F.

[1] *Moments of the output of linear random systems*, J. Acoust. Soc. Amer. **34** (1962), 1063-1066.

Brockett, R.W. and Willems, J.C.

[1]* *Average value criteria for stochastic stability*, Lecture Notes in Mathematics **294** (1972).

Bucy, R.S.

[1] *Stability and positive supermartingales*, J. Differential Equations **1** (1965), 151-155. MR **32** #8414.

Bunke, H.

[1] *Über das asymptotische Verhalten von Lösungen linearer stochastischer Differentialgleichungssysteme*, Z. Angew. Math. Mech. **45** (1965), 1-9. MR **33** #2902.

[2] *Über die fast sichere Stabilität linearer stochastischer Systeme*, Z. Angew. Math. Mech. **43** (1963), 533-535. MR **30** #635.

[3] *Stabilität bei stochastischen Differentialgleichungssysteme*, Z. Angew. Math. Mech. **43** (1963), 63-70. MR **27** #1985.

[1]* *Gewöhnliche Differentialgleichungen mit zufälligen Parametern.* Akademie Verlag, Berlin 1972.

Car'kov, E.F.

[1]* *Asymptotic exponential stability in the mean sqaure of the trivial solution of stochastic functional differential equations.* (Russian) Teor. Verojatnost. i Primenen. **21** (1976), 871-875.

Caughey, T.K. and Dienes, J.K.

[1] *The behavior of linear systems with random parametric excitation*, J. Mathematical Phys. **41** (1962), 300-318. MR **25** #5549.

Caughey, T.K. and Gray, A.H., Jr.

[1] *On the almost sure stability of linear dynamic systems with stochastic coefficients*, Trans. ASME Ser. E. J. Appl. Mech. **32** (1965), 365-372. MR **32** #4903.

[2] *A controversy in problems involving random parametric excitation*, J. Mathematical Phys. **44** (1965), 288-296.

Čelpanov, I.B.

[1] *Vibration of a second-order system with a randomly varying parameter*, Prikl. Mat. Meh. **26** (1962), 762-766 = J. Appl.

Čelpanov, I.B. (Contd)

Math. Mech. **26** (1962), 1145-1152. MR **26** #3983.

Četaev, N.G.

[1] *Stability of motion*, 3rd edn., "Nauka", Moscow, 1965; English transl. of 2nd edn., Pergamon Press, Oxford, 1961. MR **32** #4334; **42** #6370.

Chan, S.Y. and Chuang, K.

[1] *A study of linear, time-varying systems subject to stochastic disturbances*, Automatica **4** (1966), no. 1, 31-48.

Chung, K.L.

[1] *Contributions to the theory of Markov chains. II*, Trans. Amer. Math. Soc. **76** (1954), 397-419. MR **16** #419.

Darkovskiĭ, B.S. and Leĭbovič, V.S.

[1]* *Statistical stability and output moments of a certain class of systems with at random varying structure*. (Russian) Avtom. i Telemeh **10** (1971), 36-43.

Demidovič, B.P.

[1] *On the dissipativity of a certain non-linear system of differential equations. I,II*, Vestnik Moskov. Univ. Ser. I Mat. Meh. **1961**, no. 6, 19-27; ibid. **1962**, no. 1, 3-8. (Russian) MR **25** #4178; **26** #1567.

Doob, J.L.

[1] *Asymptotic properties of Markoff transition probabilities*, Trans. Amer. Math. Soc. **63** (1948), 393-421. MR **9** 598.

[2] *Martingales and one-dimensional diffusion*, Trans. Amer. Math. Soc. **78** (1955), 168-208. MR **17** #50.

[3] *Stochastic processes*, Wiley, New York; Chapman & Hall, 1953; Russian transl., IL, Moscow, 1956. MR **15** #445; **19** #71.

Dorogovcev, A. Ja.

[1] *Some remarks on differential equations perturbed by periodic random processes*, Ukrain. Mat. Z. **14** (1962), no. 2, 119-128; English transl., Selected Transl. Math. Statist. and Probability, vol. 5, Amer. Math. Soc., Providence, R.I., 1965, pp. 259-269. MR **25** #5247.

Driml, M. and Hans. O.

[1] *Continuous stochastic approximations*, Trans. Second Prague Conf. on Information Theory, Publ. House Czech. Akad. Sci.,

Prague, 1960; Academic Press, New York, 1961, pp. 113-122. MR **24** #A562.

Driml, M. and Nedoma, J.

[1] *Stochastic approximations for continuous random processes*, Trans. Second Prague Conf. on Information Theory, Publ. House Czech. Akad. Sci., Prague, 1960; Academic Press, New York, 1961, pp. 145-158. MR **23** #A4171.

Dym, H.

[1] *Stationary measures for the flow of a linear differential equation driven by white noise*, Trans. Amer. Math. Soc. **123** (1966), 130-164. MR **33** #6696.

Dynkin, E.B.

[1] *Infinistesimal operators of Markov processes*, Teor. Verojatnost. i Primenen. **1** (1956), 38-60 = Theor. Probability Appl. **1** (1956), 34-54. MR **19** #691.

[2] *Foundations of the theory of Markov processes*, Fizmatgiz, Moscow, 1959; English transl., Prentice Hall, Englewood Cliffs, N.J.; Pergamon Press, Oxford, 1961. MR **24** #A1745; # 1747.

[3] *Markov processes*, Fizmatgiz, Moscow, 1963; English transl., Die Grundlehren der math. Wissenschaften, Bände 121, 122, Academic Press, New York; Springer-Verlag, Berlin, 1965. MR **33** #1886; #1887.

Dynkin, E.B. and Juškevič, A.A.

[1] *Strong Markov processes*, Teor. Verojatnost. i Primenen. **1** (1956), 149-155 = Theor. Probability Appl. **1** (1956), 134-139. MR **19** #469.

Eĭdel'man, S.D.

[1] *Parabolic systems*, "Nauka", Moscow, 1964; English transl., Noordhoff, Groningen; North-Holland, Amsterdam, 1969. MR **29** #4998; **40** #6023.

Eskin, L.D.

[1] *On the asymptotic behavior of solutions of parabolic equations as $t \to \infty$*, Izv. Vysš. Učebn. Zaved. Matematika **1966**, no. 5 (54), 154-164; English transl., Amer. Math. Soc. Transl. (2) **87** (1969), 87-98. MR **34** #1671; **41** #3202.

[2] *On the asymptotic behavior of solutions of the Cauchy problem*, Uspehi Mat. Nauk **23** (1968), no. 2 (140), 207-208. (Russian).

Fabian, V.

[1] *A survey of deterministic and stochastic approximation methods for the minimization of functions*, Kibernetika (Prague) **1** (1965), 499-523. (Czech.) MR **32** #5378.

[2] *Stochastic approximation method*, Czechoslovak Math. J. **10** (**85**) (1960), 123-159. MR **22** #4091.

[3] *A new one-dimensional stochastic approximation method for finding a local minimum of a function*, Trans. Third Prague Conf. on Information Theory, Statist. Decision Functions, Random Processes (Liblice, 1962), Publ. House Czech. Akad. Sci., Prague, 1964, pp. 85-105. MR **29** #6569.

Feller, W.K.

[1] *Fluctuation theory of recurrent events*, Trans. Amer. Math. Soc. **67** (1949), 98-119. MR **11** #255.

[2] *The parabolic equations and the associated semi-groups of transformations*, Ann. of Math. (2) **55** (1952), 468-519; Russian transl., Matematika **1** (1957), no. 4, 105-153. MR **13** # 948.

[3] *Diffusion processes in one dimension*, Trans. Amer. Math. Soc. **77** (1954), 1-31. MR **16** #150.

[4] *An introduction to probability theory and its applications. Vol. II*, Wiley, New York, 1966; Russian transl., "Mir", Moscow, 1967. MR **35** #1048; **39** #3535.

Fleming, W.H.

[1] *Duality and a priori estimates in Markovian optimization problems*, J. Math. Anal. Appl. **16** (1966), 254-279. MR **33** #8222.

Friedman, A.

[1] *Interior estimates for parabolic systems of partial differential equations*, J. Math. Mech. **7** (1958), 393-417. MR **21** #7362.

[1]* *Stability and angular behaviour of solutions of stochastic differential equations*. Lecture Notes in Mathematics 294 (1972), 14-20.

Friedman, A. and Pinsky, M.A.

[1]* *Asymptotic behavior of solutions of linear stochastic differential systems*. Trans. Amer. Math. Soc. **181** (1973), 1-22.

[2]* *Asymptotic behavior and spiraling properties of stochastic equations*. Trans. Amer. Math. Soc. **186** (1973), 331-358.

[3]* *Dirichlet problem for degenerate elliptic equations*. Trans. Amer. Math. Soc. **186** (1973), 359-383.

Frisch, U.

[1] *Sur la résolution des équations différentielles stochastiques à coefficients morkoviens*, C.R. Acad. Sci. Paris Sér. A-B **262** (1966), A762-A765. MR **33** /2904.

Furstenberg, H.

[1] *Noncommuting random products*, Trans. Amer. Math. Soc. **108** (1963), 377-428. MR **29** #648.

Gabasov, R.

[1] *On the stability of stochastic systems with a small parameter multiplying the derivatives*, Uspehi Mat. Nauk **20** (1965), no. 1 (121), 189-196. (Russian) MR **30** # 3277.

Gantmaher, F.R.

[1] *The theory of matrices*, 2nd edn., "Nauka", Moscow, 1966; English tranl. of 1st edn., Vols. 1, 2, Chelsea, New York, 1959. MR **21** #6372c; **34** #2585.

Gel'fand, I.M.

[1] *Lectures on linear algebra*, 2nd edn., GITTL, Moscow, 1951; English transl., Pure and Appl. Math., no. 9, Interscience, New York, 1961. MR **13** #99; **23** #A152.

[2] *Generalized random processes*, Dokl. Akad. Nauk SSSR **100** (1955), 853-856. (Russian) MR **16** #938.

Germaidze, V.E. and Kac, I.Ja.

[1] *A problem of stability in the first approximation for stochastic systems*, Ural. Politehn. Inst. Sb. **139** (1964), 27-35. (Russian) MR **32** #6008.

Gihman, I.I.

[1] *On the theory of differential equations for stochastic processes. I*, Ukrain. Mat. Ž. **2** (1950), no. 3, 37-63; English transl., Amer. Math. Soc. Transl. (2) **1** (1955), 111-137. MR **14** #61; **17** #502.

[2] *Differential equations with random functions*, Winter School in Theory of Probability and Math. Statist. (Užgorod, 1964), Izdat. Akad. Nauk Ukrain. SSR, Kiev. 1964, pp. 41-86. (Russian) MR **31** #6037.

[3] *Stability of solutoins of stochastic differential equations*, Limit Theorems Statist. Inference, Izdat. "Fan", Tashkent, 1966, pp. 14-45; English transl., Selected Transl. Statist. and Probability, vol. 12, Amer. Math. Soc., Providence, R.I. 1973, pp. 125-154. MR **40** #944.

Gihman, I.I. and Dorogovčev, A. Ja.

[1] *On stability of solutions of stochastic differential equations*, Ukrain. Mat. Ž. **17** (1965), no. 6, 3-21; English transl., Amer. Math. Soc. Transl. (2) **72** (1968), 229-250. MR **43** #6996.

Gihman, I.I. and Skorohod, A.V.

[1] *Introduction to the theory of random processes*, "Nauka , Moscow, 1965; English transl., Saunders, Philadelphia, Pa., 1969. MR **33** #6689; **40** #923.

Gladyšev, E.G.

[1] *On stochastic approximation*, Teor. Verojatnost. i Primenen. **10** (1965), 297-300 = Theor. Probability Appl. **10** (1965), 275-278. MR **32** #3184.

Genedenko, B.V.

[1] *A course in the theory of probability*, 3rd edn., Fizmatgiz, Moscow, 1961; English transl., Chelsea, New York, 1962. MR **25** # 2622.

Grenander, U.

[1] *Probabilities on algebraic structures*, Wiley, New York, 1963; Russian transl., "Mir", Moscow, 1965. MR **34** # 6810; **36** #7165.

Halmos, P.

[1] *Measure theory*, Van Nostrand, Princeton, N.J., 1950; Russian transl., IL, Moscow, 1953. MR **11** #504; **16** #22.

Has'minskiĭ, R.Z.

[1] *Ergodic properties of recurrent diffusion processes and stabilization of the solution of the Cauchy problem for parabolic equations*, Teor. Verojatnost. i Primenen. **5** (1960), 196-214 = Theor. Probability Appl. **5** (1960), 179-196. MR **24** #A3695

[2] *On the stability of the trajectory of Markov processes*, Prikl. Mat. Meh. **26** (1962), 1025-1032 = J. Appl. Math. Mech. **26** (1962), 1554-1565. MR **28** #5470.

[3] *The averaging principle for parabolic and elliptic differential equations and Markov processes with small diffusion*, Teor. Verojatnost. I Primenen. **8** (1963), 3-25 = Theor. Probability Appl. **8** (1963), 1-21. MR **28** #4253.

[4] *The behavior of a self-oscillating system acted upon by slight noise*, Prikl. Mat. Meh. **27** (1963), 683-688 = J. Appl. Math. Mech. **27** (1963), 1035-1044. MR **28** #5234.

[5] *On the dissipativeness of random processes defined by differential equations*, Problemy Peredači Informacii 1 (1965), vyp. 1, 88-104 = Problems of Information Transmission **1** (1965), no. 1, 63-77. MR **32** #4747.

[6] *On equations with random perturbations*, Teor. Verojatnost. i Primenen. **10** (1965), 394-397 = Theor. Probability Appl. **10** (1965), 361-364.

[7] *On the stability of systems under persistent random perturbations*, Theory of Information Transmission. Pattern Recognition, "Nauka", Moscow, 1965, pp. 74-87. (Russian).

[8] *A limit theorem for the solution of differential equations with random right-hand side*, Teor. Verojatnost. i Primenen. **11** (1966), 444-462 = Theor. Probability Appl. **11** (1966), 390-406. MR **34** #3637.

[9] *On the stability of nonlinear stochastic systems*, Prikl. Mat. Meh. **30** (1966), 915-921 = J. Appl. Math. Mech. **30** (1966), 1082-1089. MR **37** #530.

[10] *Remark on the article of M.G. Šur "On linear differential equations with randomly perturbed parameters"*, Izv. Akad. Nauk SSSR Ser. Mat. **30** (1966), 1311-1314. (Russian) MR **35** #3762.

[11] *Necessary and sufficient conditions for the asymptotic stability of linear stochastic systems*, Teor. Verojatnost. i Primenen. **12** (1967), 167-172 = Theor. Probability Appl. **12** (1967), 144-147. MR **35** #2345.

[12] *Stability in the first approximation for stochastic systems*, Prikl. Mat. Meh. **31** (1967), 1021-1027 = J. Appl. Math. Mech. **31** (1967), 1025-1030. MR **37** #6561.

[13] *The limit distribution of additive functionals of diffusion processes*, Mat. Zametki **4** (1968), 599-610. (Russian) MR **39** #1017.

Has'minskiĭ R.Z. and Nevel'son, M.B.

[4]* *The stability of the solutions of one-dimensional stochastic equations*. (Russian) Dokl. Akad. Nauk SSSR **200** (1971), 785-788. MR **44** # 6069.

[5]* *Stochastic approximation and recursive estimation*. Translations of Mathematical Monographs **47** (1976).

Hinčin, A.Ja.

[1] *Asymptotische Gesetze der Wahrscheinlichkeitsrechnung*, Springer, Berlin, 1933; Russian transl., ONTI, Moscow, 1936.

Hopf, E.

[1] *Ergodentheorie*, Ergebnisse der Mathematik und ihrer Grenzgebiete, Band 5, Springer-Verlag, Berlin, 1937; reprint,

Hopf, E. (Contd)

Chelsea, New York, 1948; Russian transl., Uspehi Mat. Nauk **4** (1949), no. 1 (29), 113-182. MR **10** #549.

Ibragimov, I.A. and Linnik, Ju.V.

[1] *Independent and stationarily connected variables*, "Nauka", Moscow, 1965; English transl., Noordhoff, Groningen, 1971. MR **34** #2049.

Il'in, A.M. and Has'minskiĭ, R.Z.

[1] *Asymptotic behavior of solutions of parabolic equations and an ergodic property of nonhomogeneous diffusion processes*, Mat. Sb. (**60**) (**102**) (1963), 366-392; English transl., Amer. Math. Soc. Transl. (2) **49** (1965), 241-268. MR **26** #6608.

Il'in, A.M., Kalašinikov, A.S. and Oleĭnik, O.A.

[1] *Second-order linear equations of parabolic type*, Uspehi Mat. Nauk **17** (1962), no. 3 (105), 3-146 = Russian Math. Surveys **17** (1962), no. 3, 1-143. MR **25** #2328.

Itô, K.

[1] *On stochastic differential equations*, Mem. Amer. Math. Soc. No. 4 (1951); Russian transl., Matematika **1** (1957), no. **1**, 78-116. MR **12** #724.

[2] *On a formula concerning stochastic differentials*, Nagoya Math. J. **3** (1951), 55-65; Russian transl., Matematika **3** (1959), no. 5, 131-141. MR **13** #363.

[3] *Stationary random distributions*, Mem. Coll. Sci. Univ. Kyoto Ser. A. Math. **28** (1954), 209-223. MR **16** #378.

Itô, K. and Nisio, M.

[1] *On stationary solutoins of a stochastic differential equation*, J. Math. Kyoto Univ. **4** (1964), 1-75. MR **31** #1719.

Jakubovič, V.A. and Levit, M.V.

[1]* *Algebraic criteria for stochastic stability of linear systems with parametric control of the type of white noise.* (Russian) Prikl. Mat. Meh. **1**, **36** (1972), 142-147.

Jaglom, A.M.

[1] *On the statistical reversibility of Brownian motion*, Mat. Sb. **24** (**66**) (1949), 457-492; German transl., Abhandlungen aus der Sowjetischen Physik, Folge III, Verlag Kultur und Fortschritt, Berlin, 1952, 7-42. MR **11** #41; **17** #51.

[2] *An introduction to the theory of stationary random functions*, Uspehi Mat. Nauk **7** (1952), no. 5 (51), 3-168; English transl.,

rev. edn., Prentice-Hall, Englewood Cliffs, N.J., 1962. MR **14** #485; **32** #1762.

Kac, I.Ja.

[1] *On stability in the first approximation of systems with random parameters*, Ural. Gos. Univ. Mat. Zap. **3** (1962), no. 2, 30-37. (Russian) RŽMat. **1964** #7B59.

[2] *On the stability of stochastic systems in the large*, Prikl. Mat. Meh. **28** (1964), 366-372 = J. Appl. Math. Mech. **28** (1964), 449-456. MR **31** #6038.

[3] *Converse of the theorem on the asymptotic stability of stochastic systems*, Ural. Gos. Univ. Mat. Zap. **5** (1965), tetrad' 2, 43-51. (Russian) MR **33** #770.

[4] *On the stability in the first approximation of systems with random lag*, Prikl. Mat. Meh. **31** (1967), 447-452 = J. Appl. Math. Mech. **31** (1967), 478-482. MR **38** #4784.

Kac, I.Ja and Krasovskiĭ, N.N.

[1] *On stability of systems with random parameters*, Prikl. Mat. Meh. **24** (1960), 809-823 = J. Appl. Math. Mech. **24** (1960), 1225-1246.

Kalman, R.E.

[1] *Control of randomly varying linear dynamical systems*, Proc. Sympos. Appl. Math., vol. 13, Amer. Math. Soc., Providence, R.I., 1962, 287-298. MR **27** #2379.

Kamke, E.

[1] *Differentialgleichungen. Lösuingsmethoden and Lösungen. Teil I. Gëwohnliche Differentialgleichungen*, 3rd edn., Math. und ihrer Anwendungen in Physik und Technik, Band 18, Geest & Portig, Leipzig, 1944; Russian transl., IL, Moscow, 1951. MR 9 #33.

Kampé de Fériet, J.

[1] *La mecanique statistique des milieux continus*, Congrès Internat. de Philosophie des Sciences (Paris, 1949), vol. III, Actualités Sci, Indust., no. 1137, Hermann, Paris, 1951, 129-142; Russian transl., "Mir", Moscow, 1964. MR **15** #477.

Kesten, J. and Furstenberg, H.

[1] *Products of random matrices*, Ann. Math. Statist. **31** (1960), 457-469.

Kolmanovskiĭ, V.B.

[1] *Stationary solutions of equations with delay*, Problemy Peredači Informacii **3** (1967), vyp. 1, 64-72 = Problems of Information Transmission **3** (1967), no. 1, 50-57. MR **44** #571.

Kolmanovskiĭ, V.B. (Contd)

[1]* *The stability of non-linear systems with lag.* (Russian) Mat. Zametki **7** (1970), 743-751. MR **41** #878

Kolmogorov, A.N.

[1] *Über die analytischen Methoden in der Wahrscheinlichkeitsrechnung*, Math. Ann. **104** (1931), 415-458; Russian transl., Uspehi Mat. Nauk **5** (1938), 5-41.

[2] *Markov chains with countably many possible states*, Bjull. Moskov. Gos. Univ. Sekc. A **1** (1937), no. 3, 1-16. (Russian)

Kolomīēć, V.G.

[1] *Parametric action of a random force on a non-linear oscillatory system*, Ukrain. Mat. Ž. **14** (1962), 211-214. (Russian) MR **26** #425.

[2] *Parametric random effect on linear and non-linear oscillatory systems*, Ukrain. Mat. Z. **15** (1963), 199-205. (Russian MR **28** #3477.

[3] *Parametric random oscillations in linear and nonlinear systems*, Sb. Dokl. Taškent. Politehn. Inst. **1964**, no. 6, 49-59. (Russian) RŽMat. **1966** #4B51.

Kozin, F.

[1] *On almost sure stability of linear systems with random coefficients*, J. Math. Phys. **42** (1963), 59-67. MR **26** #2688.

[2] *On relations between moment properties and almost sure Lyapunov stability for linear stochastic systems*, J. Math. Anal. Appl. **10** (1965), 342-353. MR **30** #4630.

[3] *On almost sure asymptotic sample properties of diffusion processes defined by stochastic differential equations*, J. Math. Kyoto Univ. **4** (1964/65), 515-528. MR **31** #5239.

[1]* *Stability of the linear stochastic system*, Lecture Notes in Mathematics, **294** (1972), 186-229.

Krasnosel'skiĭ, M.A.

[1] *The operator of translation along the trajectories of differential equations*, "Nauka", Moscow, 1966; English transl., Transl. Math. Monographs, vol. 19, Amer. Math. Soc. Providdence, R.I., 1968. MR **34** #3012; **36** #6688.

Krasovskiĭ, A.A.

[1] *Sufficient conditions for statistical stability of motion*, Izv. Akad. Nauk USSR Tehn. Kibernet. **1966**, no. 2, 107-113. (Russian) MR **34** # 1087.

Krasovskiĭ, N.N.

[1] *Certain problems in the theory of stability of motion*, Fizmatgiz. Moscow, 1959; English transl., Stanford Univ. Press, Standford, Calif., 1963. MR **21** #5047; **26** #5258.

[2] *Problems in stabilization of controlled motion*, Appendix IV to I.G. Malkin [2], 2nd edn., 475-514. (Russian)

[3] *The stabilization of systems in which the noise depends on the magnitude of the action of the controller*, Izv. Akad. Nauk USSR Tehn. Kibernet. **1965**, no. 2 102-109. (Russian) MR **32** #5429.

Krasovskiĭ, N.N. and Lidskiĭ, È.A.

[1] *Analytic design of controllers in systems with random attributes. I,II,III*, Avtomat. i Telemeh. **22** (1961), 1145-1150; 1273-1278; 1425-1431 = Automat. Remote Control **22** (1961), 1021-1025; 1145-1146; 1289-1294. MR **27** #2375; #2377.

Krasulina, T.P.

[1] *The Robbins-Monro process in the case of several roots*, Teor. Verojatnost. i Primenen. 12 (1967), 386-390 = Theor. Probability Appl. **12** (1967), 333-337. MR **35** #5085.

Kushner, H.J.

[1] *On the stability of stochastic dynamical systems*, Proc. Nat. Acad. Sci. U.S.A. 53 (1964), 8-12. MR **31** #1138.

[2] *On the construction of stochastic Liapunov functions*, IEEE Trans. Automatic Control **AC-10** (1965), 477-478.

[3] *Finite time stochastic stability and the analysis of tracking systems*, IEEE Trans. Automatic Control **AC-11** (1966), 219-227. MR **35** #2676.

[4] *Stochastic stability and control*, Math. in Sci. and Engineering, vol. 33, Academic Press, New York, 1967. MR **35** #7723.

[5] *Converse theorems for stochastic Liapunov functions*, SIAM J. Control **5** (1967), 228-233. MR **35** #4048.

[6]* *The concept of invariant set for stochastic dynamical systems and applications to stochastic stability*, "Stochastic Optimization and Control", John Wiley and Sons, New York, 1968.

[7]* *Stochastic stability*, Lecture Notes in Mathematics **294** (1972), 97-124.

[8]* *Stochastic stability and control*, New York, Academic Press, 1967.

[9]* *On the stability of processes defined by stochastic difference-differential equations*, J. Differential Equations **4** (1968), 424-443.

Langevin, P.

[1] *Sur la théorie du mouvement brownien*, C.R. Acad. Sci. Paris **146** (1908), 530-533.

Laning, J.H. and Battin, R.H.

[1] *Random processes in automatic control*, McGraw-Hill, New York, 1956. MR **18** #74.

LaSalle, J.P. and Lefschetz, S.

[1] *Stability by Liapunov's direct method, with applications*, New York, 1961; Russian transl., "Mir", Moscow, 1964. MR **24** #A2712; 34 #6051.

Lavrenti'ev, M.A. and Šabat, B.V.

[1] *Methods of the theory of functions of a complex variable*, 2nd rev. edn., Fizmatgiz, Moscow, 1958; German transl., Mathematik fur Naturwiss. und Technik, Bank 13, VEB Deutscher Verlag, Berlin, 1967. MR **21** #116; **35** #345.

Leibowitz, M.A.

[1] *Statistical behavior of linear systems with randomly varying parameters*, J. Mathematical Phys. **4** (1963), 852-858. MR **27** #903.

Levit, M.V.

[1]* *A partial criterion for absolute stochastic stability of non-linear systems of differential equations.* (Russian) Uspiehi Mat. Nauk **4** (1961), **27** (1972).

Lidskiĭ, E.A.

[1] *On the stability of solutions of a stochastic system*, Proc. Interuniversity Conf. on Applications of Stability Theory, Kazan', 1964, 99-102. (Russian)

Ljapunov, A.M.

[1] *Problème général de la stabilité du mouvement*, Kharkov, 1892; reprint, GITTL, Moscow, 1950.

Loève, M.

[1] *Probability theory. Foundations. Random sequences*, Van Nostrand, Princeton, N.J., 1960; Russian transl., IL, Moscow, 1962. MR **23** #A670; **25** #3551.

Lur'e, A.I.

[1] *Minimal quadratic performance index of a controlled system*, Izv. Akad. Nauk SSSR Tehn. Kibernet. **4** (1963), 140-146. (Russian)

Malahov, A.N.

[1] *Statistical stability of motion*, Izv. Vysš. Učebn. Zaved. Radiofizika **6** (1963), no. 1, 42-53. (Russian)

Malkin, I.G.

[1] *Das Existenzproblem von Liapounoffschen Funktionen*, Izv. Kazan. Fiz.-Mat. Obšč. **5** (1931), 63-84.

[2] *Theory of stability of motion*, 2nd rev. edn., "Nauka", Moscow, 1966; English transl., Atomic Energy Commission Transl. #3352. MR **21** #2791; **34** #6388.

Maruyama, G. and Tanaka, H.

[1] *Some properties of one-dimensional diffusion processes*, Mem. Fac. Sci. Kyushu Univ. Ser. A. Math. **11** (1957), 117-141. MR **20** #3607.

[2] *Ergodic property of N-dimensional recurrent Markov processes*, Mem. Fac. Sci. Kyushu Univ. Ser. A. Math. **13** (1959), 157-172. MR **22** #3030.

McKenna J. and Morrison, J.

[1]* *Moments of solutions of a class of stochastic differential equations*, Journ. Math. Phys. **12**, 10 (1971), 2126-2135.

Mihaĭlov, V.P.

[1] *A solution of a mixed problem for a parabolic system using the method of potentials*, Dokl. Akad. Nauk USSR **132** (1960), 291-294 = Soviet Math. Dokl. **1** (1960), 556-559. MR **22** #12307.

Miranda, C.

[1] *Equazioni alle derivate parziali di tipo ellittico*, Ergebnisse der Mathematik und ihrer Grenzgebiete, Heft 2, Spinger-Verlag, Berlin, 1955; Russian transl., IL, Moscow, 1957; English transl., Springer-Verlag, Berlin, 1970. MR **19** #421.

Moiseev, N.N.

[1] *On a probability treatment of the concept "stability of motion"*, Rostov. Gos. Univ. Uč. Zap. Fiz-Mat. Fak. **18** (1953), no. 3, 79-82. (Russian) MR **17** #738.

Morozan, T.

[1] *La stabilité des solutions des systèmes d'équations différentielles aux paramètres aléatoires*, Rev. Roumaine Math. Pures Appl. **11** (1966), 211-238. MR **34** #450.

[2] *Stability of some linear stochastic systems*, J. Differential Equations **3** (1967), 153-169. MR **34** #7911

Morozan, T. (Contd)

[3] *Stability of controlled systems with random parameters*, Rev. Roumaine Math. Pures Appl. **12** (1967), 545-552. (Russian) MR **37** #6125.

[4] *Sur l'approximation stochastique*, C.R. Acad. Sci. Paris Sér. A-B **264** (1967), A633-A635. MR **35** #5045.

[5] *Stability in the case of small stochastic perturbations*, Rev. Roumaine Math. Pures Appl. **13** (1968), 71-74. (Russian) MR **38** #788.

[6] *Stability of stochastic discrete systems*, J. Math. Anal. Appl. **23** (1968), 1-9. MR **37** #4864.

[7]* *Stabilitatea sistemelor cu parametri aleatori*, Edit. Acad. R.S.R., Bucareşti, 1969.

[8]* *Boundedness properties for stochastic systems*. Lecture Notes in Mathematics **294** (1972) 21-34.

Nagaev, S.V.

[1] *Ergodic theorems for discrete-time Markov processes*, Sibirsk. Mat. Ž. **6** (1965), 413-432. (Russian) MR **31** #1715.

Nevel'son, M.B.

[1] *Stability in the large of a trajectory of the Markov processes of diffusion-type*, Differencial'nye Uravnenija **2** (1966), 1052-1060 = Differential Equations **2** (1966), 544-548. MR **34** #1632.

[2] *Some remarks on the stability of a linear stochastic system*, Prikl. Mat. Meh. **30** (1966), 1124-1127 = J. Appl. Math. Mech. **30** (1966), 1332-1335. MR **36** #3598.

[3] *Behavior of a linear system under small random excitation of its parameters*, Prikl. Mat. Meh. **31** (1967), 527-530 = J. Appl. Math. Mech. **31** (1967), 552-555. MR **38** #3933.

[4] *On the question of the recurrence and ergodicity of a Gaussian Markov process*, Teor. Verojatnost. i Primenen. **14** (1969), 35-43 = Theor. Probability Appl. **14** (1969), 35-43. MR **40** #6632.

[5] *Criterion of existence of an optimal control for a class of linear stochastic systems*, Prikl. Mat. Meh. **33** (1969), 573-577 = J. Appl. Math. Mech. **33** (1969), 561-565. MR **41** #1416.

Nevel'son, M.B. and Has'minskiĭ, R.Z.

[1] *Stability of a linear system with random disturbances of its parameters*, Prikl. Mat. Mech **30** (1966), 487-493. MR **36** #1134.

[2] *Stability of stochastic systems*, Problemy Peredači Informacii **2** (1966), vyp. 3, 76-91 = Problems of Information Transmission **2** (1966), no. 3, 61-74. MR **34** #7233.

[3] *Stability and stabilization of stochastic differential equations*, Proc. Sixth Math. Summer School on Probability Theory and Math. Statist. (Kaciveli, 1968), Izdat. Akad. Nauk Ukrain, SSR., Kiev, 1969, 59-122. (Russian) MR **42** #8583.

Petrovskiĭ, I.G.

[1] *Lectures on partial differential equations*, GITTL, Moscow, 1950; English transl., Interscience, New York, 1954. MR **13** #241; **16** #478.

Pliss, V.A.

[1] *Non-local problems in the theory of oscillations*, "Nauka", Moscow, 1964; English transl., Academic Press, New York, 1966. MR 30 #2188; 33 #4391.

Pontrjagin, L.S., Andronov, A.A. and Vitt, A.A.

[1] *On the statistical investigation of dynamic systems*, Ž. Èksper. Teoret. Fiz. **3** (1933), no. 3, 165-180. (Russian)

Pontrjagin, L.S., Boltjanskiĭ, V.G., Gamkrelidze, R.V. and Miščenko, E.F.

[1] *The mathematical theory of optimal processes*, Fizmatgiz, Moscow, 1961; English transl., Wiley, New York, 1962; Macmillan, New York, 1964. MR **29** #3316a,b.

Pinsky, M.A.

[1]* *Stochastic stability and Dirichlet problem*, Comm. on Pure and Appl. Math. **27** (1974), 311-350.

Popov, V.M.

[1]* *Hiperstabilitatea sistemelor automate*. Edit. Acad. R.S.R., Bucareşti, 1966.

Prohorov, Ju.V.

[1] *Convergence of random processes and limit theorems in probability theory*, Teor. Verojatnost, i Primenen. **1** (1956), 177-238 = Theor. Probability Appl. **1** (1956), 106-134. MR **18** #943.

Pugačev, V.S.

[1] *The theory of random functions and its application to control problems*, Fizmatgiz, Moscow, 1960; English transl., Internat. Series of Monographs on Automation and Automatic Control, vol. 5, Pergamon Press, Oxford; Addison-Wesley, Reading Mass., 1965. MR **23** #B2119; **31** #2091.

Rabotnikov, Ju.L.

[1] *On the impossibility of stabilizing a system in the mean-*

Rabotnikov, Ju.L. (Contd)

square by random perturbations of its parameters, Prikl. Mat. Meh. **28** (1964), 935-940 = J. Appl. Math. Mech. **28**, (1964), 1131-1136. MR **32** #7303.

[2] *Boundedness of solutions of differential equations with random coefficients whose averages are constant*, Zap. Meh.-M t. Fak. Har'kov. Gos. Univ. i Har'kov. Mat. Obšč. (4) **30** 1964), 75-84. (Russian) MR **36** #2916.

Robbins, H. and Monro, S.

[1] *A stochastic approximation method*, Ann. Math. Statist. **22** (1951), 400-407. MR **13** #144.

Romanovskiĭ, Ju.M.

[1] *Parametric random perturbations in some aeroelasticity problems*, Izv. Akad. Nauk USSR Ser. Meh. Maš. **4** (1960), 133-135. (Russian).

Rosenbloom, A.

[1] *Analysis of linear systems with randomly time-varying parameters*, Proc. Sympos. on Information Networks (New York, 1954), Polytechnic Institute of Brooklyn, Brooklyn, N.Y., 1955, 145-153. MR **17** #1100.

Rozanov, Ju.A.

[1] *Stationary random processes*, Fizmatgiz, Moscow, 1963; English transl., Holden-Day, San Francisco, Calif., 1967. MR **28** #2580; **35** #4985.

Rytov, S.M.

[1] *Introduction to statistical radiophysics*, "Nauka", Moscow, 1966. (Russian)

Sagirov, P.

[1]* *Stochastic methods in the dynamics of Satellites*. Intern. Center for Mech. SIAM Udine, Course Lectures 57, 1970.

Šaĭhet, L.E.

[1]* *Stability investigations of stochastic systems with lag by the method of Ljapunov functionals*. (Russian) Problemy Peredači Informacii **4**, 11 (1975), 70-75.

Samuels, J.C.

[1] *On the stability of random systems and the stabilization of deterministic systems*, J. Acoust. Soc. Amer. **32** (1960), 594-601. MR **22** #1945.

Samuels, J.C. and Eringen, A.C.

[1] *On stochastic linear systems*, J. Mathematical Phys. **38** (1959/60), 83-103. MR **21** #6041.

Sazonov, V.V. and Tutubalin, V.N.

[1] *Probability distributions on topological groups*, Teor. Verojatnost. i Primenen. **11** (1966), 3-55) = Theor. Probability Appl. **11** (1966), 1-45.

Šefl, O.

[1] *On stability of a randomized linear system*, Sci. Sinica **7** (1958), 1027-1034. MR **21** #644.

Serrin, J.

[1] *On the Harnack inequality for linear elliptic equations*, J. Analyse Math. **4** (1955/56), 292-308; Russian transl., Matematika **2** (1958), no. 6, 49-62. MR **18** #398.

Širjaev, A.N.

[1] *On stochastic equations in the theory of conditional Markov processes*, Teor. Verojatnost. i Primenen. **11** (1966), 200-206. = Theor. Probability Appl. **11** (1966), 179-184.

Skorohod, A.V.

[1] *Studies in the theory of random processes*, Izdat. Kiev. Univ. Kiev, 1961; English transl., Addison-Wesley, Reading, Mass., 1965. MR **32** #3082a,b.

[2] *Stochastic processes with independent increments*, "Nauka", Moscow, 1964. (Russian) MR **31** #6280.

Stepanov, V.V.

[1] *Zur Definition der Stabilitätswahrscheinlichkeit*, C.R. (Dokl.) Acad. Sci. URSS **18** (1938), 151-154.

Stratonovič, R.L.

[1] *Topics in the theory of random noise. Vol. 1: General theory of random processes. Nonlinear transformations of signals and noise*, "Sovetskoe Radio", Moscow, 1961; English transl., Gordon and Breach, New York, 1963. MR **28** #1660.

[2] *A new form of representing stochastic integrals and equations*, Vestnik Moskov. Univ. Ser. I Mat. Meh. **1964**, no. 1, 3-12. Russian) MR **28** #3476.

[3] *Conditional Markov processes and their application to the theory of optimal control*, Izdat. Moskov. Univ. Moscow, 1966; English transl., American Elsevier, New York, 1967. MR **33** #5391.

Stratonovič, R.L. and Romanovskiĭ, Ju.M.

[1] *Parametric representation of linear and nonlinear oscillatory systems by a random force*, Naučn. Dokl. Vysš. Školy. Fiz.-Mat. Nauki **1958**, no. 3, 221-224. (Russian) RŽMat. **1961** #8B224.

Šur, M.G.

[1] *Linear differential equations with randomly perturbed parameters*, Izv. Akad. Nauk USSR Ser. Mat. **29** (1965), 783-806; English transl., Amer. Math. Soc. Transl. (2) 72 (1968), 251-276. MR **33** #1556.

Tihonov, V.I.

[1] *Effect of fluctuation action on the simplest parametric systems*, Avtomat. i Teleheh. **19** (1958), 717-724 = Automat. Remote Control **19** (1958), 705-711. MR **20** #1599.

Tutubalin, V.N.

[1] *Limit theorems for a product of random matrices*, Teor. Verojatnost. i Primenen. **10** (1965), 19-32 = Theor. Probability Appl. **10** (1965), 15-27. MR **30** #5354.

Ueno, T.

[1] *The diffusion satisfying Wentzell's boundary condition and Markov process on the boundary. I,II*, Proc. Japan Acad. **36** (1960), 533-538; 625-629. MR **26** #1926.

Ventcel', A.D.

[1] *Semigroups of operators that correspond to a generalized differential operator of second-order*, Dokl. Akad. Nauk USSR **111** (1956), 269-272. (Russian) MR **19** #1060.

[2] *On lateral conditions for multidimensional diffusion processes*, Teor. Verojatnost. i Primenen. **4** (1959), 172-185 = Theor. Probability Appl. **4** (1959), 164-177. MR **21** #5246.

Vorovič, I.I.

[1] *On stability of motion for random disturbances*, Izv. Akad. Nauk USSR Ser. Mat. **20** (1956), 17-32. (Russian) MR **17** #977.

Wang, P.K.C.

[1] *On the almost sure stability of linear time-lag systems with stochastic parameters*, Internat. J. Control (1) **2** (1965), 433-440.

Watson, G.N.

[1] *A treatise on the theory of Bessel functions*, 2nd edn., Cambridge, Univ. Press, Cambridge; Macmillan, New York, 1944;

Russian transl., IL, Moscow, 1947. MR **6** #64.

Willems, J.L.

[1]* *Mean-square stability criteria for linear white noise stochastic systems*, Problems of Control and Information Theory 2, 3-4 (1973), 199-217.

[2]* *Lyapunov functions and global frequency domain stability criteria for a class of stochastic feed back systems*. Lecture Notes in Mathematics **294** (1972).

Wong, E. and Zakai, M.

[1] *On the relation between ordinary and stochastic differential equations*, Internat. J. Engrg. Sci. **3** (1965), 213-229. MR **32** #505.

[2] *On the convergence of ordinary integrals to stochastic integrals*, Ann. Math. Statist. **36** (1965), 1560-1564. MR **33** #3345.

Wonham, W.M.

[1] *Liapunov criteria for weak stochastic stability*, J. Differential Equations **2** (1966), 195-207. MR **33** #2906.

[2] *A Liapunov method for the estimation of statistical averages*, J. Differential Equations **2** (1966), 365-377. MR **34** #6257.

[3] *Lecture notes on stochastic control. Part 1*, Brown University, Providence, R.I., 1967.

[4] *Optimal stationary control of a linear system with state-dependent noise*, SIAM J. Control **5** (1967), 486-500. MR **36** #2421.

[1]* *Random differential equations in control theory*. Reprinted from Probabilistic methods in applied mathematics, Academic Press Inc., New York **2** (1970), 131-212.

Yoshizawa, T.

[1] *Liapunov's function and boundedness of solutions*, Funkcial. Ekvac. **2** (1959), 95-142. MR **22** #5789.

GENERAL INDEX

Typesetter: *Academic Industrial Epistemology*
Printer: *Samsom Sijthoff Grafische Bedrijven*
Binder: *Abbringh*